Molecular Symmetry and Group Theory: An Introduction

Molecular Symmetry and Group Theory: An Introduction

Sara Ruth

STATES
ACADEMIC PRESS
www.statesacademicpress.com

Molecular Symmetry and Group Theory: An Introduction
Sara Ruth
ISBN: 978-1-63989-366-9 (Hardback)

STATES
ACADEMIC PRESS

Published by States Academic Press,
109 South 5th Street,
Brooklyn, NY 11249, USA

Cataloging-in-Publication Data

Molecular symmetry and group theory : an introduction / Sara Ruth.
p. cm.
Includes bibliographical references and index.
ISBN 978-1-63989-366-9
1. Molecular theory. 2. Symmetry (Physics). 3. Group theory. I. Ruth, Sara.
QD461 .M65 2022
541.22--dc23

Table of Contents

Preface

The fundamental concept of chemistry which refers to the symmetry present in the molecular structures and the classification of these molecules on the basis of their symmetry is called molecular symmetry. It attempts to explain the chemical properties of the molecules and their spectroscopic transitions. The basic framework for studying molecular symmetry is provided by group theory. The subject of group theory can be classified into finite group theory, Lie theory, and combinatorial and geometric group theory. Molecular symmetry classifies the different states of molecules by using the irreducible representations from the character table. X-ray crystallography and other forms of spectroscopy are used extensively for the assessment of molecular symmetry. Such selected concepts that redefine molecular symmetry and group theory have been presented in this book. Different approaches, evaluations and methodologies on molecular symmetry and group theory have been included herein. This book is a complete source of knowledge on the present status of this important field.

A detailed account of the significant topics covered in this book is provided below:

Chapter 1- The symmetry which is present in molecules as well as their classification is described through molecular symmetry. The study of algebraic structures known as groups is called group theory. This chapter has been carefully written to provide an easy introduction to the varied facets of molecular symmetry and group theory.

Chapter 2- The mathematical function which describes the wave like behavior as well as the location of an electron within a molecule is known as a molecular orbital. A few of its major aspects are symmetry properties of orbitals, symmetry-adapted wavefunctions, and orbital symmetry rules and reactions. This chapter discusses in detail these aspects of molecular orbitals.

Chapter 3- The different symmetry operations which can be performed on a molecule which result in a conformation that is identical to the original are described by a point group. The topics elaborated in this chapter will help in gaining a better perspective about the representation of point groups as well as matrix representation of groups and irreducible representations.

Chapter 4- The molecular transformations which are engaged in chemical reactions depend on the symmetries of the orbitals involved. The possibility of a particular chemical reaction mechanism can be determined through the symmetry properties of molecular orbitals and reaction coordinates. This chapter discusses in detail the symmetry rules for chemical reactions.

Chapter 5- The spectroscopy which focuses on the infrared region of the electromagnetic spectrum is termed as vibrational spectroscopy. It is based on absorption spectroscopy. This chapter closely examines the key concepts of group theory and vibrational spectroscopy such as molecular vibrations and vibration modes to provide an extensive understanding of the subject.

I would like to make a special mention of my publisher who considered me worthy of this opportunity and also supported me throughout the process. I would also like to thank the editing team at the back-end who extended their help whenever required.

Sara Ruth

1

Introduction to Molecular Symmetry and Group Theory

The symmetry which is present in molecules as well as their classification is described through molecular symmetry. The study of algebraic structures known as groups is called group theory. This chapter has been carefully written to provide an easy introduction to the varied facets of molecular symmetry and group theory.

Symmetry

Symmetry is an important concept in chemistry. Firm understanding of this concept of the symmetry and group theory is extremely useful in dealing with molecular structures, stereochemistry, spectroscopy and reaction mechanisms. The concept of symmetry is important to almost every aspect of life in our universe. The "principle of symmetry" plays an important role in modern science. The idea of symmetry principle originally resulted from the study of geometrical forms and observations of natural objects. There are countless examples of symmetry in nature. Symmetry principle finds lot of applications in natural science i.e. physics, chemistry where geometrical properties are generally taken into account. The Oxford English dictionary gives the following two definitions for symmetry:

- Mutual relation of the parts of something in respect of magnitude and position; relative measurement and arrangement of parts; proportion.
- Due or just proportion, harmony of parts with each other and the whole; fitting, regular or balanced arrangement and relation of parts of elements; the condition or quality of being well proportion or balanced.

The first symmetry definition clearly has more scientific importance. The concept of symmetry sounds simple and familiar, yet symmetry is far more complex and difficult to apply in practice than one might think. Let us introduce the concept of symmetry in molecular systems in more quantitative manner. The symmetry of the molecule is determined by the existence of symmetry elements about which symmetry operations can be performed.

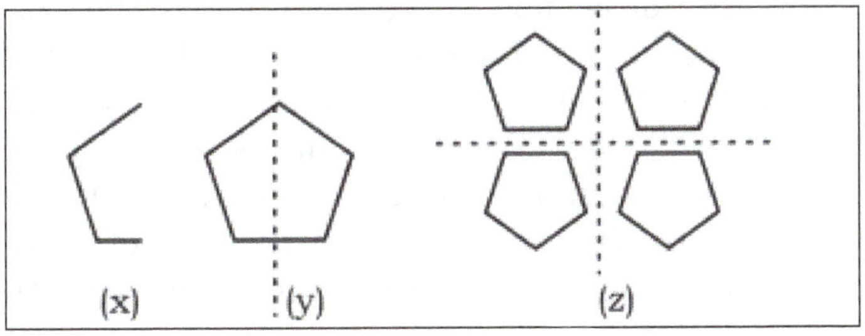

Let us examine these objects to understand the concept of symmetry in terms of symmetry operations (certain movements) that can be performed on these objects. If one performs a symmetry operation on the object then object should look the same before after the symmetry operation is carried out i.e. object before and after the symmetry operation should have equivalent configuration or is indistinguishable. Configurations may not be identical as some parts may have interchanged when symmetry operation is carried out. What are the characteristics of these objects that lead us to say that object in part (x) has no symmetry and but objects in parts (y) and (z) have symmetry? One can even say that the object in part (z) has more symmetry than the object in (y). Dotted lines in parts (y) and (z) indicate that the objects are symmetrical about these lines. Let us explain it further by taking the example of unlabeled square piece of card and rotate this by 0°, 90°, 180°, 270° and 360° successively about dotted line perpendicular to the plane of the card. Figure gives some of these results of rotations.

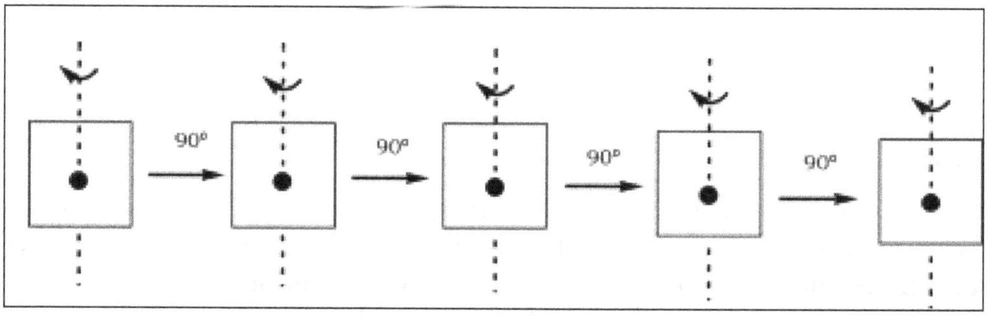

Rotations of square planar card without labeling.

All resulting configurations look the same and cannot be distinguished from each other. Thus these rotations/movements are the symmetry operations. However if one labels the corners of the square piece of card and then perform similar rotations about the dotted line one gets the result as shown in figure.

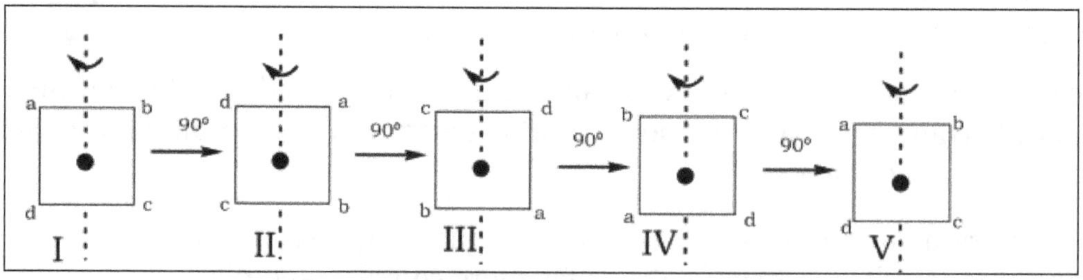

Rotations of square planar card with labeling.

If we label the four corners, as a, b, c and d then it is possible to observe the effects of these movements/rotations. If we carry out four successive 90° rotations about dotted line, the configuration V (the final configuration) is now strictly identical to the configuration I (the original configuration). The labels a, b, c, and d help distinguish the effects of rotation. If we have labeled or tagged corners of the square card then the intermediate 90° rotations (configurations II, III and IV) do not lead to equivalent configurations and these can be distinguished from each other. Thus these movements/rotations when the square card corners are labeled are not truly symmetry operations. Let us now differentiate between closely related terms symmetry elements and symmetry operations. For this let us consider the Figure and label the arms and perform the rotation through 180°.

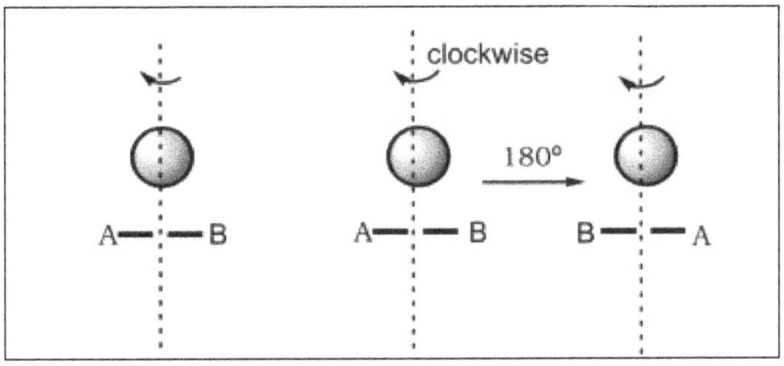

180° rotation of the object about the dotted line.

Now rotate the figure around the vertical dotted line by 180° (half a complete cycle, a clock wise rotation). Because the hands are identified by letters, the two figures can be differentiated after rotation and thus this rotation/movement is not a symmetry rotation. But if we remove the labeling, the figures will be identical/equivalent/ indistinguishable. The concept of symmetry is familiar one and generally one speaks of a shape as being, "symmetrical", "unsymmetrical" or even "more symmetrical than some other shape". For scientific purposes, one needs to specify concept of symmetry in a more quantitative way. Let us consider two triangles A and B as shown in figure.

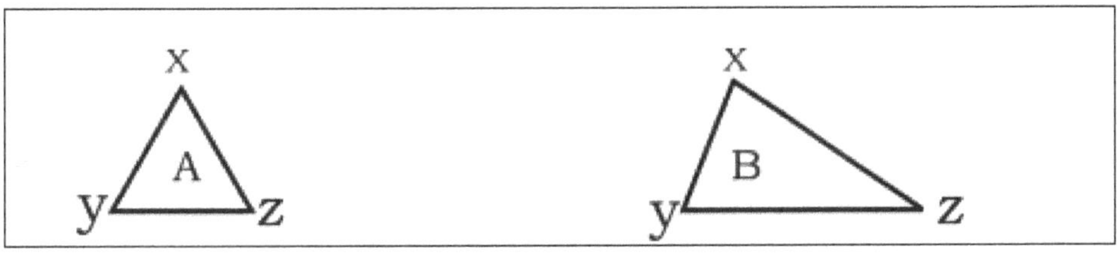

Shapes of the triangles A and B.

Find the triangle which is more symmetrical. Your answer will be that triangle A is more symmetrical than triangle B, which is correct. Why triangle A is more symmetrical than triangle B? Let us examine it. Label the corners of triangles as x y z. Draw a dotted line in the plane of the paper through vertex x such that it bisects the side yz at p as shown in figure.

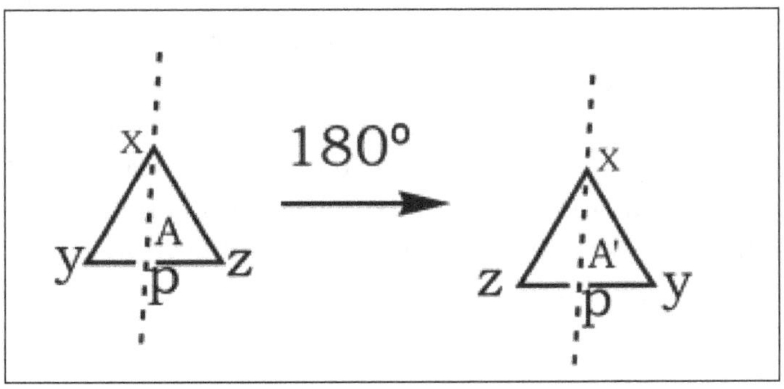

The 180° rotation about dotted line in triangle A.

The 180° rotation about dotted line in triangle A Triangle A has been divided into two halves which are equivalent i.e. triangle xzp and triangle xpy are equivalent to each other. Now rotate triangle A through 180° about dotted line (in plane). Triangle A goes to triangle A′. Triangles A and A′ are distinguishable hence rotation is not a symmetry rotations. If the corners of triangle A are not labeled and 180° rotation is performed about the dotted line you will not be able to distinguish the original triangle A and the triangle obtained after rotation i.e. triangle A′. In that case rotation is a symmetry rotation. Can you carry out similar rotation process in triangle B i.e. rotate about dotted line through 180° and get the indistinguishable figure B′? Your answer will be 'No'.

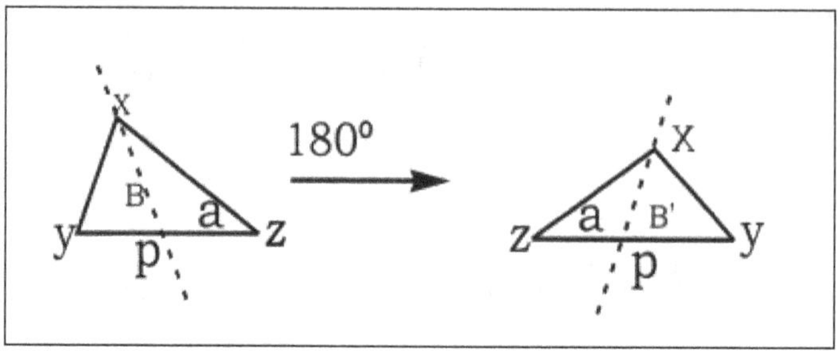

The 180° rotation in triangle B.

The smaller angle a is pointing in opposite direction. You can distinguish between triangles B and B′. Thus 180° rotation on triangle B does not give similar results as were obtained in case of triangle A. Hence triangle A is more symmetrical than triangle B about dotted line in plane of the paper. Let us again take triangle A and pass a dotted line perpendicular to the plane of the triangle through the midpoint of this triangle A and perform three successive 120° clockwise (negative rotation) rotations about this dotted line. The results of these rotations are shown in figure.

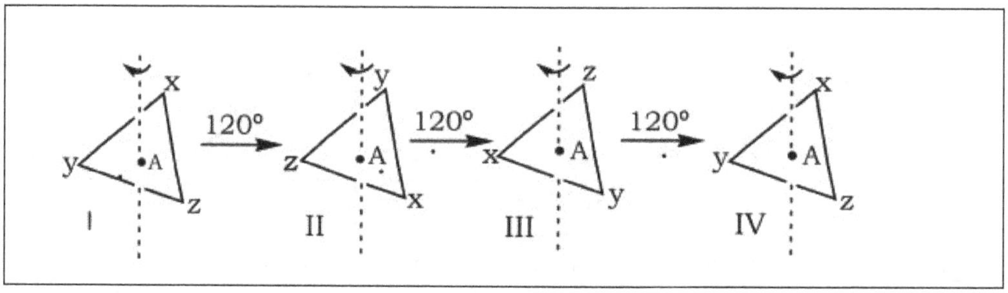

Effect of successive 120° rotations on triangle A.

Configurations I and IV of the triangle are indistinguishable even when the vertices are labeled. These three successive 120° rotations bring back the triangle A to original position. If vertices are not labeled then you will not be able to distinguish configurations I, II, III, IV from each other. In the triangle A, so far, we have given two types of movements or rotations about dotted line (a line or an axis) in plane and a line perpendicular to plane of the paper. After these physical movements/rotations (or operations) you find the new triangle A′. The configurations II, III, IV are equivalent or indistinguishable from A i.e. from configuration I. In these types of movements/rotations/operations on triangle A we had imaginary lines in plane and perpendicular to the triangle plane about which these movements/rotations/operations were performed. Since we are getting indistinguishable configurations, these imaginary lines and physical movements/rotations/operations are special.

Symmetry Elements and Symmetry Operation

A symmetry operation is an action that leaves an object looking the same after it has been carried out. For example, if we take a molecule of water and rotate it by 180° about an axis passing through the central O atom (between the two H atoms) it will look the same as before. It will also look the same if we reflect it through either of two mirror planes, as shown in the figure below:

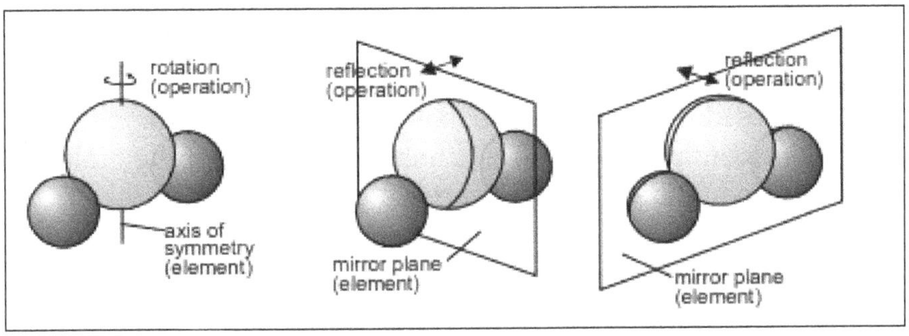

Each symmetry operation has a corresponding symmetry element, which is the axis, plane, line or point with respect to which the symmetry operation is carried out. The symmetry element consists of all the points that stay in the same place when the symmetry operation is performed. In a rotation, the lines of points that stay in the same place constitute a symmetry axis; in a reflection the points that remain unchanged make up a plane of symmetry. The symmetry elements that a molecule may possess are:

- E: The identity - The identity operation consists of doing nothing, and the corresponding symmetry element is the entire molecule. Every molecule has at least this element.
- C_n: An n-fold axis of rotation. Rotation by 360°/n leaves the molecule unchanged. The H_2O molecule above has a C_2 axis. Some molecules have more than one C_n axis, in which case the one with the highest value of n is called the principal axis. Note that by convention rotations are counterclockwise about the axis.
- σ: A plane of symmetry. Reflection in the plane leaves the molecule looking the same. In a molecule that also has an axis of symmetry, a mirror plane that includes the axis is called a vertical mirror plane and is labelled σ_v, while one perpendicular to the axis is called a horizontal mirror plane and is labelled σ_h. A vertical mirror plane that bisects the angle between two C_2 axes is called a dihedral mirror plane, σ_d.
- i: A centre of symmetry. Inversion through the centre of symmetry leaves the molecule unchanged. Inversion consists of passing each point through the centre of inversion and out to the same distance on the other side of the molecule. An example of a molecule with a centre of inversion is shown below:

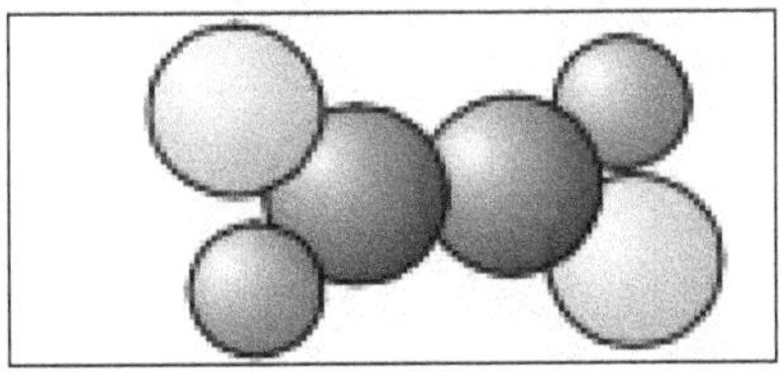

- S_n: An n-fold improper rotation axis (also called a rotary-reflection axis). The rotary reflection operation consists of rotating through an angle 360°/n about the axis, followed by reflecting in a plane perpendicular to the axis. S_1 is the same as reflection and S_2 is the same as inversion. The molecule shown above has two S_2 axes.

The identity E and rotations C_n are symmetry operations that could actually be carried out on a molecule. For this reason they are called proper symmetry operations. Reflections, inversions and improper rotations can only be imagined (it is not actually possible to turn a molecule into its mirror image or to invert it without some fairly drastic rearrangement of chemical bonds) and as such, are termed improper symmetry operations. Conventionally, when imposing a set of Cartesian axes on a molecule, the z axis lies along the principal axis of the molecule, the x axis lies in the plane of the molecule (or in a plane containing the largest number of atoms if the molecule is non-planar), and the y axis makes up a right handed axis system.

Symmetry Classification of Molecules – Point Groups

It is only possible for certain combinations of symmetry elements to be present in a molecule (or any other object). As a result, we may group together molecules that possess the same symmetry elements and classify molecules according to their symmetry. These groups of symmetry elements are called point groups (due to the fact that there is at least one point in space that remains unchanged no matter which symmetry operation from the group is applied). There are two systems of notation for labelling symmetry groups, called the Schoenflies and Hermann-Mauguin (or International) systems. The symmetry of individual molecules is usually described using the Schoenflies notation.

Some of the point groups share their names with symmetry operations, so be careful you don't mix up the two. It is usually clear from the context which one is being referred to. The molecular point groups are listed below:

- C_1: Contains only the identity (a C_1 rotation is a rotation by 360° and is the same as the identity operation E) e.g. CHDFCl.

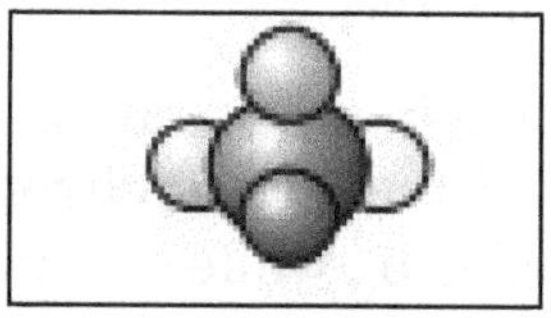

- C_i: Contains the identity E and a centre of inversion i.

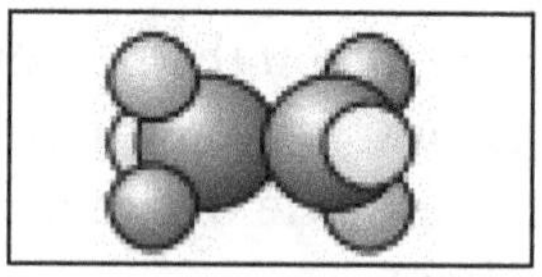

- C_S: Contains the identity E and a plane of reflection σ.

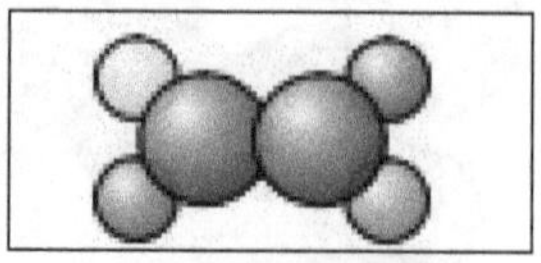

- C_n: Contains the identity and an n-fold axis of rotation.

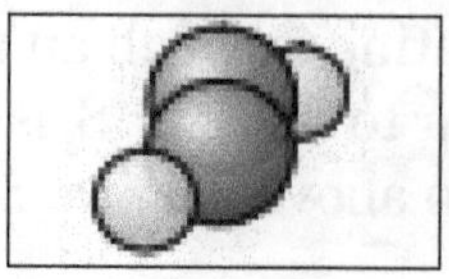

- C_{nv}: Contains the identity, an n-fold axis of rotation, and n vertical mirror planes σ_v.

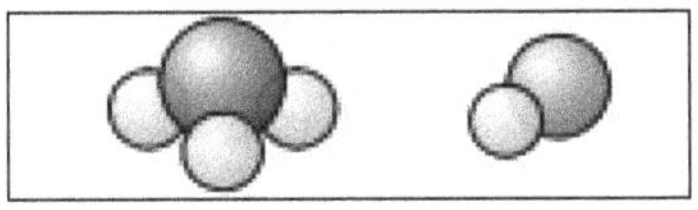

- C_{nh}: Contains the identity, an n-fold axis of rotation, and a horizontal reflection plane σ_h (note that in C_2h this combination of symmetry elements automatically implies a centre of inversion).

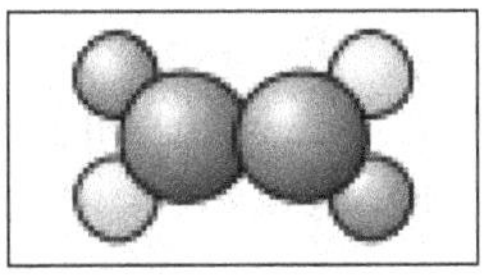

- D_n: Contains the identity, an n-fold axis of rotation, and n 2-fold rotations about axes perpendicular to the principal axis.
- D_{nh}: Contains the same symmetry elements as D_n with the addition of a horizontal mirror plane.

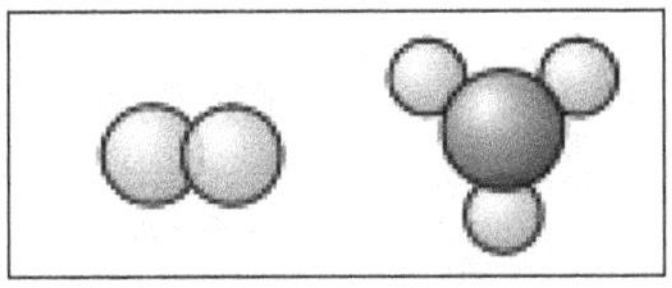

- D_{nd}: Contains the same symmetry elements as D_n with the addition of n dihedral mirror planes.

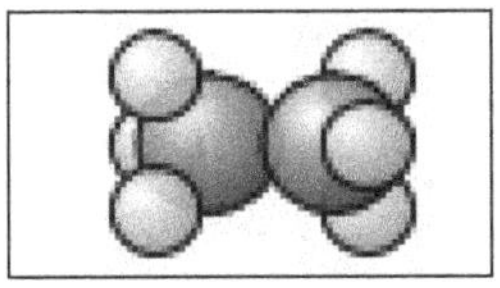

- S_n: Contains the identity and one S_n axis. Molecules only belong to S_n if they have not already been classified in terms of one of the preceding point groups (e.g. S_2 is the same as C_i, and a molecule with this symmetry would already have been classified).

The following groups are the cubic groups, which contain more than one principal axis. They separate into the tetrahedral groups (T_d, T_h and T) and the octahedral groups (O and O_h). The icosahedral group also exists but is not included below:

- T_d: Contains all the symmetry elements of a regular tetrahedron, including the identity, 4 C_3 axes, 3 C_2 axes, 6 dihedral mirror planes, and 3 S_4 axes e.g. CH_4.

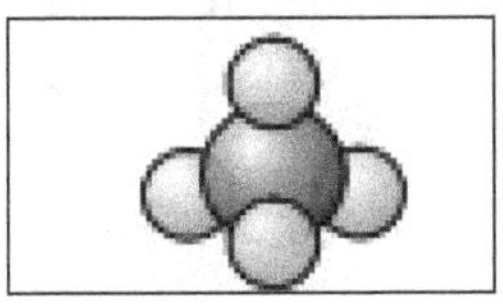

- T: As for T_d but no planes of reflection.
- T_h: As for T but contains a centre of inversion.

- O_h: The group of the regular octahedron e.g. SF_6.

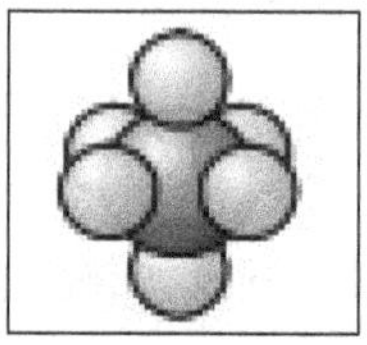

- O: As for O_h but with no planes of reflection.

The final group is the full rotation group R_3, which consists of an infinite number of C_n axes with all possible values of n and describes the symmetry of a sphere. Atoms (but no molecules) belong to R_3, and the group has important applications in atomic quantum mechanics. Here the symmetry elements and point groups described above after this, you will find it quite straightforward to classify a molecule in terms of its point group. In the meantime, the flowchart shown below provides a step-by-step approach to the problem.

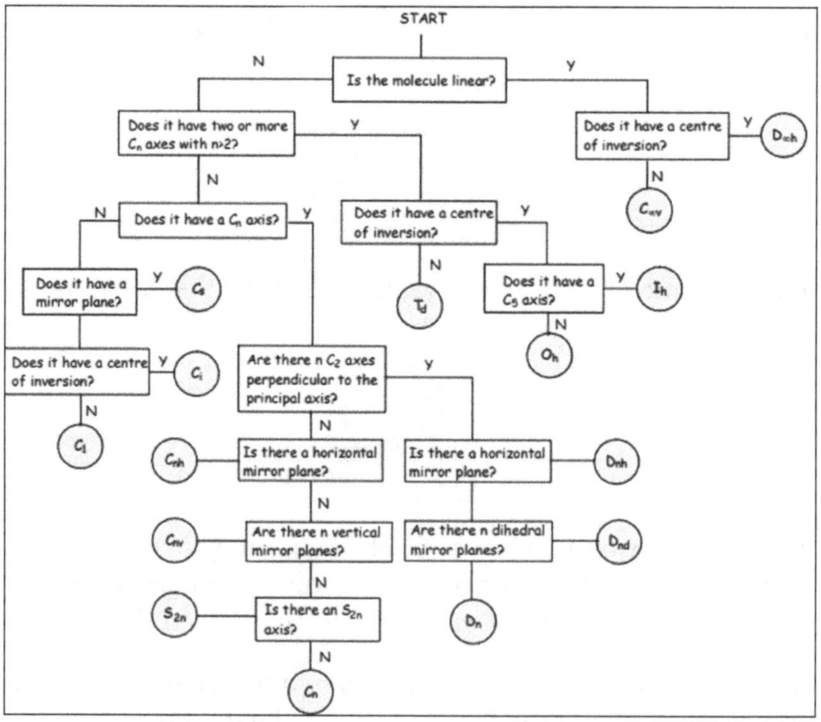

Symmetry and Physical Properties

Carrying out a symmetry operation on a molecule must not change any of its physical properties. It turns out that this has some interesting consequences, allowing us to predict whether or not a molecule may be chiral or polar on the basis of its point group.

Polarity

For a molecule to have a permanent dipole moment, it must have an asymmetric charge distribution. The point group of the molecule not only determines whether the molecule may have a dipole moment, but also in which direction(s) it may point. If a molecule has a C_n axis with n>1, it cannot have a dipole moment perpendicular to the axis of rotation (for example, a C_2 rotation would interchange the ends of such a dipole moment and reverse the polarity, which is not allowed – rotations with higher values of n would also change the direction in which the dipole points). Any dipole must lie parallel to a C_n axis.

Also, if the point group of the molecule contains any symmetry operation that would interchange the two ends of the molecule, such as a σ_h mirror plane or a C_2 rotation perpendicular to the principal axis, then there cannot be a dipole moment along the axis. The only groups compatible with a dipole moment are C_n, C_{nv} and C_s. In molecules belonging to C_n or C_{nv} the dipole must lie along the axis of rotation.

Chirality

One example of symmetry in chemistry that you will already have come across is found in the isomeric pairs of molecules called enantiomers. Enantiomers are non-superimposable mirror images of each other, and one consequence of this symmetrical relationship is that they rotate the plane of polarised light passing through them in opposite directions. Such molecules are said to be chiral, meaning that they cannot be superimposed on their mirror image. Formally, the symmetry element that precludes a molecule from being chiral is a rotation-reflection axis S_n. Such an axis is often implied by other symmetry elements present in a group. For example, a point group that has C_n and σh as elements will also have S_n. Similarly, a centre of inversion is equivalent to S_2. As a rule of thumb, a molecule definitely cannot have be chiral if it has a centre of inversion or a mirror plane of any type (σ_h, σ_v or σ_d), but if these symmetry elements are absent the molecule should be checked carefully for an S_n axis before it is assumed to be chiral.

Chirality and Optical Activity

The term "chiral" was derived from the Greek word for hand, as they are non-superimposable mirror images of each other. This property of asymmetry is termed as chirality. On the contrary, molecules identical to their mirror images are known as achiral. Most chiral molecules lack symmetry elements like a plane of symmetry (a plane that divides the molecule in two halves such that one half is mirror image of other) or a center of symmetry (a point from which any atom in the molecule lies diametrically opposite to identical atom at equal distance). Chirality is more often observed in molecules having at least one carbon atom which is attached to four different groups. To understand the difference between chiral and achiral molecules let us compare two molecules; 1,1-dichloroethane and 1,1-bromochloroethane.

We can clearly see that 1,1-dichloroethane has a plane of symmetry passing through CH_3-C-H whereas no plane of symmetry is present in 1,1-bromochloroethane. Further, to test the superimposability of both 1,1-dichloroethane and 1,1-bromochloroethane with their respective mirror images, we will rotate the along the axis passing through C-H bond. The spatial arrangement of mirror image after rotation and 1,1-dichloroethane is identical to original molecule but no similar observation can be made in case of 1,1-bromochloroethane. Thus, 1,1-dichloroethane is achiral whereas 1,1-bromochloroethane is a chiral molecule.

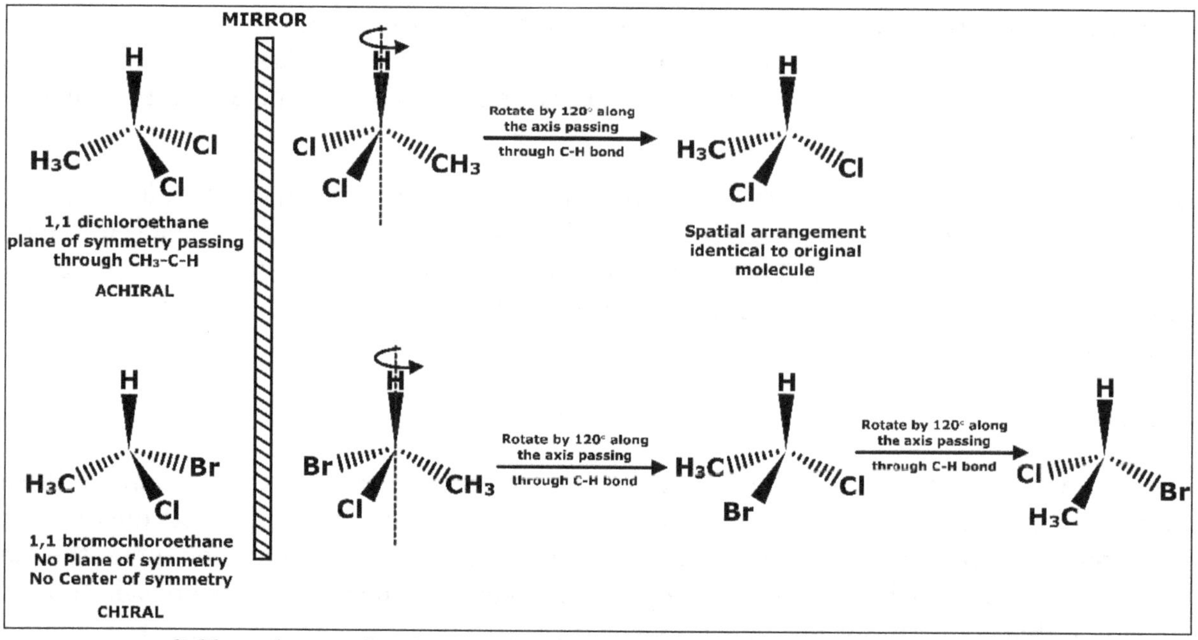

1,1-dichloroethane and 1,1-bromochloroethane with their mirror image representations.

A carbon atom bonded to four different groups is termed as a stereogenic center, chiral carbon or asymmetric carbon. The mirror images of 1,1-bromochloroethane have the same constitution and sequence of bonded atoms, however, they differ in the spatial arrangement of atoms; thus they are known as stereoisomers. Stereoisomers which are non-superpimposable mirror images of each other are termed as enantiomers.

It is pretty interesting that enantiomers have identical chemical (except when interacting with chiral reagents or in a chiral environment) and usual physical properties (like solubility, density, boiling point etc.) but have striking difference in their ability to rotate the plane of polarized light as well as the properties that depend on the spatial arrangement of the atoms. Further, a molecule can have more than one stereocenter. More than two stereoisomers are possible for such molecules. For 'n' number of stereogenic centers present in a molecule, there will be a maximum of '2^n' stereoisomers. The number of stereoisomers increases with increasing stereocenters.

Table: Relationship between number of stereocenters and maximum number of stereoisomers possible.

n No. of stereocenters	2^n Maximum number of stereoisomers possible
1	2
2	4
3	8
4	16
5	32

Carbohydrates are a very good example to explain '2^n' rule as they have more than one stereocenter. Let us take an example of Erythrose. It has two stereocenters and thus has four stereoisomers. There are two sets of stereoisomers that are mirror images (A & B, C & D) and therefore they are enantiomers. All the other sets of stereoisomers which are not mirror images (A & C, A & D, B & C, B & D) are known as diastereomers. Diastereomers are stereoisomers which are not mirror images of each other. Unlike enantiomers, diastereomers show different physical and chemical properties.

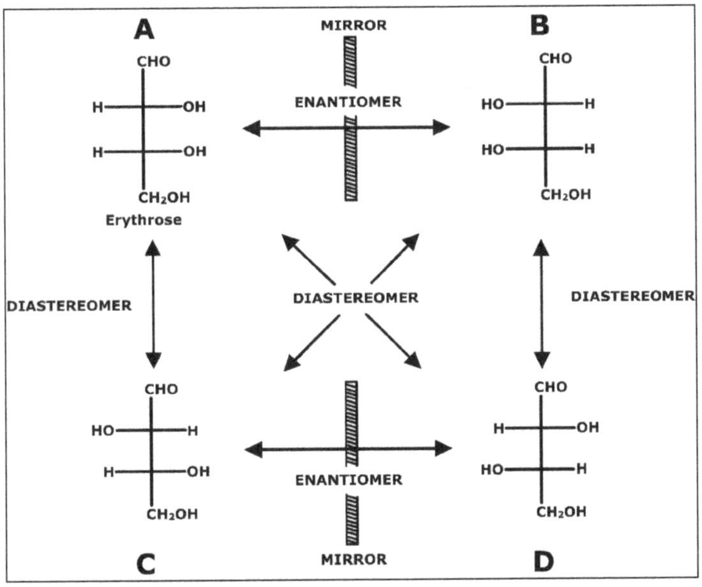

Relationship between four stereoisomers of erythrose (molecule with two stereocenters).

A molecule with one stereogenic center will always be chiral. However, this is not necessarily the case with a molecule having multiple stereocenters. A molecule with two or more stereocenters which is superimposable on its mirror image is known as meso compounds. This is due to presence of plane of symmetry in the molecule. Thus they are optically inactive. To understand this concept more clearly let us take the example of tartaric acid. Here you can see that A & B are non-superimposable mirror image of tartaric acid and therefore are enantiomers, whereas, C & D are identical due to the presence of plane of symmetry and therefore is optically inactive form i.e. meso form. So tartaric acid has just three stereoisomers. For molecules having plane of symmetry, number of meso compounds can be calculated using following formula:

- For 'n' odd number of stereocenters:
    - No. of meso forms = $2^{\frac{n}{2}-\frac{1}{2}}$.
    - No. of optically active forms = 2^{n-1}.
    - Total no. of stereoisomers = $2^{n-1}+2^{\frac{n}{2}-\frac{1}{2}}$.
- For 'n' even number of stereocenters:
    - No. of meso forms = $2^{\frac{n}{2}-1}$.
    - No. of optically active forms = 2^{n-1}.
    - Total no. of stereoisomers = $2^{n-1}+2^{\frac{n}{2}-\frac{1}{2}}$.

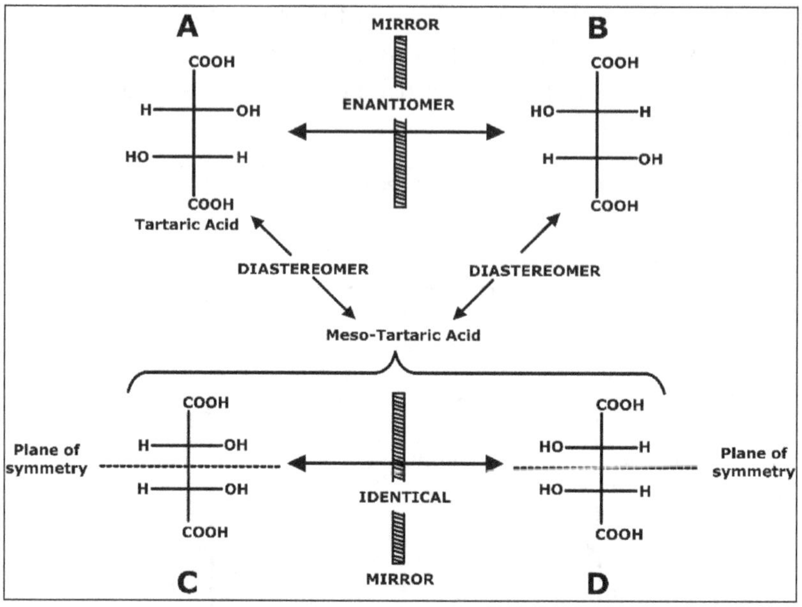

Representation of stereoisomers of tartaric acid.

Optical Activity

Optical activity is an ability of a chiral molecule to rotate the plane polarized light passing through it in a clockwise or anti-clockwise direction. Enantiomers will rotate the plane of polarisation in exactly equal amounts (same magnitude) but in opposite directions. An enantiomer that rotates plane-polarized light in clockwise direction is called dextrorotary [(+), or d-], while the enantiomer that rotates the light in anti-clockwise direction is called levorotary [(-), or l-]. When both d- and l- isomers are present in equal amounts, the mixture is called a racemic mixture. Achiral compounds and racemic mixture do not show optical activity.

Optical activity is measured using a device called a polarimeter. The observed rotation 'α' measured in degrees by polarimeter and is influenced by several factors like the distance travelled by light through the sample i.e. pathlength 'l', wavelength 'λ' of the light passing through the sample, temperature, the concentration 'c' of the sample etc. The standard measurement for rotation for a particular compound is called the specific rotation and is denoted by [α]λ. It is defined as an angle of rotation of plane polarized light measured at a path length of 1 decimeter and a concentration of 1g/ml,

$$[\alpha]_{\lambda}^{T} = \frac{\alpha}{l \cdot c}$$

where,

- [α] is specific rotation in degrees $cm^3\ dm^{-1}\ g^{-1}$.
- T is temperature.
- λ is wavelength of plane polarized light.
- α is measured rotation in degrees.
- l is pathlength in decimeter.

- c is concentration of sample in g/ml.

Optical Purity or Enantiomeric Excess

For sample which is not optically pure (containing only one enantiomer) but consists of a mixture of enantiomers, a part of the effect of enantiomer present in major amount is cancelled by the enantiomer present in minor amount. We can easily calculate the ratio of both the enantiomers present in such mixtures by using the concept of enantiomeric excess and optical purity. Optical purity is numerically equal to enantiomeric excess for chemically pure sample. From the observed specific rotation, we can calculate the optical purity as follows:

$$\text{Optical Purity} = \frac{\text{Specific Rotation of Mixture}}{\text{Specific Rotation of Pure Enantiomer}} \times 100$$

The excess of one enantiomer over the other in a mixture of enantiomers is termed as enantiomeric excess. It can be expressed mathematically as follows:

$$\text{Enantiomeric excess} = \%\text{ of Major Enantiomer} - \%\text{ of Minor Enantiomer}$$

Enantiomeric excess can also be calculated from the concentration of major enantiomer and minor enantiomer,

$$\text{Enantiomeric excess} = \frac{[\text{Major Enantiomer}] - [\text{Minor Enantiomer}]}{[\text{Major Enantiomer}] + [\text{Minor Enantiomer}]} \times 100$$

Where,

$$[\text{Major Enantiomer}] = \text{Concentration of major enantiomer}$$

$$[\text{Minor Enantiomer}] = \text{Concentration of minor enantiomer}$$

Configuration of Chiral Molecules

Specification of the four different groups present on a stereogenic center is termed as configuration. The precise arrangement of substituents at a stereogenic center is known as the absolute configuration of the molecule. However, prior to 1951, when it was not possible to assign absolute configuration, the arrangement of atoms in thousands of optically active compounds was determined relative to one another on the basis of chemical interconversions. This method of specifying configuration is known as relative configuration.

Relative Configuration – D and L System

In carbohydrate chemistry, configuration is assigned to various monosaccharides with respect to glyceraldehyde since they can be synthesized from one of the enantiomeric form of glyceraldehyde by a complex series of reactions. If in the fischer projection of glyceraldehyde, the -OH is present on the right hand side, it is assigned D-configuration and if -OH is present on the left hand side, it is assigned L-configuration. It should be noted that D/L configuration is independent of dextro- and laevo-designation.

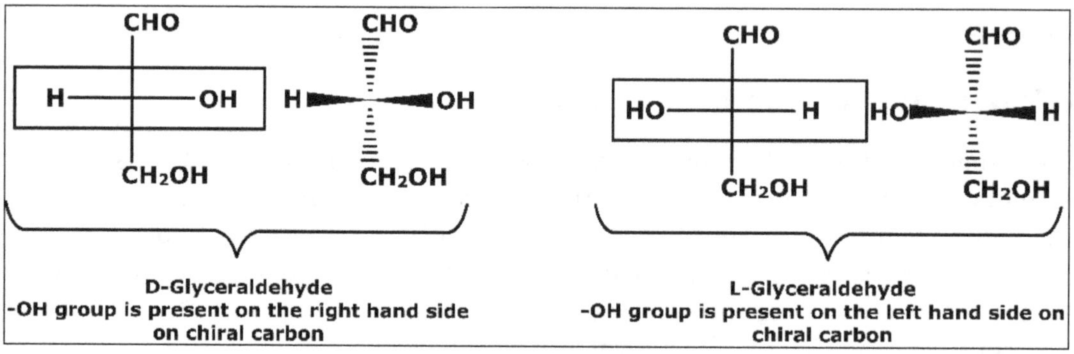

Enantiomers of glyceraldehyde.

So for monosaccharides, position of -OH group on the chiral carbon farthest from the anomeric carbon will determine its relative configuration. In figure, enantiomers of glucose are illustrated. When -OH group on the chiral carbon in red box (it is farthest from anomeric carbon) is present on right, it is assigned D-Glucose and when -OH group is present on left, then as L-Glucose. This system is not used much these days except for certain classes of compounds like carbohydrates and amino acids.

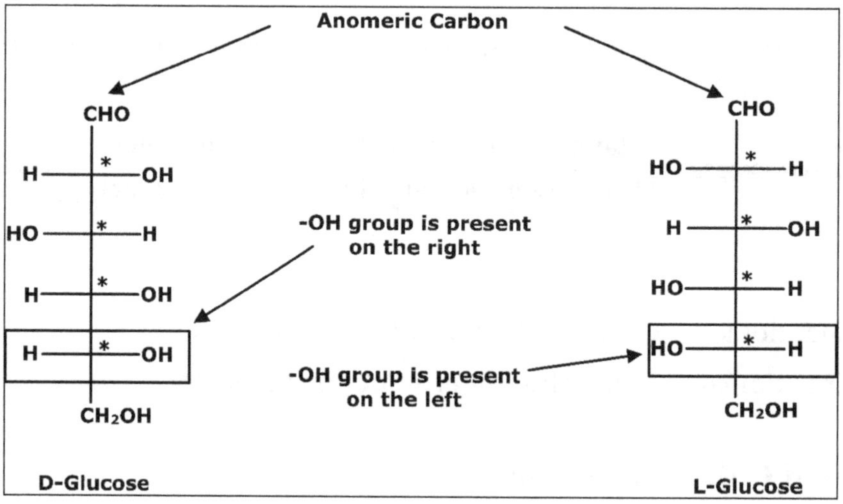

Enantiomers of glucose.

Absolute Configuration – R and S System (Cahn-Ingold-Prelog System)

Absolute configuration is assigned according to a set of rules devised by three European chemists: R. S. Cahn, C. K. Ingold and V. Prelog and is also often called the Cahn-Ingold-Prelog system or CIP system. The rules given by them are known as sequence rules and they assign the nomenclature of stereoisomers by prefixing the name by R (from the Latin word 'rectus' meaning right) or S (from the Latin word 'sinister' meaning left) depending upon the configuration of chiral center. The rules for CIP system of stereochemical nomenclature are given below:

1. Assign the priorities to the four substituents bonded to the chiral carbon according to following rules:

- Higher is the atomic number of atom directly attached to chiral center, higher will be the priority e.g. I > Br > Cl > S > F > O > N > C > H.

- When dealing with isotopes, higher is the atomic mass of isotope, higher will be the priority e.g. ^3H (T) > ^2H (D) > ^1H.
- When we can't decide the priority on the basis of atom directly attached to the chiral center, then the groups are prioritize on the basis of next set of atoms attached to the atom directly attached to the chiral center. Next set of atoms are evaluated until the first point of difference is obtained. e.g.

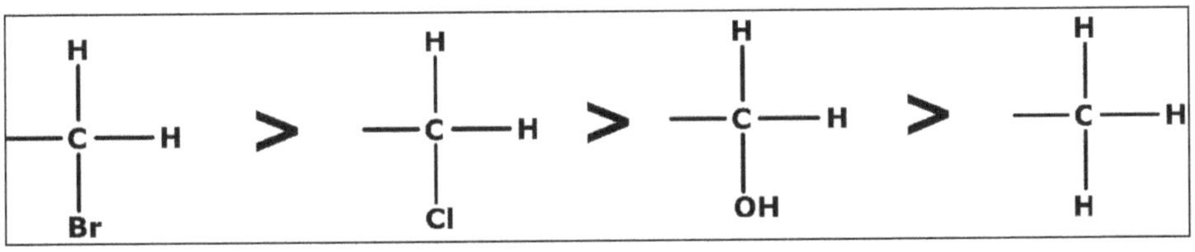

- Multiply bonded atoms in substituents are considered to be singly bonded to an equivalent number of similar atoms. e.g.

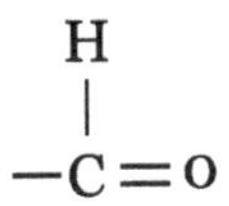 becomes

—C≡N becomes

2. Orient the molecule in such a way that lowest priority substituent points away from you.

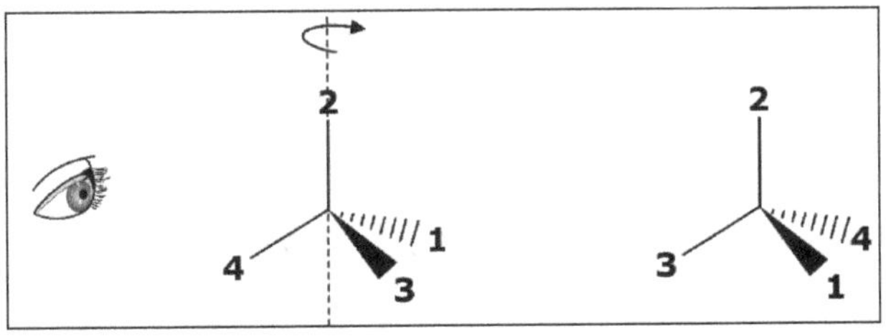

In case of Fischer projection formula of a compound, the lowest priority group is always kept in the top or bottom position, that is, on the vertical line to find the absolute configuration of the stereocenter. If the lowest priority group is not on the vertical line all we have to do is to exchange any two substituents bonded to chiral center twice, such that lowest priority group is on a vertical line. It should be noted that a single exchange between any two groups will result in opposite configuration.

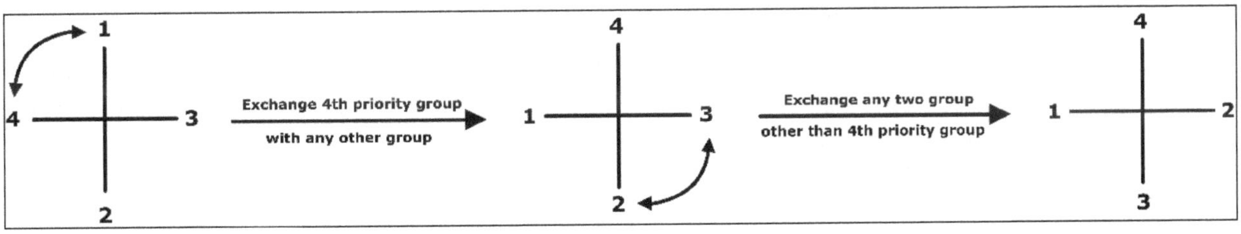

3. Trace the direction of curved arrow starting from highest priority to second highest priority to the third highest priority substituent. (i.e. 1 → 2 → 3). If the direction of curved arrow is clockwise, then the chiral center has 'R' configuration and if it is anti-clockwise, then the chiral center has 'S' configuration.

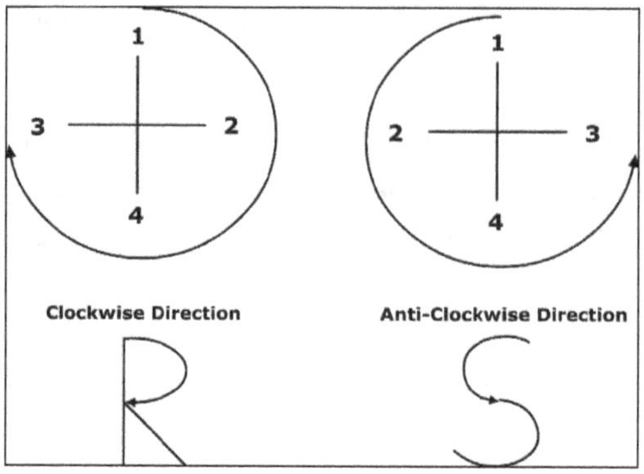

Group and its Characteristics

A group is a set/collection of elements, which are combined with certain operation, such that:

- The group contains an identity.
- The group contains inverses.
- The operation is associative.
- The group is closed under the operation.

In mathematical sense a group may be defined as, "collection/set of elements/numbers having certain properties in common i.e. these elements are bound by certain conditions, known as group properties/postulates". The elements do not need to have some physical significance.

The number of elements h is called the order of the group. For our purposes, the elements are the symmetry operations of a molecule and the rule for combining them is the sequential application of symmetry operations investigated in the previous section. The elements of the group and the rule for combining them must satisfy the following criteria:

- The group must include the identity E, for which $Eg_i = g_iE = g_i$ for all the elements of the group.
- The elements must satisfy the group property that the combination of any pair of elements is also an element of the group.
- Each element gi must have an inverse g_i^{-1}, which is also an element of the group, such that $g_ig_i^{-1} = g_i^{-1}g_i = E$ (e.g. in $C3_v$ the inverse of $C3^+$ is $C3^-$, the inverse of σv is σv; the inverse g_i^{-1} 'undoes' the effect of the symmetry operation g_i).
- The rule of combination must be associative i.e. $g_i(g_jg_k) = (g_ig_j)g_k$

The above definition does not require the elements to commute (which would require $g_ig_k = g_kg_i$). In many groups the outcome of consecutive application of two symmetry operations depends on the order in which the operations are applied. Groups for which the elements do not commute are called non-Abelian groups; those for which they elements do commute are Abelian.

Group theory is an important area in mathematics, and luckily for chemists the mathematicians have already done most of the work for us. Along with the formal definition of a group comes a comprehensive mathematical framework that allows us to carry out a rigorous treatment of symmetry in molecular systems and learn about its consequences. Many problems involving operators or operations (such as those found in quantum mechanics or group theory) may be reformulated in terms of matrices. Any of you who have come across transformation matrices before will know that symmetry operations such as rotations and reflections may be represented by matrices. It turns out that the set of matrices representing the symmetry operations in a group obey all the conditions laid out above in the mathematical definition of a group, and using matrix representations of symmetry operations simplifies carrying out calculations in group theory.

Basic Properties of a Group

There are four basic properties/conditions/postulates/characteristics of a group. The elements belonging to the group should follow/adhere to these conditions in a true sense. Let us state these conditions and explain these by taking suitable examples. Let the elements of the group be [A, B, C, D, E----X-]. The four conditions are:

- In the collection (i.e. group) there must an element such when it combines/multiplies with each and every other elements of the group, it leaves them unchanged. This element is the identity element symbolized as 'E' and this property can be expressed as: AE=A, BE=B, CE=C, DE=D and so on. Further this type of combination with E is commutative (order of combination/product is immaterial) i.e. when order of combination is reversed the results, EA=A, EB=B, EC=C, ED=D are also true. Here mode of combination may be multiplication, addition, subtraction in mathematical sense or in symmetry sense one symmetry operation followed by another symmetry operation. There is only one identity element for every group.
- Each and every element in the group must have an inverse, say (X), which is also the member of the group i.e. AX=XA=E. Here, X is the inverse of A and A is the inverse of X i.e. AX-=XA=E or $A=X^{-1}$, $A^{-1}=X$ i.e. $AA^{-1}=E$. Inverses are unique. It is to be noted that there exists only one identity for every single element in the group but each element in the group has a different inverse.
- Although the combination of two elements may or may not be commutative, but it must be associative. In group {A,B,C,D----X---} following relations are valid A(BC)=(AB)C=ABC. Provided the order of combination is not changed. This associative property can be extended to any number of elements of the group. This associative result of elements must be an element of the group.
- Closure: Results of combination (multiplication, addition, subtraction) of two or more elements or square of the element (the element is combined with itself) must be equivalent to an element of the group, which is also the member of the group. The group is closed under the given combination.

In group [A, B, C, D---X-] these type of combinations and their results may be AB=C, AC=D, BC= another element of the group only. It is not necessary that combination AB= combination BA i.e. combination may be commutative or non-commutative. Order of combination matters in a group.

This order of combination is very significant. This is true in case of symmetry product of symmetry operations.

Examples of Group and Verification of the Characteristics

Example: Show that a group [0] and mode of combination as addition constitutes a group. Group contains a single element 0 in it. There should be an element in the group such that when it combines with each and every element of the group leaves it unchanged. Since group contains only 0 as the element, such element will be 0 itself as 0+0=0.

Element 0 is unchanged. Each element in the group must have its unique inverse. Here inverse of 0 is 0 itself, as $0+0^{-1}=0$. Other properties are easy to verify as there is a single element in the group.

Example: Show that the integers [------ -3,-2,-1, 0,+1,+2,+3-----] under the mode of combination as addition and zero as the identity element constitutes a group of infinite order. Here, the identity element is 0. You add 0 to any element of the group and it will remain unchanged. 3+0=3, 5+0=5, -6+0=-6 and so on. Addition of 0 to the right or left to the number does not matter, i.e. 4+0=4 or 0+4+4. According to previous property each element in the group has its unique inverse. Let inverse of integer n is n^{-1}. Since the mode of combination is addition in this example the inverse of +n will -n as (n)+(-n)=0(identity). Element when combined with its inverse gives identity. For example -3+3=0, (+4)+(-4)=0 and so on.

You can combine any number of elements the net result will be an element of the group. Order of combination of elements should not be changed. In this example changing of order does not matter. But when we deal with symmetry elements of the group it matters a lot. For example (-3)+(-2)+(-1) = [-3+(-2)]+(-1) = (-3) + [(-2) +(-1)] = -6. Identity is zero. A combination of two or more elements or square of the element (addition of number to itself) must give a number that belongs to group.

- -2+(-1) = -3, and -3 belongs to group (because it is of infinite order).
- -2+-2 = -4 (square of the element),-4 belongs to group.
- -2 +-3 +3 = -2 (more elements combined).-2 belongs to group.

This group under the mode of combination as multiplication does not constitute a group as the identity element 0 when combined (here multiplied) with other elements, changes them i.e. 4x0=0, thus 4 changes to 0. The set of natural numbers 0, 1, 2, 3,----- under addition is not a group as there is no inverse of the elements except that of zero.

Example: Show that the group [-1,1] under the mode of multiplication constitutes a group. First let us find the identity element in the group. There are three possibilities -1 is the inverse, +1 is the inverse or there is no inverse. Let us take the following products: 1x1=1, -1x1=-1, -1x-1=1, number 1 does not change 1 or -1 when these are combined (here multiplication) with 1.-1 is not the identity as it change the element when it is combined with other element. Now we need to find the inverse of the element of the group and it should be such that 1x1-1=1, -1x-1-1=1.

Here, 1x1=1 so 1 is the inverse of 1 and -1x-1=1 so the inverse of -1 is -1 itself as these products give identity 1 as the element. Since in the group there are only two elements you combine in any way you will get the same result and net result will be an element of the group. 1x1=1, -1x-1=1,

-1x1=-1(element is combined with itself or square of the element) or 1x-1=-1 etc. The group [-1, 1] under the mode of addition does not constitute a group. As 1+-1=0 and 0 is not the element of the group and so on you can verify other properties.

Example: Let us explain the group properties by taking an example of face movements of a soldier in exercise drill.

L stands for Left Turn command R stands for Right Turn command A stands for About Turn command E means no command or stand at ease	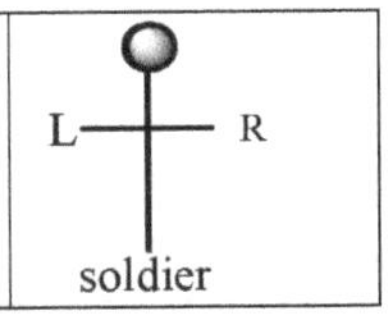

These four face movements L,R,A,E of soldier constitute a group with E (stand at ease) as identity and mode of combination is one face movement on command followed by the other face movement on command i.e. these four face movements on command constitute a group [L,R,A,E]. Let us now verify the four properties of this group. There must be an element into the group which on combination with other element does not change the element. This is E here i.e. no command is given to the soldier AE=A, means E movement followed A (about turn) movement i.e. soldier follows the face movements in the order as; stand at ease followed by about turn by 180°. This combined movement is equivalent to A. Also EA=A .Similarly, LE=EL=L and RE=ER=R etc. Here order of face movement matters. The product or combination of any two or more elements or square of an element must be an element of the group.

$$\overleftarrow{LR}=E \quad \overleftarrow{AR}=L \quad \overleftarrow{AA}=E \quad \overleftarrow{LAR}=A$$

Direction of arrow indicates the order of face movement to be carried i.e. move from right to left and must be followed in all cases. Results of some face movements are shown in Figure.

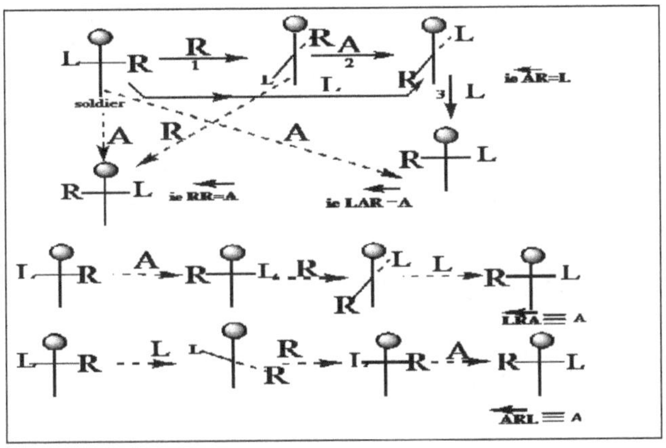

Theorem of Reciprocals

In a group the reciprocal of the product of two or more than two elements is equal to the product of the reciprocals of elements taken in reverse order. Let the group be [A, B, C, D, E, ---------X]. Then according to the theorem reciprocals,

$$(ABCD----------)^{-1}=(D^{-1}C^{-1}B^{-1}A^{-1}--------)$$

Let us prove this. If A, B, C, D,---- are the elements of the group then the product (ABCD) must be an element of the group and (ABCD) product is equivalent another element X of the group such that, (ABCD)=X.

Right multiply both sides with ($D^{-1}C^{-1}B^{-1}A^{-1}$ --------).

$$(ABCD)(D^{-1}C^{-1}B^{-1}A^{-1}--------)=X(D^{-1}C^{-1}B^{-1}A^{-1}--------)$$

Taking help of law of associative multiplication we can work out the left hand side of this relation as:

$$(ABC)(D)(D^{-1})(C^{-1}B^{-1}A^{-1}--------)=X(D^{-1}C^{-1}B^{-1}A^{-1}--------)$$

$$(ABC)(DD^{-1})(C^{-1}B^{-1}A^{-1}--------)=X(D^{-1}C^{-1}B^{-1}A^{-1}--------)$$

$$(ABC)(E)(C^{-1}B^{-1}A^{-1}--------)=X(D^{-1}C^{-1}B^{-1}A^{-1}--------)$$

Element and its reciprocal when combined gives identity E and identity E when combined with element leaves it unchanged,

$$(ABC)(E)(C^{-1}B^{-1}A^{-1}--------)=X(D^{-1}C^{-1}B^{-1}A^{-1}--------)$$

$$(ABC)(C^{-1}B^{-1}A^{-1}--------)=X(D^{-1}C^{-1}B^{-1}A^{-1}--------)$$

$$(AB)(CC^{-1})(B^{-1}A^{-1}--------)=X(D^{-1}C^{-1}B^{-1}A^{-1}--------)$$

$$(AB)(E)(B^{-1}A^{-1}--------)=X(D^{-1}C^{-1}B^{-1}A^{-1}--------)$$

$$(AB)(B^{-1}A^{-1}--------)=X(D^{-1}C^{-1}B^{-1}A^{-1}--------)$$

$$(A)(B\,B^{-1})(A^{-1})--------)=X(D^{-1}C^{-1}B^{-1}A^{-1}--------)$$

$$(A)(E)(A^{-1})--------)=X(D^{-1}C^{-1}B^{-1}A^{-1}--------)$$

$$(A)(A^{-1})--------)=X(D^{-1}C^{-1}B^{-1}A^{-1}--------)$$

$$(E)=X(D^{-1}C^{-1}B^{-1}A^{-1}--------)$$

$$E=X(D^{-1}C^{-1}B^{-1}A^{-1}--------)$$

Since X combined with ($D^{-1}C^{-1}B^{-1}A^{-1}$ --------) gives identity E, therefore, X and ($D^{-1}C^{-1}B^{-1}A^{-1}$ --------) are inverse to each other or,

$$X^{-1}=(D^{-1}C^{-1}B^{-1}A^{-1}--------)$$

and X=ABCD-----as defined earlier. Putting the value of X one gets,

$$(ABCD----)^{-1} = (D^{-1}C^{-1}B^{-1}A^{-1}--------)$$

Hence the theorem is proved.

Some Common Types of Groups

Abelian Group

In case of integers and addition as the mode of combination the order of combination does not matter but in symmetry operations as group elements it matters a lot. An abelian group is that group in which all elements when combine in either way give the same result. In a group [A, B, C, D----] all combination products, AB=BA, AC=CA, BC=CB-------- are true i.e. the elements commute with each other. The integers [------ -3, -2, -1, 0, +1, +2, +3-----] under the mode of combination as addition and zero as the identity element constitutes an abelian group of infinite order.

Cyclic Group

In this group the elements are generated by taking the power of one group element. Let the element be X that generates all other group elements. Then the group elements are X^1, X^2, X^3, X^4 ---------X^n=E. There are n elements in the group and thus the order of the group is n. Cyclic group is always an abelian group as powers are additive and their order of combination does not matter.

$$X^1 + X^2 = X^2 + X^1 = X^3$$

Finite and Infinite Groups

The group which has infinite number of elements in it is called infinite group. All positive and negative integers with zero as the identity and addition as the mode of combination constitute an infinite order group i.e. [------4, 3, 2, 1 0,-1,-1,-3,-4-------] with addition as mode is a n infinite group. In finite group the number of elements is finite. The group (1,-1, i,-i) is a finite group of order four.

Total Symmetry Elements and Total Operations Generated

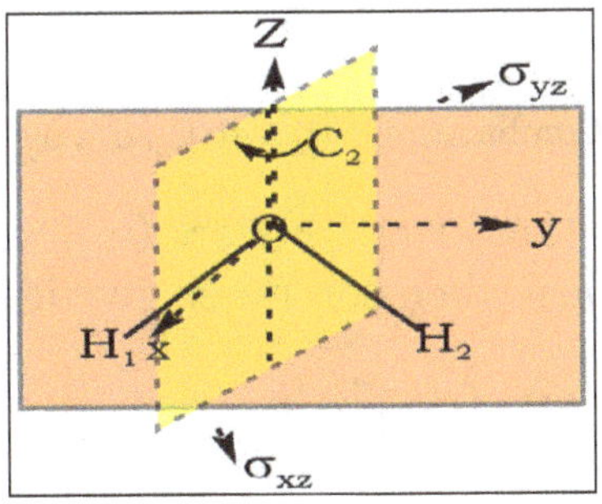

Various symmetry elements in H_2O molecule.

Let us take the example of water molecule and find total symmetry elements in it and total symmetry operations generated by these symmetry elements. Figure shows the various symmetry elements in structure of water molecule.

There are four symmetry elements σ_{xz}, σ_{yz}, C_2 and E in H_2O molecule. Total symmetry elements are four. Let us find the symmetry operations generated by each of these:

- E generates only one operation and it is E.
- σ_{xz} generates σ_{xz}^1 and σ_{xz}^2 = E symmetry operations and these are σ_{xz} and E only.
- σ_{yz} generates σ_{yz}^1 and σ_{yz}^2 = E symmetry operations and these are σ_{yz} and E only.
- C_2 generates C_2^1 and C_2^2 = E symmetry operations and these are C_2 and E only.

Thus total symmetry operations generated here: σ_{xz}, σ_{yz}, C_2 and E in H_2O i.e. only four symmetry operations are generated by four symmetry elements. In case of H_2O molecule number of symmetry elements and symmetry operations are same. It is not same always. These four symmetry operations constitute a group ie [σ_{xz}, σ_{yz}, C_2, E] is a group of order four. Number of elements in a group is its order. Now let us verify the four group properties:

1. Here E is the identity, EE=E, $EC_2=C_2E=C_2$; $E\sigma_{xz}=\sigma_{xz}E=\sigma_{xz}$; $E\sigma_{yz}=\sigma_{yz}E=\sigma_{yz}$, thus group element remains unchanged. Figure below shows all these results. Symmetry operations are always carried out from right to left.

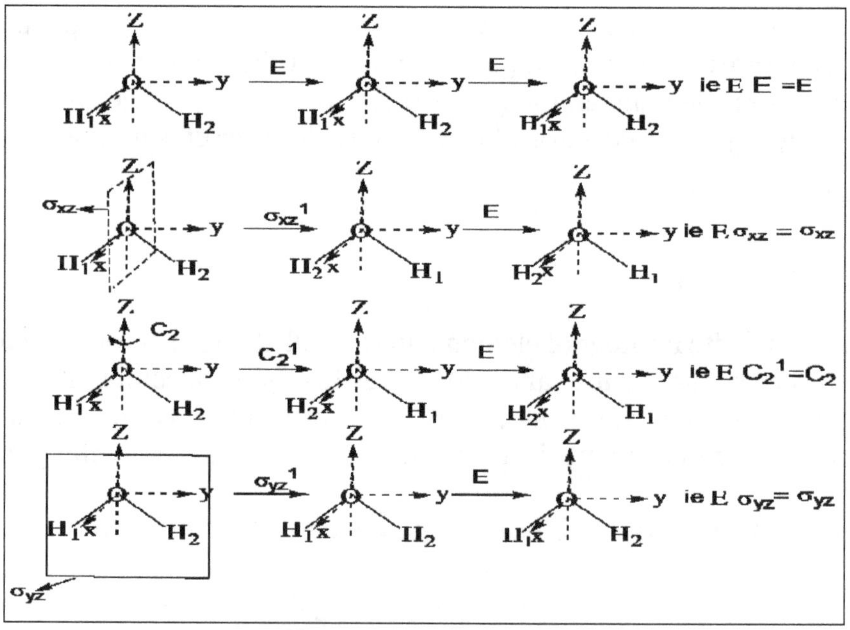

Combination of E with C_2, σ_{xz}, σ_{yz}.

2. Every element has its reciprocal (i.e. inverse): Here $(\sigma_{xz})^{-1}$ is (σ_{xz}); $(\sigma_{yz})^{-1} = \sigma_{yz}$ and $(C_2)^{-1}$ is C_2^1. Symmetry operation and its reciprocal when combined give identity i.e.

$$\sigma_{xz}\left(\sigma_{xz}\right)^{-1}=E;\ \sigma_{yz}\left(\sigma_{yz}\right)^{-1}=E;\ C_2^1\left(C_2^1\right)^{-1}=E$$

In Figure below the symmetry product $C_2.C_2$ =E, it means C_2 is its own inverse i.e. $(C_2)^{-1}=C_2$. Similarly σ_{xz}=E, therefore, σ_{xz} is the inverse of itself and similarly inverse of σ_{yz} is σ_{yz}. Thus in the group each element has a unique inverse. Here each and every group element (symmetry operation) is the inverse of itself.

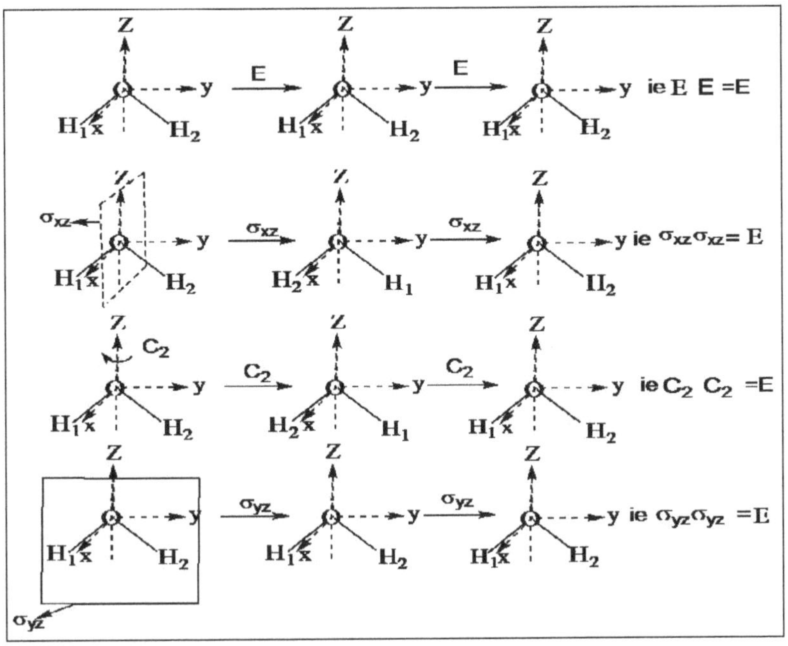

Combination of C_2 with C_2, σ_{xz} with σ_{xz}, σ_{yz} with σ_{yz}.

3. Law of associative multiplication (order of multiplication should not be changed). $(C_2^1\,\sigma_{xz})\,\sigma_{yz} = C_2^1\,(\sigma_{xz}\,\sigma_{yz})$. These products are shown in the Figure.

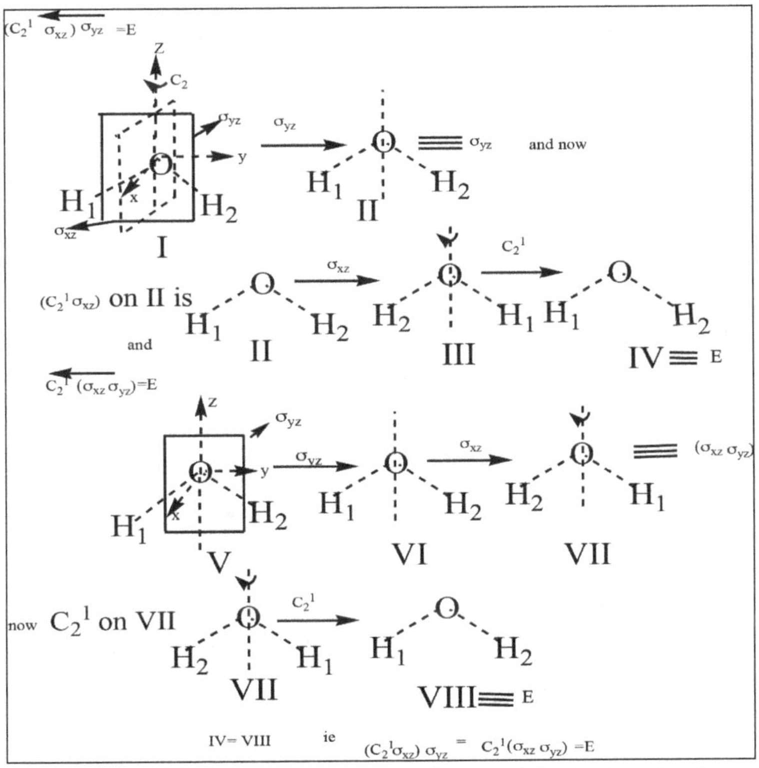

Associative products $(C_2\,\sigma_{xz})\,\sigma_{yz} = C_2^1\,(\sigma_{xz}\,\sigma_{yz})$ for H_2O molecule.

4. Product of two or more elements or square of the element must be an element of the group. The combinations of various symmetry operations are shown in figure.

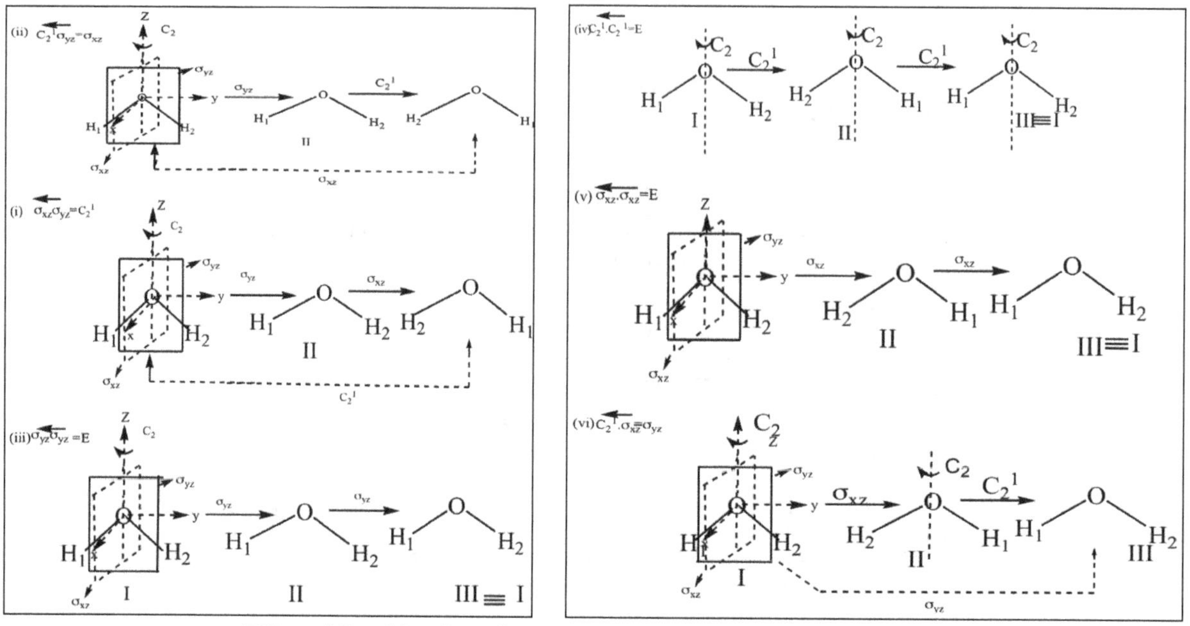

Effect of Combination of Symmetry Operations on H_2O Molecule.

Thus all the four properties of a group [σ_{xz}, σ_{yz}, C_2, E] have been verified. Let us take another example of C_3 axis in an equilateral triangle and see that operations generated by C_3 axis constitute a group. C_3 axis generates the following symmetry operations: C_3^1, C_3^2, C_3^3 = E i.e. in all three symmetry operations are generated. Various symmetry products are shown in figure below.

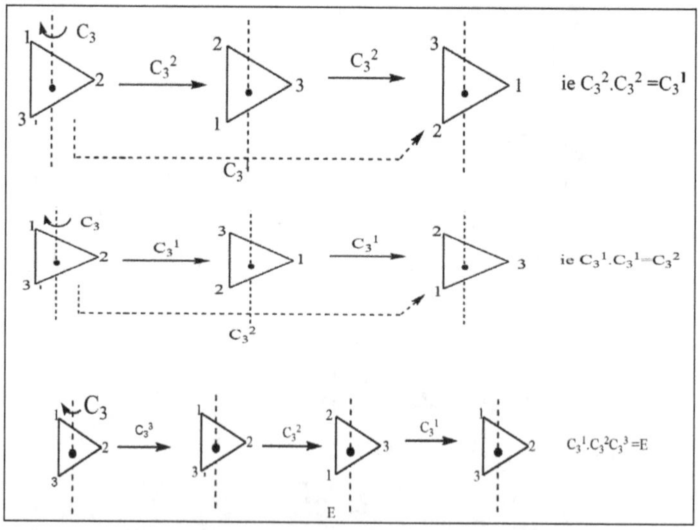

Various symmetry products of Total symmetry elements present in the given molecule symmetry operations C_3^1, C_3^2, C_3^3.

Similarly other symmetry products can be worked out and all the four properties of the group can be verified. Let us now find the following products or inverses.

$$C_3^1 . C_3^2 = ? \ ; \ \left(C_3^1\right)^{-1} = ?$$

$$C_3^2 . C_3^2 = ? \ C_3^1 . C_3^1 = ? \ ; \ \left(C_3^1\right)\left[\left(C_3^2\right)\left(C_3^3\right)\right] = \left[\left(C_3^1\right)\left(C_3^2\right)\right]\left(C_3^3\right)$$

Inverse of $C_3^1)^{-1} = C_3^2$ so now show that $C_3^1.C_3^2 = E$.

Subgroup

It is defined as a part of the main group which satisfies all the properties of the group i.e. within the main group of order' h' there may be smaller groups of lower order then 'h'.

- Sub group always contains identity element E.
- E itself constitutes a group of order one and it is trival.
- The sub group of order two also contains identity element E.
- Groups of order two can be E,A; E,B; E,C etc. in the main group A,B,C,E.
- Sub groups always follow the group conditions of the main group.
- Order of the sub group is an integral factor of the order of the main group.
- If the order of sub group is' g' then h/g is an integer and not in fraction.
- If the order of the main group is six, then sub groups can have orders 1,2,3 only(6/6=1,6/2=3,6/3=2) only and 6/4 or 6/5 are fractions.

Let us now try to find some sub groups in main group by taking suitable examples. [1,-1,I,-I] is a group of order four under the mode of combination as multiplication. Since the order of the group is four, the sub groups order can be 4/2=2,4/4=1. There will be no sub group of order 4/3=1.33. Since identity always constitutes a sub group order one. Here, identity is 1. So a sub group of order one is [1] .It satisfies all four group properties. The other sub groups of order two in it can be [1,-1], [I,-I], [-1,i] [1,I], [1,-I], [-1,-I]. In these groups of order two, only group [1-1] satisfies all four group properties under the mode of multiplication and 1 as the identity. Here we write some products of this sub group as: 1x1=1, 1x-1=-1, -1x-1=1. Multiplication with 1 does not change the number. Number multiplied with itself also gives the element belonging to the sub group. Thus [1,-1] constitutes a sub group of order two and there are only two sub groups [1] of order 1 and [1,-1] of order two in the main group [1,-1,I,-i] of order four.

Let us take another example of face movements of a soldier during exercise. Here L stands for left turn, R stands for right turn, A stands for about turn and E stands for stand at ease or do nothing or stand still. Thus main group is [L, R, A, E].

<table>
<tr><td>L stands for Left Turn command
R stands for Right Turn command
A stands for About Turn command
E means no command or stand at ease</td><td>
</td></tr>
</table>

Let us find sub groups in the main group [L, R, A, E]. Here E is trivial group of order one and non-trivial sub groups of order of two can be[E,A], [E,L], [E,R] ,[L,R],[A,L],[A,R] (because 4/2= 2 is integer) and so on. Out of [E,A], [E,L], [E,R] sub groups only [E,A]sub group satisfies all the group condition and this is the only subgroup of order two of the main group (L,R,A,E). EA=AE=A, AA=E, AAA=A, $(A)^{-1}$=A and E^{-1} is E and so on.

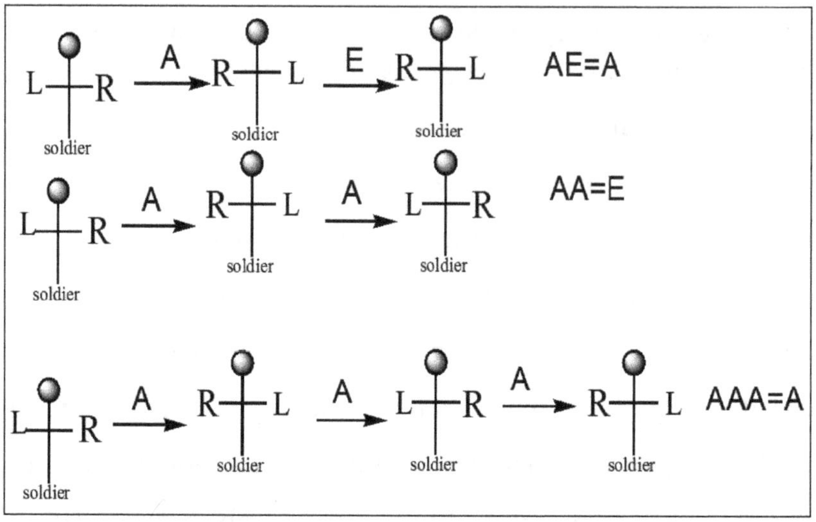

The remaining groups of order of order two [E,L], [E,R], [L,R], [A,L], [A,R] etc. do not constitute groups To verify that the [E,L] and [E,R] etc. do not constitute sub groups is left an exercise for the reader. All sub groups of order g (h/g=integer) need not be sub groups in true sense and that may not follow the conditions of the group.

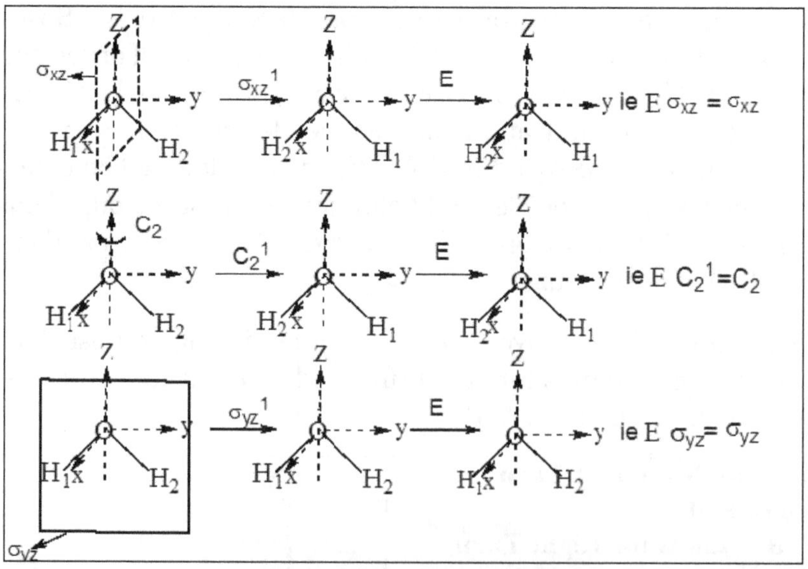

Combination of E with C_2, σ_{xz} and σ_{yz}.

Symmetry operations $\sigma_x z$, σ_{yz}, C_2 and E in H_2O molecules constitute a group. Let us now try to find the sub group in the group [σ_{xz}, σ_{yz}, C_2, E]. E is always a sub group of order one. The other sub groups of order two may be [E, C_2], [E, σ_{yz}], [E, σ_{xz}], [σ_{xz}, σ_{yz}], [C_2, σ_{yz}], [C_2, σ_{xz}] and so on. Here [E, C_2], [E, σ_{yz}] and [E, σ_{xz}] are the sub groups of order two and [σ_{xz}, σ_{yz}], [C_2, σ_{yz}] and [C_2, σ_{xz}] are not the sub groups of order two. Let us see some of the products by taking example of water molecule.

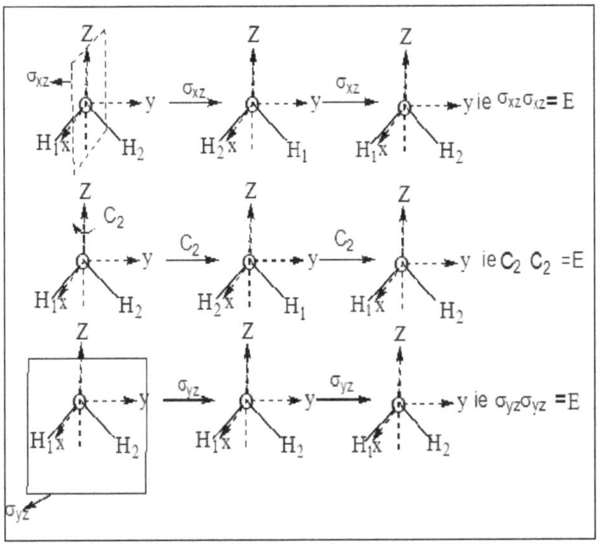

Combination of C_2 with C_2, σ_{xz}, with σ_{xz} and σ_{yz} and σ_{yz}.

These results clearly show that E, [E, C_2], [E, σ_{yz}] and [E, σ_{xz}] are sub groups of the main group [σ_{xz}, σ_{yz}, C_2, E].

The elements of a subgroup should obey the following conditions:

If g is the order of the group & s is the order of the subgroup, then g/s is a natural number. Example: Water molecule has symmetry elements- E, C_2,σ_v, σ_v^1.

Group: E,C_2,σ_v, σ_v^1.

Subgroups: E

E,C_2

E,σ_v,

E, σ_v^1

Classes: This is the subdivision of a group.

Two elements A & B in a group form a class if they are conjugate to each other. Conjugate elements are related by the equation,

$$X-1AX=B$$

Where X is similarity transformation element .It is used to find whether a set of elements form a class. Example- water molecule has symmetry elements- E,C_2,σ_v, σ_v^1.

Group: E,C_2,σ_v, σ_v^1.

Classes: $E^{-1}C_2E = C_2$

$$\sigma_v^{-1}C_2\sigma_v = C_2$$

$$\sigma_v^{-1}C_2\sigma_v^1 = C_2$$

$$C_2^{-1}C_2C_2 = C_2$$

Order: The order of a class of a group must be an integral factor of the order of a group and the number of elements is called the group order of the group.

Method to Find the Class

- A symmetry operation which commutes with all symmetry operations forms a class. E, σ_h, I belong to separate class.
- Rotation operation & its inverse form a class like C_2^{-1} & C_2.
- Improper axis & inverse forms a class $S_1 S_1^{-1}$.
- Two rotations about different axis form a class if there is a third operation which interchanges the points of the axis.
- Two reflections about different planes belong to the same class if there is a third operation which interchange points on the two planes.

Example: Square Planar AB_4 molecule has Symmetry operations: 16- E, i, σ_h, C_2^1 C_4^1 C_4^3 S_4^1 S_4^3 $4C_2^1$ 4σv. No of Elements: 13.

Classes:

- E, i, σ_h, C_2,
- C_4^1 C_4^3,
- S_4^1 S_4^3,
- 2 C_2^1 operations about C_2 axis (reflection),
- 2 C_2^1 operations about C_2' axis (reflection),
- 2 reflection operations in two σ_v planes,
- 2 reflection operations in two σ'_v planes.

Relation Between Orders of a Finite Group and its Subgroup

If there are a finite number of elements, the group is called a finite group and the number of elements is called the group order of the group.

- A subset of a group that is closed under the group operation and the inverse operation is called a subgroup. Subgroups are also groups, and many commonly encountered groups are in fact special subgroups of some more general larger group.
- A finite group is a group having finite group order. Examples of finite groups are the modulo multiplication groups, point groups, cyclic groups, dihedral groups, symmetric groups, alternating groups, and so on.
- The finite (cyclic) group forms the "Finite Simple Group of Order 2".
- A basic example of a finite group is the symmetric group, which is the group S_n, of permutations (or "under permutation") of n objects.

Class

This is another way of subdividing a group into smaller collection of elements. It is defined as the collection of all conjugate elements of the group. It has different bases for the division of the group. All conjugate elements are related by similarity transformation. The order of a class is the integral factor of the order of the main group.

Conjugacy Relation and Class

A complete set of mutually conjugate group elements. Each element in a group belongs to exactly one class, and the identity element (I = 1) is always in its own class. The conjugacy class orders of all classes must be integral factors of the group order of the group.

- A group of prime order has one class for each element.
- In an Abelian group, each element is in a conjugacy class by itself.
- Two operations belong to the same class when one may be replaced by the other in a new coordinate system which is accessible by a symmetry operation. These sets correspond directly to the sets of equivalent operations.
- Two elements A & B in a group form a class if they are conjugate to each other. Conjugate elements are related by the equation, $X^{-1}AX = B$:
    - Where X is similarity transformation element. It is used to find whether a set of elements form a class.
- Conjugacy is an equivalence relation. Also note that conjugate elements have the same order. The set of all elements conjugate to a is called the class of a.
- To find conjugacy class similarity transformations $X^{-1}AX = X^{-1}(AX)$ on A. Applying a similarity transformation gives:
    - $A^{-1}DA = E$
    - $B^{-1}DB = E$
    - $C^{-1}DC = E$
    - $D^{-1}DD = D$
    - $E^{-1}DE = D$

  So {D, E} form a conjugacy class.

Point Symmetry Groups

Each molecule has a set of symmetry operations that describes the molecule's overall symmetry. This set of operations defines the point group of the molecule. Since all the elements of symmetry present in the molecule intersect at a common point & this point remains fixed under all symmetry operations of the molecule and is known as point symmetry groups.

Schoenflies Notation

The point groups are denoted by their component symmetries. There are a few standard notations used by crystallographers. The Schoenflies notation or Schönflies notation, named after the German mathematician Arthur Moritz Schoenflies, is one of two conventions commonly used to describe crystallographic point groups. This notation is used in spectroscopy. The other convention is the Hermann-Mauguin notation, also known as the International notation. A point group in the Schoenflies convention is completely adequate to describe the symmetry of a molecule; this is sufficient for spectroscopy. The Hermann-Maunguin notation is able to describe the space group of a crystal lattice, while the Schoenflies notation isn't. Thus the Hermann-Mauguin notation is used in crystallography.

Schönflies Notation

In Schönflies notation, point groups are denoted by a letter symbol with a subscript. The symbols used in crystallography mean the following:

- The letter O (for octahedron) indicates that the group has the symmetry of an octahedron (or cube), with (O_h) or without (O) improper operations.
- The letter T (for tetrahedron) indicates that the group has the symmetry of a tetrahedron. Td includes improper operations, T excludes improper operations, and T_h is T with the addition of an inversion.
- The letter I (for icosahedron) indicates that the group has the symmetry of an icosahedron (or dodecahedron), either with (I_h) or without (I) improper operations.
- C_n (for cyclic) indicates that the group has an n-fold rotation axis. C_{nh} is C_n with the addition of a mirror (reflection) plane perpendicular to the axis of rotation. C_{nv} is C_n with the addition of a mirror plane parallel to the axis of rotation.
- S_n (for Spiegel, German for mirror) denotes a group that contains only an n-fold rotation-reflection axis.
- D_n (for dihedral, or two-sided) indicates that the group has an n-fold rotation axis plus a twofold axis perpendicular to that axis. D_{nh} has, in addition, a mirror plane perpendicular to the n-fold axis. D_{nv} has, in addition to the elements of D_n, mirror planes parallel to the n-fold axis.

Point Groups

- Low Symmetry Groups:

C_1: only E

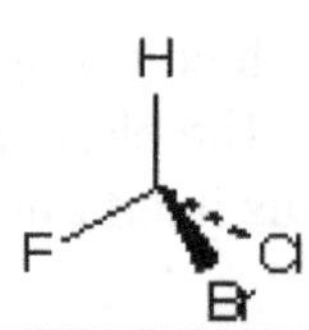

C_s: E and σ only

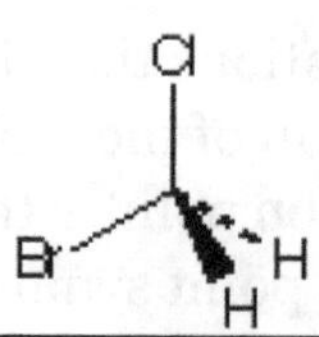

C_i: E and i only

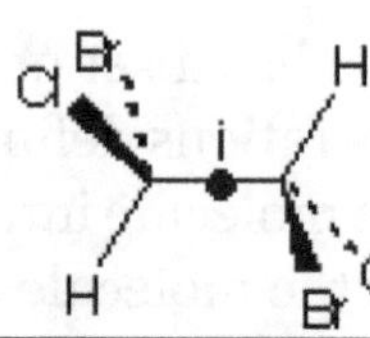

- C_n, C_{nv}, C_{nh} Groups:

C_n: E and C_n only

C_2:

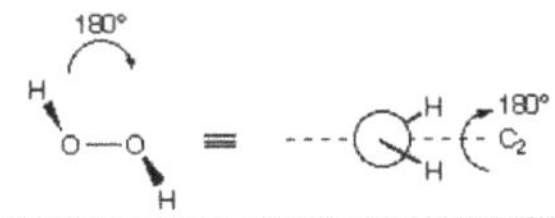

C_3:

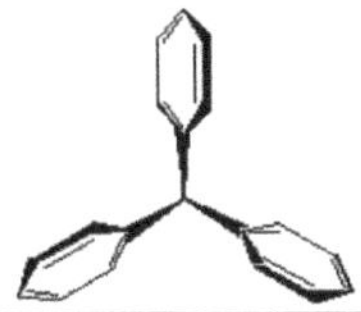

C_{nv}: E and C_n and n σ_v's
C_{2v}: E, C_2, 2 σ_v H_2O C_{3v}: E, C_3, 3 σ_v NH_3 Co_v: E, C, σ_v HF, HCN
C_{nh}: E and C_n and σ_h (and others as well)
C_{2h}: E, C_2, σ_h, I

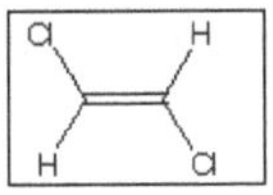

- D_n, D_{nv}, D_{nh} Groups:

D_n: E, C_n, n C_2 axes to C_n

D_3: E, C_3, 3 C_2

$[Co(en)_3]^{3+}$

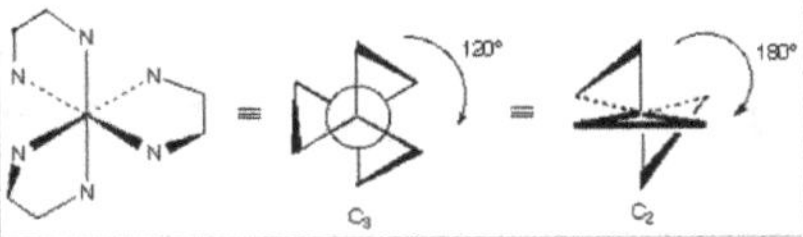

D_{nh}: E, C_n, n C_2 axes , σ_h

D_{3h}: E, C_3, 3 C_2, σ_h

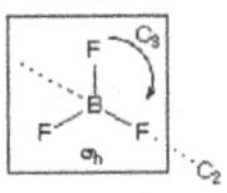

D_{3h}: E, C_3, 3 C_2, σ_h

eclipsed ethane

D_{6h}: E, C_6, 6 C_2, σ_h

D_{h}: E, C , C_2, σ_h

O=C=O

H_2

D_{nd}: E, C_n, n C_2 axes ⊥ to C_n,
D_{3d}: E, C_3, 3 C_2, 3 σ_d staggered ethane

- S_n Group:

S_{2n}: E, C_n, S_{2n} (no mirror planes)
S_4, S_6, S_8, etc. (Note: never S_3, S_5, etc.)
S_4: E, C_2, S_4 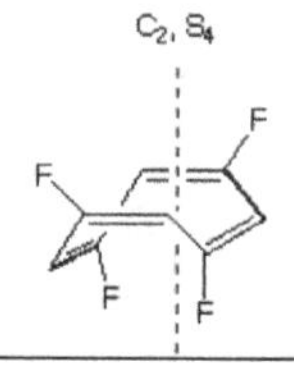

- High Symmetry Cubic Groups, T_d, O_h, I_h:

T_d: E, 8 C_3, 3 C_2, 6 S_4, 6 σ_d
Tetrahedral structures No need to identify all the symmetry elements - simply recognize T_d shape. methane, CH_4
O_h: E, 8 C_3, 6 C_2, 6 C_4, i, 6 S_4, 8 S_6, 3 σ_h, 6 σ_d
Octahedral structures No need to identify all the symmetry elements - simply recognize O_h shape. 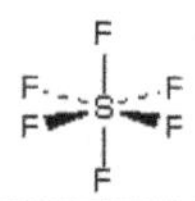
I_h: E, 12 C_5, 20 C_3, 15 C_2, i, 12 S_{10}, 20 S_6, 15 σ Icosahedron
Other rare high symmetry groups are **T**, **T_h**, **O**, and **I**

Method of Determination of Point Group of Molecules

The process used to assign a molecule to a point group is straightforward with a few exceptions. It is a procedure. Here are set of steps to quickly guide you:

- Look at the molecule and see if it seems to be very symmetric or very unsymmetric. If so, it probably belongs to one of the special groups (low symmetry: C_1, C_s, C_i or linear $C_{\infty v}$, $D_{\infty h}$) or high symmetry (T_d, O_h, I_h).
- For all other molecules find the rotation axis with the highest n, the highest order C_n axis of the molecule.
- Does the molecule have any C_2 axes perpendicular to the C_n axis? If it does, there will be n of such C_2 axes, and the molecule is in one of D point groups. If not, it will be in one of C or S point groups.

Similarity Transformation and Conjugate

A and B are two elements in a group, X is any elements in this group. If,

$$X^{-1}AX = B$$

Then we can say the relationship of A and B is similarity transformation. A and B are conjugate. Conjugate elements have three properties.

All the symmetry operations of a molecule as a group can be written in the form of group multiplication table and they obey all the properties of a group. This group is called symmetry point group. It is called point group for two reasons. First reason is that this group has all the properties of a group. Second reason is that all the symmetry operations are related to a fixed point in the molecule, which is not necessarily to be an atom of the molecule. According to the symmetry of molecules, they can be classified as symmetry point groups. To determine the symmetry point group of a molecule is very important, because all symmetry related properties are dependent on the symmetry point group of the molecule. Symmetry point groups can be divided into 5 classes which are summarized below and they are described in details here:

Point groups		Symmetry Elements	Order	Example
Nonaxial	C_1	E	1	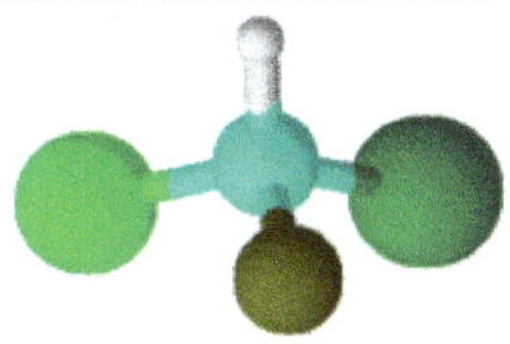 HCFBrCl
	C_i	E, i	2	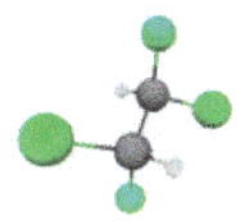 $C_2H_2F_2Cl_2$
	C_s	E, ?	2	 CH_2BrCl
Cyclic	C_n	E, C_n	n	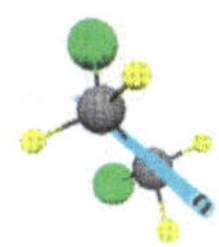 $C_2H_4Cl_2$
	C_{nh}	E, C_n, $n\sigma_v$	2n	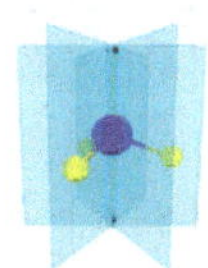 NH_3
	C_{nv}	E, C_n, σ_h, S_n	2n	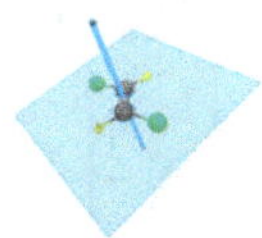 $C_2H_2F_2$

	S_n	E, S_n	n	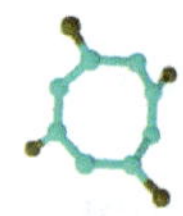 1,3,5,7-tetrafluoracyclooctatetrane
Dihedral	D_n	E, C_n, nC_2 (?C_n)	2n	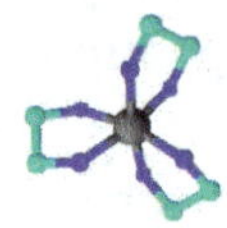 $[Co(en)_3]^{3+}$
	D_{nh}	E, C_n, σ_h, nC_2 (?C_n)	4n	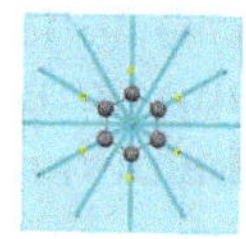 Benzene
	D_{nd}	E, C_n, σ_d, nC_2 (?C_n)	4n	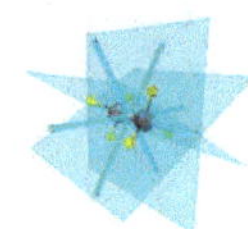 C_2H_6
Polyhedral	T_d	E, $3C_2$, $4C_3$, $3S_4$, $6\sigma_d$	24	 CCl_4
	O_h	E, $3S_4$, $3C_4$, $6C_2$, $4S_6$, $4C_3$, $3\sigma_h$, $6\sigma_d$, i	48	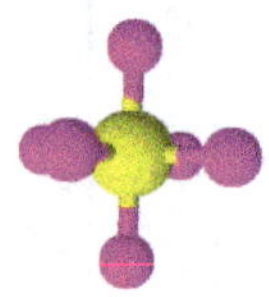 SF_6
	I_h	E, $6S_{10}$, $10S_6$, $6C_5$, $10C_3$, $15C_2$, 15?	120	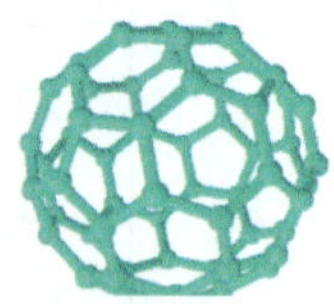 C_{60}
Linear	$C_{\sigma v}$	E, $C_?$, σ_v	?	 HCl
	$D_{\sigma h}$	E, $C_?$ σ_v, σ_h, i, ?C_2	?	 O_2

Group Multiplication

Now we will investigate what happens when we apply two symmetry operations in sequence. As an example, consider the NH_3 molecule, which belongs to the C_{3v} point group. Consider what happens if we apply a C_3 rotation followed by a σ_v reflection. We write this combined operation $\sigma_v C_3$.

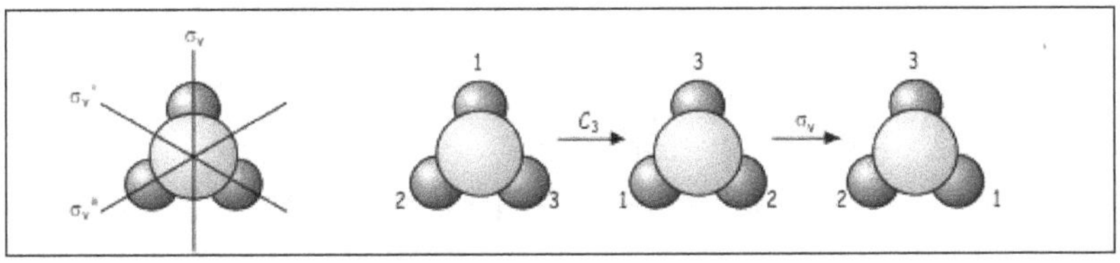

The combined operation $\sigma_v C_3$ is equivalent to σ_v", which is also a symmetry operation of the C_{3v} point group. Now let's see what happens if we apply the operators in the reverse order i.e. $C_3\sigma_v$ (σ_v followed by C_3).

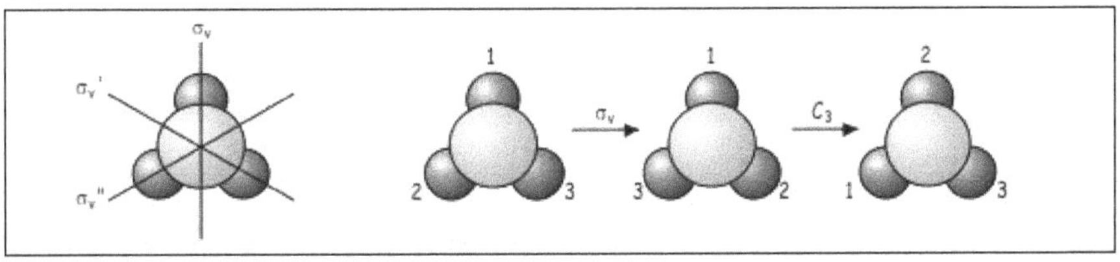

Again, the combined operation $C_3\sigma_v$ is equivalent to another operation of the point group, this time σ_v'. There are two important points that are illustrated by this example:

- The order in which two operations are applied is important. For two symmetry operations A and B, AB is not necessarily the same as BA, i.e. symmetry operations do not in general commute. In some groups the symmetry elements do commute; such groups are said to be Abelian.
- If two operations from the same point group are applied in sequence, the result will be equivalent to another operation from the point group. Symmetry operations that are related to each other by other symmetry operations of the group are said to belong to the same class. In NH_3, the three mirror planes σ_v, σ_v' and σ_v" belong to the same class (related to each other through a C_3 rotation), as do the rotations C_3^+ and C_3^- (anticlockwise and clockwise rotations about the principal axis, related to each other by a vertical mirror plane).

The effects of applying two symmetry operations in sequence within a given point group are summarized in group multiplication tables. As an example, the complete group multiplication table for C_{3v} using the symmetry operations as defined in the figures above is shown below. The operations written along the first row of the table are carried out first, followed by those written in the first column (note that the table would change if we chose to name σ_v, σ_v' and σ_v" in some different order).

C_{3v}	E	C_3^+	C_3^-	σ_v	σ_v'	σ_v"
E	E	C_3^+	C_3^-	σ_v	σ_v'	σ_v"
C_3^+	C_3^+	C_3^-	E	σ_v'	σ_v"	σ_v
C_3^-	C_3^-	E	C_3^+	σ_v"	σ_v	σ_v'
σ_v	σ_v	σ_v"	σ_v'	E	C_3^-	C_3^+
σ_v'	σ_v'	σ_v	σ_v"	C_3^+	E	C_3^-
σ_v"	σ_v"	σ_v'	σ_v	C_3^-	C_3^+	E

Constructing Higher Groups from Simpler Groups

A group that contains a large number of symmetry elements may often be constructed from simpler groups. This is probably best illustrated using an example. Consider the point groups C_2 and C_S. C_2 contains the elements E and C_2, and has order 2, while C_S contains E and σ and also has order 2. We can use these two groups to construct the group C_{2v} by applying the symmetry operations of C_2 and C_S in sequence.

C_2 operation	E	E	C_2	C_2
C_S operation	E	$\sigma(xz)$	E	$\sigma(xz)$
Result	E	$\sigma_v(xz)$	C_2	$\sigma_v'(xz)$

C_{2v} has order 4, which is the product of the orders of the two lower-order groups. C_{2v} may be described as a direct product group of C_2 and C_S.

Matrix

An n × m matrix is a two dimensional array of numbers with n rows and m columns. The integers n and m are called the dimensions of the matrix. If n = m then the matrix is square. The numbers in the matrix are known as matrix elements (or just elements) and are usually given subscripts to signify their position in the matrix e.g. an element a_{ij} would occupy the ith row and j^{th} column of the matrix. For example:

$$M = \begin{pmatrix} 1 & 2 & 3 \\ 4 & 5 & 6 \\ 7 & 8 & 9 \end{pmatrix}$$ is a 3 × 3 matrix with a11 = 1, a12 = 2, a13 = 3, a21 = 4 etc.

In a square matrix, diagonal elements are those for which i = j (the numbers 1, 5 and 9 in the above example). Off-diagonal elements are those for which i ≠ j (2, 3, 4, 6, 7, and 8 in the above example). If all the off-diagonal elements are equal to zero then we have a diagonal matrix. We will see later that diagonal matrices are of considerable importance in group theory.

A unit matrix or identity matrix (usually given the symbol I) is a diagonal matrix in which all the diagonal elements are equal to 1. A unit matrix acting on another matrix has no effect – it is the same as the identity operation in group theory and is analogous to multiplying a number by 1 in everyday arithmetic.

The transpose A^T of a matrix A is the matrix that results from interchanging all the rows and columns. A symmetric matrix is the same as its transpose ($A^T = A$ i.e. $a_{ij} = a_{ji}$ for all values of i and j). The transpose of matrix M above (which is not symmetric) is,

$$M^T = \begin{pmatrix} 1 & 4 & 7 \\ 2 & 5 & 8 \\ 3 & 6 & 9 \end{pmatrix}$$

The sum of the diagonal elements in a square matrix is called the trace (or character) of the matrix (for the above matrix, the trace is $\chi = 1 + 5 + 9 = 15$). The traces of matrices representing symmetry operations will turn out to be of great importance in group theory.

A vector is just a special case of a matrix in which one of the dimensions is equal to 1. An n × 1 matrix is a column vector; a 1 × m matrix is a row vector. The components of a vector are usually only labelled with one index. A unit vector has one element equal to 1 and the others equal to zero (it is the same as one row or column of an identity matrix). We can extend the idea further to say that a single number is a matrix (or vector) of dimension 1 × 1.

Matrix Algebra

1. Two matrices with the same dimensions may be added or subtracted by adding or subtracting the elements occupying the same position in each matrix. e.g.

$$A = \begin{pmatrix} 1 & 0 & 2 \\ 2 & 2 & 1 \\ 3 & 2 & 0 \end{pmatrix} \quad B = \begin{pmatrix} 2 & 0 & -2 \\ 1 & 0 & 1 \\ 1 & -1 & 0 \end{pmatrix} \quad A+B = \begin{pmatrix} 3 & 0 & 0 \\ 3 & 2 & 2 \\ 4 & 1 & 0 \end{pmatrix} \quad A-B = \begin{pmatrix} -1 & 0 & 4 \\ 1 & 2 & 0 \\ 2 & 3 & 0 \end{pmatrix}$$

2. A matrix may be multiplied by a constant by multiplying each element by the constant.

$$4B = \begin{pmatrix} 8 & 0 & -8 \\ 4 & 0 & 4 \\ 4 & -4 & 0 \end{pmatrix} \quad 3A = \begin{pmatrix} 3 & 0 & 6 \\ 6 & 6 & 3 \\ 9 & 6 & 0 \end{pmatrix}$$

3. Two matrices may be multiplied together provided that the number of columns of the first matrix is the same as the number of rows of the second matrix i.e. an n x m matrix may be multiplied by an mxl matrix. The resulting matrix will have dimensions n × l. To find the element a_{ij} in the product matrix, we take the dot product of row i of the first matrix and column j of the second matrix (i.e. we multiply consecutive elements together from row i of the first matrix and column j of the second matrix and add them together i.e. $c_{ij} = \Sigma_k a_i k_b k_j$ e.g. in the 3 x 3 matrices A and B used in the above examples, the first element in the product matrix C = AB is $c_{11} = a_{11}b_{11} + a_{12}b_{21} + a_{13}b_{31}$:

$$AB = \begin{pmatrix} 1 & 0 & 2 \\ 2 & 2 & 1 \\ 3 & 2 & 0 \end{pmatrix} \begin{pmatrix} 2 & 0 & -2 \\ 1 & 0 & 1 \\ 1 & -1 & 0 \end{pmatrix} = \begin{pmatrix} 4 & -2 & -2 \\ 7 & -1 & -2 \\ 8 & 0 & -4 \end{pmatrix}$$

An example of a matrix multiplying a vector is,

$$Av = \begin{pmatrix} 1 & 0 & 2 \\ 2 & 2 & 1 \\ 3 & 2 & 0 \end{pmatrix} \begin{pmatrix} 1 \\ 2 \\ 3 \end{pmatrix} = \begin{pmatrix} 7 \\ 9 \\ 7 \end{pmatrix}$$

Matrix multiplication is not generally commutative, a property that mirrors the behaviour found earlier for symmetry operations within a point group.

Direct Products

The direct product of two matrices (given the symbol ⊗) is a special type of matrix product that generates a matrix of higher dimensionality if both matrices have dimension greater than one. The easiest way to demonstrate how to construct a direct product of two matrices A and B is by an example:

$$
\begin{aligned}
A \otimes B &= \begin{pmatrix} a_{11} & a_{12} \\ a_{21} & a_{22} \end{pmatrix} \otimes \begin{pmatrix} b_{11} & b_{12} \\ b_{21} & b_{22} \end{pmatrix} \\
&= \begin{pmatrix} a_{11}B & a_{12}B \\ a_{21}B & a_{22}B \end{pmatrix} \\
&= \begin{pmatrix} a_{11}b_{11} & a_{12}b_{12} & a_{12}b_{11} & a_{11}b_{12} \\ a_{11}b_{21} & a_{11}b_{22} & a_{12}b_{21} & a_{12}b_{22} \\ a_{21}b_{11} & a_{21}b_{12} & a_{22}b_{11} & a_{22}b_{12} \\ a_{21}b_{21} & a_{21}b_{22} & a_{22}b_{21} & a_{22}b_{22} \end{pmatrix}
\end{aligned}
$$

Though this may seem like a somewhat strange operation to carry out, direct products crop up a great deal in group theory.

Inverse Matrices and Determinants

If two square matrices A and B multiply together to give the identity matrix I (i.e. AB = I) then B is said to be the inverse of A (written A^{-1}). If B is the inverse of A then A is also the inverse of B. Recall that one of the conditions imposed upon the symmetry operations in a group is that each operation must have an inverse. It follows by analogy that any matrices we use to represent symmetry elements must also have inverses. It turns out that a square matrix only has an inverse if its determinant is non-zero. For this reason (and others which will become apparent later on when we need to solve equations involving matrices) we need to learn a little about matrix determinants and their properties.

For every square matrix, there is a unique function of all the elements that yields a single number called the determinant. Initially it probably won't be particularly obvious why this number should be useful, but matrix determinants are of great importance both in pure mathematics and in a number of areas of science. Historically, determinants were actually around before matrices. They arose originally as a property of a system of linear equations that 'determined' whether the system had a unique solution. When such a system of equations is recast as a matrix equation this property carries over into the matrix determinant.

There are two different definitions of a determinant, one geometric and one algebraic. In the geometric interpretation, we consider the numbers across each row of an n × n matrix as coordinates in n-dimensional space. In a one-dimensional matrix (i.e. a number), there is only one coordinate, and the determinant can be interpreted as the (signed) length of a vector from the origin to this point. For a 2 × 2 matrix we have two coordinates in a plane, and the determinant is the (signed) area of the parallelogram that includes these two points and the origin. For a 3 × 3 matrix the determinant is the (signed) volume of the parallelepiped that includes the three points (in three-dimensional space) defined by the matrix and the origin. This is illustrated below. The idea extends up to higher dimensions in a similar way. In some sense then, the determinant is therefore related to the size of a matrix.

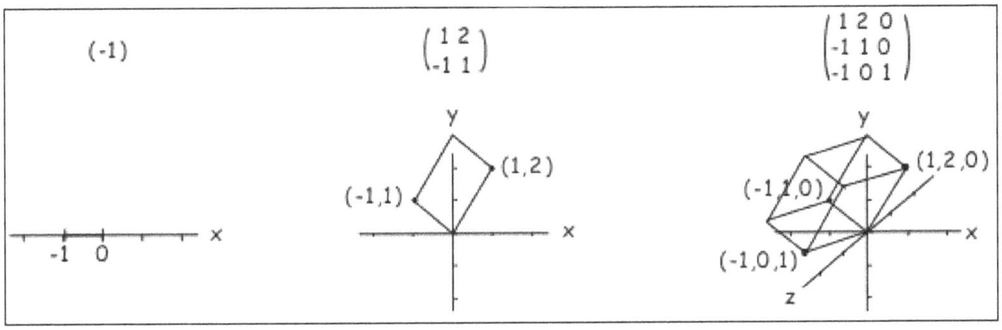

The algebraic definition of the determinant of an n x n matrix is a sum over all the possible products (permutations) of n elements taken from different rows and columns. The number of terms in the sum is n!, the number of possible permutations of n values (i.e. 2 for a 2 × 2 matrix, 6 for a 3 × 3 matrix etc.). Each term in the sum is given a positive or a negative sign depending on whether the number of permutation inversions in the product is even or odd. A permutation inversion is just a pair of elements that are out of order when described by their indices. For example, for a set of four elements (a_1, a_2, a_3, a_4), the permutation $a_1a_2a_3a_4$ has all the elements in their correct order (i.e. in order of increasing index). However, the permutation $a_2a_4a_1a_3$ contains the permutation inversions a_2a_1, a_4a_1, a_4a_3.

For example, for a two-dimensional matrix,

$$\begin{pmatrix} a_{11} & a_{12} \\ a_{21} & a_{22} \end{pmatrix}$$

where the subscripts label the row and column positions of the elements, there are 2 possible products/permutations involving elements from different rows and column, $a_{11}a_{22}$ and $a_{12}a_{21}$. In the second term, there is a permutation inversion involving the column indices 2 and 1 (permutation inversions involving the row and column indices should be looked for separately) so this term takes a negative sign, and the determinant is $a_{11}a_{22}$-$a_{12}a_{21}$.

For a 3 × 3 matrix,

$$\begin{pmatrix} a_{11} & a_{12} & a_{13} \\ a_{21} & a_{22} & a_{23} \\ a_{31} & a_{32} & a_{33} \end{pmatrix}$$

the possible combinations of elements from different rows and columns, together with the sign from the number of permutations required to put their indices in numerical order are:

- $a_{11}a_{22}a_{33}$ (0 inversions).
- $-a_{11}a_{23}a_{32}$ (1 inversion – 3>2 in the column indices).
- $-a_{12}a_{21}a_{33}$ (1 inversion – 2>1 in the column indices).
- $a_{12}a_{23}a_{31}$ (2 inversion2 – 2>1 and 3>1 in the column indices).
- $a_{13}a_{21}a_{32}$ (2 inversions – 3>1 and 3>2 in the column indices).

- $-a_{13}a_{22}a_{31}$ (3 inversions – 3>2, 3>1 and 2>1 in the column indices).

and the determinant is simply the sum of these terms.

This may all seem a little complicated, but in practice there is a fairly systematic procedure for calculating determinants. The determinant of a matrix A is usually written det(A) or |A|.

For a 2 × 2 matrix:

$$A = \begin{pmatrix} a & b \\ c & d \end{pmatrix} \qquad \det(A) = |A| = \begin{vmatrix} a & b \\ c & d \end{vmatrix} = ad - bc$$

For a 3 × 3 matrix:

$$B = \begin{pmatrix} a & b & c \\ d & e & f \\ g & h & i \end{pmatrix} \qquad \det(B) = a\begin{vmatrix} e & f \\ h & i \end{vmatrix} - b\begin{vmatrix} d & f \\ g & i \end{vmatrix} + c\begin{vmatrix} d & e \\ g & h \end{vmatrix}$$

For a 4 × 4 matrix:

$$C = \begin{pmatrix} a & b & c & d \\ e & f & g & h \\ i & j & k & l \\ m & n & o & p \end{pmatrix} \qquad \det(C) = a\begin{vmatrix} f & g & h \\ j & k & l \\ n & o & p \end{vmatrix} - b\begin{vmatrix} e & g & h \\ i & k & l \\ m & o & p \end{vmatrix} + c\begin{vmatrix} e & f & h \\ i & j & l \\ m & n & p \end{vmatrix} - d\begin{vmatrix} e & f & g \\ i & j & k \\ m & n & o \end{vmatrix}$$

and so on in higher dimensions. Note that the submatrices in the 3 × 3 example above are just the matrices formed from the original matrix B that don't include any elements from the same row or column as the premultiplying factors from the first row.

Matrix determinants have a number of important properties:

- The determinant of the identity matrix is 1.

$$\text{e.g.} \begin{vmatrix} 1 & 0 \\ 0 & 1 \end{vmatrix} = \begin{vmatrix} 1 & 0 & 0 \\ 0 & 1 & 0 \\ 0 & 0 & 1 \end{vmatrix} = 1$$

- The determinant of a matrix is the same as the determinant of its transpose i.e. det(A) = det(A^T).

$$\text{e.g.} \begin{vmatrix} a & b \\ c & d \end{vmatrix} = \begin{vmatrix} a & c \\ b & d \end{vmatrix}$$

- The determinant changes sign when any two rows or any two columns are interchanged.

$$\text{e.g.} \begin{vmatrix} a & b \\ c & d \end{vmatrix} = -\begin{vmatrix} b & a \\ d & c \end{vmatrix} = -\begin{vmatrix} c & d \\ a & b \end{vmatrix} = \begin{vmatrix} d & c \\ b & a \end{vmatrix}$$

- The determinant is zero if any row or column is entirely zero, or if any two rows or columns are equal or a multiple of one another.

$$\text{e.g.}\begin{vmatrix}1 & 2\\ 0 & 0\end{vmatrix}=0,\ \begin{vmatrix}1 & 2\\ 2 & 4\end{vmatrix}=0$$

- The determinant is unchanged by adding any linear combination of rows (or columns) to another row (or column).
- The determinant of the product of two matrices is the same as the product of the determinants of the two matrices i.e. det(AB) = det(A)det(B).

The requirement that in order for a matrix to have an inverse it must have a non-zero determinant follows from above property. The product of a matrix and its inverse yields the identity matrix. We therefore have:

$$\det(A^{-1}A)=\det(A^{-1})\det(A)=\det(I)$$
$$\det(A^{-1})=\det(I)/\det(A)=1/\det(A)$$

It follows that a matrix A can only have an inverse if its determinant is non-zero, otherwise the determinant of its inverse would be undefined.

Transformation Matrices

Matrices can be used to map one set of coordinates or functions onto another set. Matrices used for this purpose are called transformation matrices. In group theory, we can use transformation matrices to carry out the various symmetry operations considered at the beginning of the course. As a simple example, we will investigate the matrices we would use to carry out some of these symmetry operations on a vector (x, y).

1. The identity operation: The identity operation leaves the vector unchanged, and as you may already suspect, the appropriate matrix is the identity matrix.

$$(x,y)\begin{pmatrix}1 & 0\\ 0 & 1\end{pmatrix}=(x,y)$$

2. Reflection in a plane: The simplest example of a reflection matrix corresponds to reflecting the vector (x, y) in either the x or y axes. Reflection in the x axis maps y to –y, while reflection in the y axis maps x to -x. The appropriate matrix is very like the identity matrix but with a change in sign for the appropriate element. Reflection in the x axis transforms the vector (x, y) to (x,-y), and the appropriate matrix is,

$$(x,y)\begin{pmatrix}1 & 0\\ 0 & -1\end{pmatrix}=(x,-y)$$

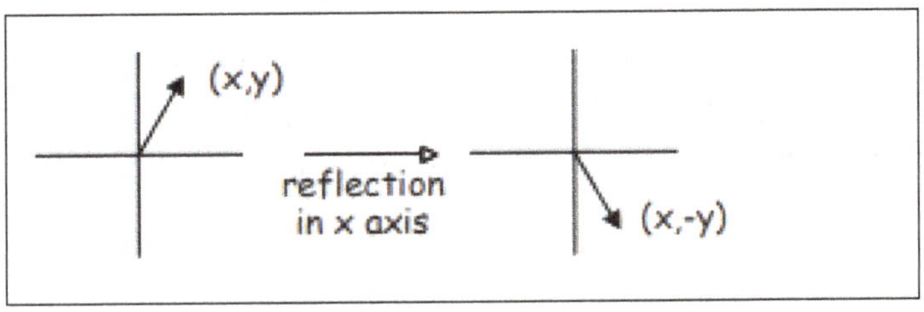

Reflection in the y axis transforms the vector (x, y) to (-x, y), and the appropriate matrix is,

$$(x,y)\begin{pmatrix} -1 & 0 \\ 0 & 1 \end{pmatrix} = (-x,y)$$

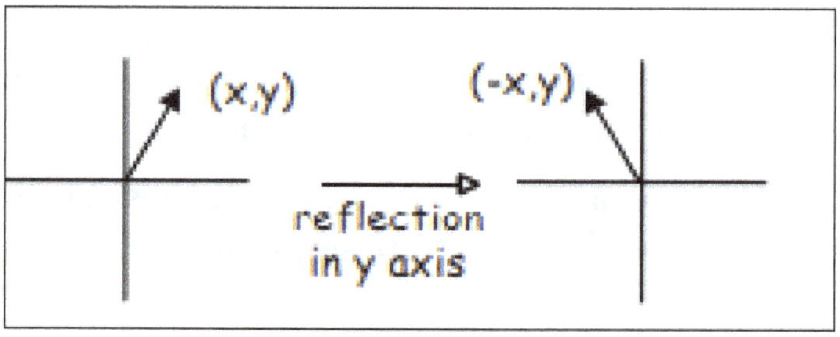

More generally, matrices can be used to represent reflections in any plane (or line in 2D). For example, reflection in the 45° axis shown below maps (x, y) onto (-y, x).

$$(x,y)\begin{pmatrix} 0 & -1 \\ -1 & 0 \end{pmatrix} = (-y,-x)$$

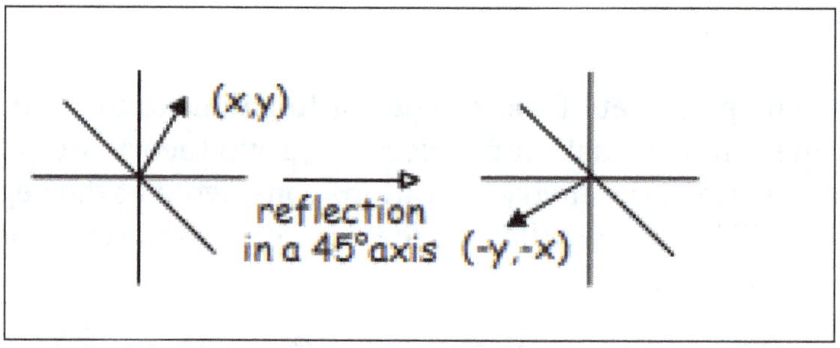

3. Rotation about an axis: In two dimensions, the appropriate matrix to represent rotation by an angle θ about the origin is,

$$R(\theta)\begin{pmatrix} \cos\theta & -\sin\theta \\ \sin\theta & \cos\theta \end{pmatrix}$$

In three dimensions, rotations about the x, y and z axes acting on a vector (x, y, z) are represented by the following matrices:

$$R_x(\theta) = \begin{pmatrix} 1 & 0 & 0 \\ 0 & \cos\theta & -\sin\theta \\ 0 & \sin\theta & \cos\theta \end{pmatrix} \quad R_y(\theta) = \begin{pmatrix} \cos\theta & 0 & -\sin\theta \\ 0 & 1 & 0 \\ \sin\theta & 0 & \cos\theta \end{pmatrix} \quad R_z(\theta) = \begin{pmatrix} \cos\theta & -\sin\theta & 0 \\ \sin\theta & \cos\theta & 0 \\ 0 & 0 & 1 \end{pmatrix}$$

One particularly useful way to describe the symmetry operations of a molecule is to write $\mathcal{R}$ in terms of a transformation matrix D($\mathcal{R}$), the 3 × 3 matrix, which determines the manner in which a set of basis vectors x are transformed into a new set x' as a result of the symmetry operation.

$$x' = D(\mathcal{R})x$$

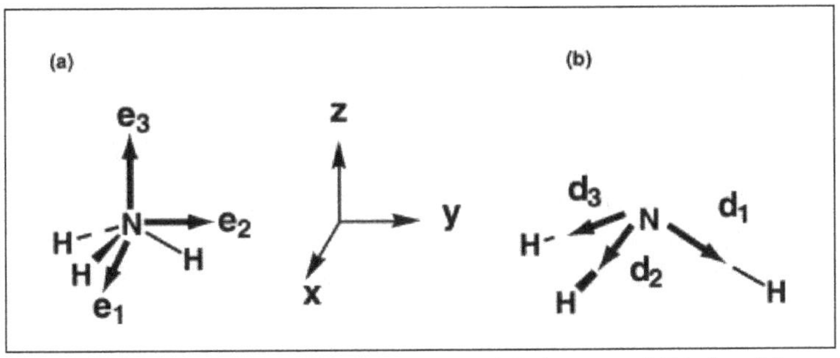

Two different bases for representations: (a) the e_i and (b) the d_i bases.

Given the set of vectors e_i (i = 1, 2, 3) which lie along the Cartesian axes of figure a, the transformation matrices for the set of operations of the C_{3v} point group are given by the following:

$$E = \begin{pmatrix} 1 & 0 & 0 \\ 0 & 1 & 0 \\ 0 & 0 & 1 \end{pmatrix} \quad C_3 = \begin{pmatrix} -\frac{1}{2} & -\frac{\sqrt{3}}{2} & 0 \\ \frac{\sqrt{3}}{2} & -\frac{1}{2} & 0 \\ 0 & 0 & 1 \end{pmatrix} \quad C_3^2 = \begin{pmatrix} -\frac{1}{2} & \frac{\sqrt{3}}{2} & 0 \\ -\frac{\sqrt{3}}{2} & -\frac{1}{2} & 0 \\ 0 & 0 & 1 \end{pmatrix}$$

$$\sigma_a = \begin{pmatrix} -1 & 0 & 0 \\ 0 & 1 & 0 \\ 0 & 0 & 1 \end{pmatrix} \quad \sigma_b = \begin{pmatrix} \frac{1}{2} & -\frac{\sqrt{3}}{2} & 0 \\ -\frac{\sqrt{3}}{2} & -\frac{1}{2} & 0 \\ 0 & 0 & 1 \end{pmatrix} \quad \sigma_c = \begin{pmatrix} \frac{1}{2} & \frac{\sqrt{3}}{2} & 0 \\ \frac{\sqrt{3}}{2} & -\frac{1}{2} & 0 \\ 0 & 0 & 1 \end{pmatrix}$$

The products of these matrices will reflect the products of the type evaluated pictorially. As an example,

$$\sigma_c\sigma_b = \begin{pmatrix} \frac{1}{2} & \frac{\sqrt{3}}{2} & 0 \\ \frac{\sqrt{3}}{2} & -\frac{1}{2} & 0 \\ 0 & 0 & 1 \end{pmatrix} \begin{pmatrix} \frac{1}{2} & -\frac{\sqrt{3}}{2} & 0 \\ -\frac{\sqrt{3}}{2} & -\frac{1}{2} & 0 \\ 0 & 0 & 1 \end{pmatrix} = \begin{pmatrix} \frac{1}{4}-\frac{3}{4}+0 & -\frac{\sqrt{3}}{4}-\frac{\sqrt{3}}{4}+0 & 0+0+0 \\ \frac{\sqrt{3}}{4}+\frac{\sqrt{3}}{4}+0 & -\frac{3}{4}+\frac{1}{4}+0 & 0+0+0 \\ 0+0+0 & 0+0+0 & 0+0+1 \end{pmatrix}$$

$$= \begin{pmatrix} -\frac{1}{2} & -\frac{\sqrt{3}}{2} & 0 \\ \frac{\sqrt{3}}{2} & -\frac{1}{2} & 0 \\ 0 & 0 & 1 \end{pmatrix} = C_3$$

Notice that since the order of operation for group theory proceeds from right to left and the rules for matrix multiplication are left to right, the product in above equation is then evaluated in inverted order. This set of matrices form a representation of the group, and the vectors $\{e_i\}$ form the basis of the representation (Γ_a). An equally valid set (Γ_b) of vectors $\{d_i\}$ (i = 1, 2, 3) which may be used to describe the atomic coordinates i are shown in figure b. The transformation matrices for this case are readily shown to be,

$$E = \begin{pmatrix} 1 & 0 & 0 \\ 0 & 1 & 0 \\ 0 & 0 & 1 \end{pmatrix} \quad C_3 = \begin{pmatrix} 0 & 1 & 0 \\ 0 & 0 & 1 \\ 1 & 0 & 0 \end{pmatrix} \quad C_3^2 = \begin{pmatrix} 0 & 0 & 1 \\ 1 & 0 & 0 \\ 0 & 1 & 0 \end{pmatrix}$$

$$\sigma_a = \begin{pmatrix} 1 & 0 & 0 \\ 0 & 0 & 1 \\ 0 & 1 & 0 \end{pmatrix} \quad \sigma_b = \begin{pmatrix} 0 & 0 & 1 \\ 0 & 1 & 0 \\ 1 & 0 & 0 \end{pmatrix} \quad \sigma_c = \begin{pmatrix} 0 & 1 & 0 \\ 1 & 0 & 0 \\ 0 & 0 & 1 \end{pmatrix}$$

It is clear that there is an infinite collection of sets of vectors (some of which are orthogonal, the e_i, and some which are not, the d_i), which may be used to define the atomic coordinates. We need to find a measure that defines the properties of these representations and which is independent of the basis. This is the character, $\gamma(\mathcal{R})$ of the group element $\mathcal{R}$. It is simply the sum of the diagonal elements (or the trace) of the transformation matrix. Table below shows how the characters of the two representations, Γ_a, Γ_b, are in fact the same even though their transformation matrices $D(\mathcal{R})$ are very different. The set of characters, one for each element of the group is termed the character of the representation. Interestingly, the $\gamma(\mathcal{R})$ are equal for all $\mathcal{R}$ in the same class $\gamma(\mathcal{K})$. Recall that $\mathcal{K}$ represents one symmetry operation in all classes, whereas $\mathcal{R}$ corresponds to all symmetry operations. The reader should be aware that traditionally representations are given by the symbol Γ and the characters of those representations by χ. We need to reserve the latter symbol, however, for atomic orbitals (along with φ for combinations of atomic orbitals and ψ for molecular orbitals), and thus the use of γ here.

Table: The Characters of the Two Representations Γ_a and Γ_b.

	Γ_a	Γ_b
E	1 + 1 + 1 = 3	1 + 1 + 1 = 3
C_3	-1/2 - 1/2 + 1 = 0	0 + 0 + 0 = 0
C_3^2	-1/2 - 1/2 + 1 = 0	0 + 0 + 0 = 0
σ_a	-1 + 1 + 1 = 1	1 + 0 + 0 = 1
σ_b	1/2 - 1/2 + 1 = 1	0 + 1 + 0 = 1
σ_c	1/2 - 1/2 + 1 = 1	0 + 0 + 1 = 1

The two representations Γ_a and Γ_b although equivalent (have the same character) are in fact different in the following sense. Notice that Γ_a, although a three-dimensional representation (there are three basis vectors), may be written in block diagonal form for all operations $\mathcal{R}$ as,

$$D(\mathcal{R}) = \begin{bmatrix} D'(\mathcal{R}) & 0 \\ 0 & D''(\mathcal{R}) \end{bmatrix}$$

where D'($\mathcal{R}$) contains elements which involve only e_1 and e_2 (x, y) and D"($\mathcal{R}$) elements which contain e_3 (z). Γ_b is thus called a reducible representation since it may be written as a direct sum of the matrices D'($\mathcal{R}$) and D"($\mathcal{R}$). We can envisage that there is some similarity transformation which is able to block-diagonalize the set of transformation matrices. For the case described here, this leads to a one-dimensional and one two-dimensional representation. Ultimately, it will be impossible to block-diagonalize the set of transformation matrices further. In this case, one has reached an irreducible representation (IR).

At this point we need to ask what the possible irreducible representations for a given point group are. Just as there is a set of rules that define the formation of a point group, so there is a pair of rules that define the possible set of IRs of a group.

1. For a group which possesses k different classes, there are k different irreducible representations, $\Gamma_1, \Gamma_2, ..., \Gamma_k$ whose dimensions $m_1, m_2, ..., m_k$ satisfy the relationship,

$$m_1^2 + m_2^2 + m_k^2 = g$$

where g is the order of the group.

2. The characters $\gamma_i(\mathcal{R})$ or $\gamma_i(\mathcal{K})$ of the irreducible representations satisfy the relationship,

$$\sum_{\mathcal{K}} g_k \gamma_i^*(\mathcal{K}) \gamma_j(\mathcal{K}) = g\delta_{ij}$$

or

$$\sum_{\mathcal{R}} \gamma_i^*(\mathcal{R}) \gamma_j(\mathcal{R}) = g\delta_{ij}$$

In short, where the sums are implied whenever there is product of the Γ_s, the above two equations are simplified as in equations below, respectively.

$$g_k \Gamma_i(\mathcal{K}) \Gamma_j(\mathcal{K}) = g\delta_{ij} \text{ and } \Gamma_i(\mathcal{R}) \Gamma_j(\mathcal{R}) = g\delta_{ij}$$

If i = j then,

$$\sum_{\mathcal{K}} g_k \left[\gamma_j(\mathcal{K})\right]^2 = g \text{ or } \sum_{\mathcal{R}} \left[\gamma_j(\mathcal{R})\right]^2 = g$$

and if i ≠ j then,

$$\sum_{\mathcal{K}} g_k \gamma_i^*(\mathcal{K}) \gamma_j(\mathcal{K}) = 0 \text{ or } \sum_{\mathcal{R}} \gamma_i^*(\mathcal{R}) \gamma_j(\mathcal{R}) = 0$$

(All of the groups we come across in this book contain real characters, and thus, the use of the γ^* notation will not be necessary.) For each point group, these rules lead to only one choice for the irreducible representations of the group. The results are depicted in a character table. Those for the C_{2v}, C_{3v}, and D_{4h} groups are shown in Tables. Notice the important result for the Abelian group C_{2v}. Since each element falls into its own class, there is same number of irreducible representations as elements, and each irreducible representation is one-dimensional. In all groups, the first entry consists of a set of figures $\gamma_i(\mathcal{K}) = +1$ for all classes. This corresponds to the totally symmetric representation since a + 1 is found for every $\gamma_i(\mathcal{R})$.

Table: The Group Table for C_{2v}.

C_{2v}	E	C_2	$\sigma_v(xz)$	$\sigma_v(yz)$	
A_1	1	1	1	1	z^a
A_2	1	1	-1	-1	
B_1	1	-1	1	-1	x
B_2	1	-1	-1	1	y

Each of the irreducible representations could be labeled in a simple way, Γ_1, Γ_2, ..., Γ_n, but there is a notation due to Mulliken, which is very useful. A and B refer to one-dimensional representations where A refers to a representation for which the character corresponding to the highest order rotation axis is +1 and B when it is -1. E and T refer to doubly and triply degenerate representations, respectively. From the definition of these characters, the entry in the character table for the identity operation, $\gamma_i(E)$ defines the order of the representation. (1 for A, B; 2 for E; 3 for T, etc.) If there is more than one representation with the same label, then they are distinguished by subscripts 1, 2, . . ., and so on.

Table: The Group Table for C_{3v}.

C_{3v}	E	$2C_3$	$3\sigma_v$	
A_1	1	1	1	z
A_2	1	1	-1	
E	2	-1	0	(x, y)

Table: The Group Table for D_{4h}.

D_{4h}	E	$2C_4$	C_2	$2C_2^{o}$	$2C_2oo$	i	$2S_4$	σ_h	$2\sigma_v$	$2\sigma_d$		
A_{1g}	1	1	1	1	1	1	1	1	1	1		$x^2 + y^2, z^2$
A_{2g}	1	1	1	-1	-1	1	1	1	-1	-1		
B_{1g}	1	-1	1	1	-1	1	-1	1	1	-1		$x^2 - y^2$
B_{2g}	1	-1	1	-1	1	1	-1	1	-1	1		xy
E_g	2	0	-2	0	0	2	0	-2	0	0		(xz, yz)
A_{1u}	1	1	1	1	1	-1	-1	-1	-1	-1		
A_{2u}	1	1	1	-1	-1	-1	-1	-1	1	1	z	
B_{1u}	1	-1	1	1	-1	-1	1	-1	-1	1		
B_{2u}	1	-1	1	-1	1	-1	1	-1	1	-1		
E_u	2	0	-2	0	0	-2	0	2	0	0	(x, y)	

For some groups, the character table may be written in a particularly simple form as,

	{1}	{2}
g or'	X	Y
u or"	X	–Y

These groups are called product groups since they may readily be derived from a simpler group (whose character table is given by X alone) by the introduction of a σ_h or i operation. The D_{4h} group appropriate for the PtH_4^{2-} ion is one of these. Here, the D_{4h} group, {1} + {2} = {E, C_4, C_4^3, C_2, $2C_2'$, $2C_2''$, i, S_4, S_4^3, σ_h, $2\sigma_v$, $2\sigma_d$} is generated by adding an i to the set of operations for the D_4 group {1} = {E, C_4, C_4^3, C_2, $2C_2'$, $2C_2''$}. Addition of the single operation i is not possible on its own. Inclusion of the larger set of operations {2} = {i, S_4, S_4^3, σ_h, $2\sigma_v$, $2\sigma_d$} is required to satisfy the rules determining the structure of the group. The two sets of operations are equal in number. Thus, we would write $D_{4h} = D_4 \otimes i$. In general, for n even, $C_{nh} = C_n \otimes i$ and $D_{nh} = D_n \otimes i$; for n odd, $D_{nd} \otimes D_n = i$ and $O_h = O \otimes i$. Representations for which $\gamma_j(i) = +1$ are given a subscript g (German; gerade = even) and those for which $\gamma_j(i) = -1$ are given a label u (ungerade = odd). A similar result applies to product groups involving sh. For example for n odd, $D_{nh} = C_n \otimes \sigma_h$. Thus, the form of the character table for D_{3h} is just as in above equation. The representations with $\gamma_j(\sigma_h) = +1$ are given by a superscript′, and those with $\gamma_j(\sigma_h) = -1$ by a superscript″.

There are special labels for the linear point groups $D_{\infty h}$ and $C_{\infty v}$. Σ refers to one-dimensional representations and Π, Δ, Φ, . . . refer to different two-dimensional ones. In spherical systems (atoms), the labels S, P, D, F, . . . refer to 1, 3, 5, 7, . . . dimensional representations, respectively. These labels (in their lowercase form) are just those used to describe the orbitals of atoms.

Insight into the meaning of these character tables can be gained when we remind ourselves that the character is the trace of the relevant transformation matrix. For one-dimensional representations, whether a + 1 or - 1 appears under a given operation $\mathcal{R}$ specifies whether a basis function described by this character set is respectively symmetric (converted into itself) or antisymmetric (converted into minus itself) as a result of that symmetry operation. For irreducible representations, which are of higher dimension, such considerations are less obvious since by definition, it is the behavior of two functions (for an E representation) or three functions (for a T representation), which are being considered.

References

- Definition-of-group-and-its-characteristics, applications-of-molecular-symmetry-and-group-theory: inflibnet.ac.in, Retrieved 11, July 2020
- Physical-and-Theoretical-Chemistry-Textbook-Maps, Supplemental-Modules-(Physical-and-Theoretical-Chemistry): libretexts.org, Retrieved 02, April 2020

2

Molecular Orbitals

The mathematical function which describes the wave like behavior as well as the location of an electron within a molecule is known as a molecular orbital. A few of its major aspects are symmetry properties of orbitals, symmetry-adapted wavefunctions, and orbital symmetry rules and reactions. This chapter discusses in detail these aspects of molecular orbitals.

Molecular orbital theory uses group theory to describe the bonding in molecules; it complements and extends. In molecular orbital theory the symmetry properties and relative energies of atomic orbitals determine how these orbitals interact to form molecular orbitals. The molecular orbitals are then occupied by the available electrons according to the same rules used for atomic orbitals. The total energy of the electrons in the molecular orbitals is compared with the initial total energy of electrons in the atomic orbitals. If the total energy of the electrons in the molecular orbitals is less than in the atomic orbitals, the molecule is stable relative to the separate atoms; if not, the molecule is unstable and predicted not to form.

Formation of Molecular Orbitals from Atomic Orbitals

Orbitals As with atomic orbitals, Schrödinger equations can be written for electrons in molecules. Approximate solutions to these molecular Schrödinger equations can be constructed from linear combinations of atomic orbitals (LCAO), the sums and differences of the atomic wave functions. For diatomic molecules such as H_2, such wave functions have the form,

$$\Psi = c_a\psi_a + c_b\psi_b$$

where Ψ is the molecular wave function, ψ_a and ψ_b are atomic wave functions for atoms a and b, and c_a and c_b are adjustable coefficients that quantify the contribution of each atomic orbital to the molecular orbital. The coefficients can be equal or unequal, positive or negative, depending on the individual orbitals and their energies. As the distance between two atoms is decreased, their orbitals overlap, with significant probability for electrons from both atoms being found in the region of overlap. As a result, molecular orbitals form. Electrons in bonding molecular orbitals have a high probability of occupying the space between the nuclei; the electrostatic forces between the electrons and the two positive nuclei hold the atoms together.

Three conditions are essential for overlap to lead to bonding. First, the symmetry of the orbitals must be such that regions with the same sign of ψ overlap. Second, the atomic orbital energies must be similar. When the energies differ greatly, the change in the energy of electrons upon formation of molecular orbitals is small, and the net reduction in energy of the electrons is too small for significant bonding. Third, the distance between the atoms must be short enough to provide good overlap of the orbitals, but not so short that repulsive forces of other electrons or the nuclei interfere. When these three conditions are met, the overall energy of the electrons in the occupied molecular orbitals is lower in energy than the overall energy of the electrons in the original atomic orbitals, and the resulting molecule have a lower total energy than the separated atoms.

Molecular Orbitals from s Orbitals

Consider the interactions between two s orbitals, as in H_2. For convenience, we label the atoms of a diatomic molecule a and b, so the atomic orbital wave functions are $\psi(1s_a)$ and $\psi(1s_b)$. We can visualize the two atoms approaching each other, until their electron clouds overlap and merge into larger molecular electron clouds. The resulting molecular orbitals are linear combinations of the atomic orbitals, the sum of the two orbitals and the difference between them.

In general terms for H_2,

$$\Psi(\sigma) = N\left[c_a\psi(1s_a) + c_b\psi(1s_b)\right] = \frac{1}{\sqrt{2}}\left[\psi(1s_a) + \psi(1s_b)\right](H_a + H_b)$$

and $$\Psi(\sigma^*) = N\left[c_a\psi(1s_a) - c_b\psi(1s_b)\right] = \frac{1}{\sqrt{2}}\left[\psi(1s_a) - \psi(1s_b)\right](H_a - H_b)$$

where,

N = normalizing factor, so $\int \Psi\Psi^*\, d\tau = 1$,

c_a and c_b = adjustable coefficients.

In this case, the two atomic orbitals are identical, and the coefficients are nearly identical as well. These orbitals are depicted in Figure. The signs of orbital lobes are indicated by shading or color. Light and dark lobes or lobes of different color indicate opposite signs of Ψ.

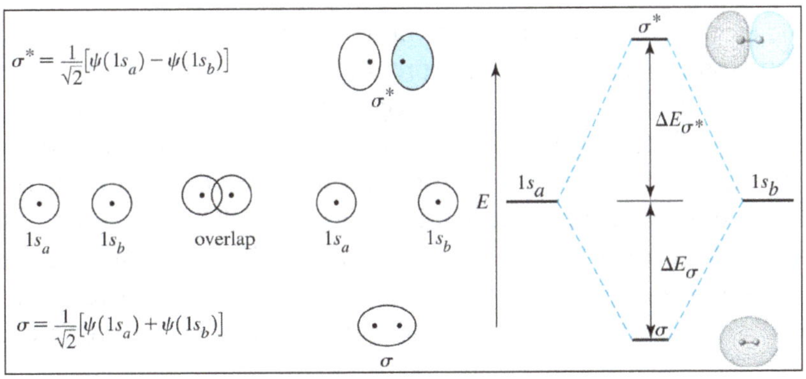

Molecular Orbitals from Hydrogen 1s Orbitals. The s molecular orbital is a bonding molecular orbital, and has a lower energy than the original atomic orbitals, since this combination of atomic orbitals results in an increased concentration of electrons between the two nuclei. The σ^* orbital is an anti-bonding orbital at higher energy since this combination of atomic orbitals results in a node with zero electron density between the nuclei.

The choice of positive and negative for specific atomic orbitals is arbitrary; what is important is how they combine to form molecular orbitals. The different colors in figure below, show opposite signs of the wave function, both in the schematic sketches on the left of the energy level diagram and in the calculated molecular orbital images on the right.

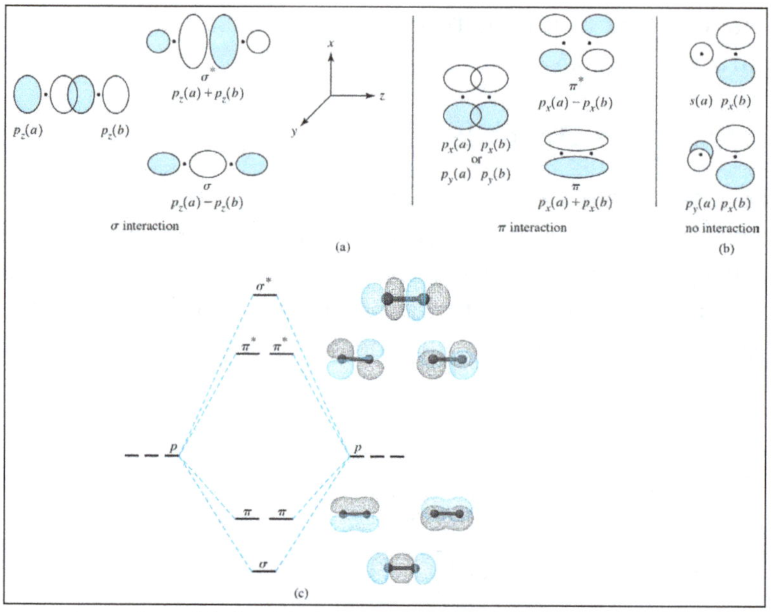

Interactions of p Orbitals: (a) Formation of molecular orbitals. (b) Orbitals that do not form molecular orbitals. (c) Energy-level diagram.

Because the σ molecular orbital is the sum of two atomic orbitals, $\frac{1}{\sqrt{2}}\left[\psi(1s_a)+\psi(1s_b)\right]$, and results in an increased concentration of electrons between the two nuclei, it is a bonding molecular orbital and has a lower energy than the original atomic orbitals. The σ* molecular orbital is the difference of the two atomic orbitals, $\frac{1}{\sqrt{2}}\left[\psi(1s_a)+\psi(1s_b)\right]$. It has a node with zero electron density between the nuclei, due to cancellation of the two wave functions, and a higher energy; it is therefore called an antibonding orbital. Electrons in bonding orbitals are concentrated between the nuclei and attract the nuclei, holding them together. Antibonding orbitals have one or more nodes between the nuclei; electrons in these orbitals are destabilized relative to the parent atomic orbitals; the electrons do not have access to the region between the nuclei where they could experience the maximum nuclear attraction. Nonbonding orbitals are also possible. The energy of a nonbonding orbital is essentially that of an atomic orbital, either because the orbital on one atom has a symmetry that does not match any orbitals on the other atom or the orbital on one atom has a severe energy mismatch with symmetry-compatible orbitals on the other atom.

The σ (sigma) notation indicates orbitals that are symmetric to rotation about the line connecting the nuclei:

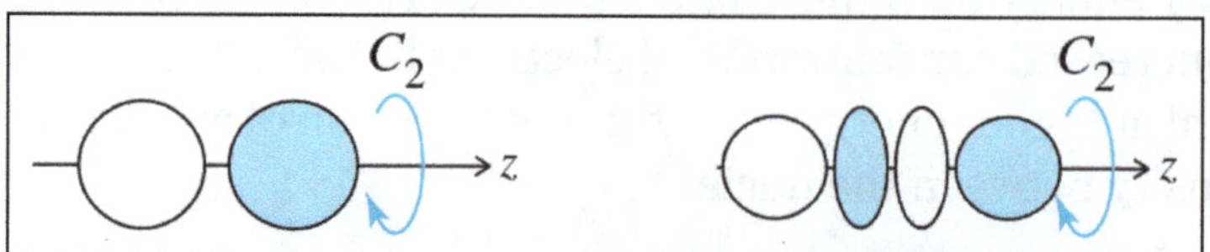

σ* from s orbital.	σ* from p_z orbital.

An asterisk is frequently used to indicate antibonding orbitals. Because the bonding, nonbonding, or antibonding nature of a molecular orbital is not always straightforward to assign in larger molecules, we will use the asterisk notation only for those molecules where bonding and antibonding orbital descriptions are unambiguous.

The pattern described for H_2 is the usual model for combining two orbitals: two atomic orbitals combine to form two molecular orbitals, one bonding orbital with a lower energy and one antibonding orbital with a higher energy. Regardless of the number of orbitals, the number of resulting molecular orbitals is always the same as the initial number of atomic orbitals; the total number of orbitals is always conserved.

Molecular Orbitals from p Orbitals

Molecular orbitals formed from p orbitals are more complex since each p orbital contains separate regions with opposite signs of the wave function. When two orbitals overlap, and the overlapping regions have the same sign, the sum of the two orbitals has an increased electron probability in the overlap region. When two regions of opposite sign overlap, the combination has a decreased electron probability in the overlap region. Figure shows this effect for the 1s orbitals of H_2; similar effects result from overlapping lobes of p orbitals with their alternating signs. For convenience, we will choose a common z axis connecting the nuclei and assign x and y axes as shown in the figure.

When we draw the z axes for the two atoms pointing in the same direction, the p_z orbitals subtract to form s and add to form σ* orbitals, both of which are symmetric to rotation about the z axis, with nodes perpendicular to the line that connects the nuclei. Interactions between p_x and p_y orbitals lead to π and π* orbitals. The π (pi) notation indicates a change in sign of the wave function with C_2 rotation about the bond axis:

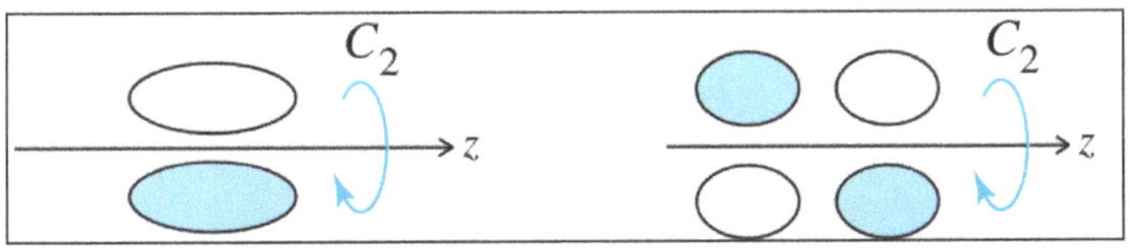

As with the s orbitals, the overlap of two regions with the same sign leads to an increased concentration of electrons, and the overlap of two regions of opposite signs leads to a node of zero electron density. In addition, the nodes of the atomic orbitals become the nodes of the resulting molecular orbitals. In the π* antibonding case, four lobes result that are similar in appearance to a d orbital.

The p_x, p_y, and p_z orbital pairs need to be considered separately. Because the z axis was chosen as the internuclear axis, the orbitals derived from the p_z orbitals are symmetric to rotation around the bond axis and are labeled σ and σ* for the bonding and antibonding orbitals, respectively. Similar combinations of the p_y orbitals form orbitals whose wave functions change sign with C_2 rotation about the bond axis; they are labeled π and π*. In the same way, the p_x orbitals also form π and π* orbitals.

It is common for s and p atomic orbitals on different atoms to be sufficiently similar in energy for their combinations to be considered. However, if the symmetry properties of the orbitals do not match, no combination is possible. For example, when orbitals overlap equally with both the same and opposite signs, as in the s + p_x example in figure, the bonding and antibonding effects cancel, and no molecular orbital results. If the symmetry of an atomic orbital does not match any orbital of the other atom, it is called a nonbonding orbital. Homonuclear diatomic molecules have only bonding and antibonding molecular orbitals.

Molecular Orbitals from d Orbitals

In the heavier elements, particularly the transition metals, d orbitals can be involved in bonding. Figure shows the possible combinations. When the z axes are collinear, two d_z2 orbitals can combine end-on for σ bonding. The d_{xz} and d_{yz} orbitals form π orbitals. When atomic orbitals meet from two parallel planes and combine side to side, as do the $d_{x^2-y^2}$ and dxy orbitals with collinear z axes, they form (δ) delta orbitals. (The δ notation indicates sign changes on C_4 rotation about the bond axis). Sigma orbitals have no nodes that include the line connecting the nuclei, pi orbitals have one node that includes the line connecting the nuclei, and delta orbitals have two nodes that include the line connecting the nuclei. Again, some orbital interactions are forbidden on the basis of symmetry; for example, p_z and d_{xz} have zero net overlap if the z axis is chosen as the bond axis since the p_z would approach the d_{xz} orbital along a d_{xz} node. It is noteworthy in this case that p_x and d_{xz} would be eligible to interact in a π fashion on the basis of the assigned coordinate system. This example emphasizes the importance of maintaining a consistent coordinate system when assessing orbital interactions.

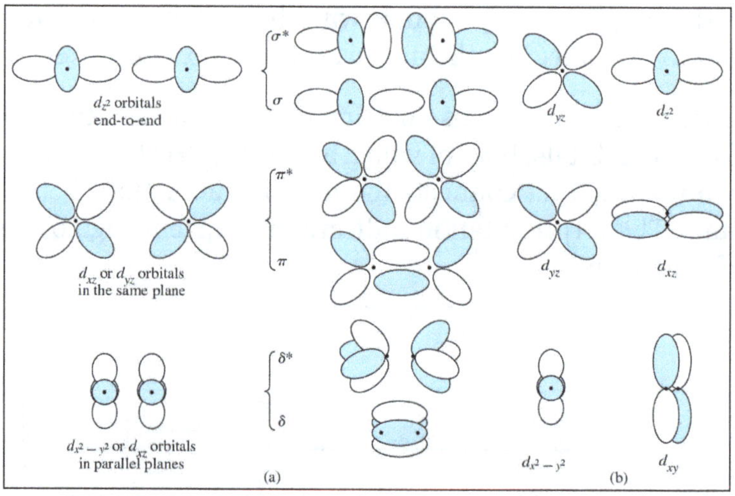

Interactions of d Orbitals: (a) Formation of molecular orbitals. (b) Atomic orbital combinations that do not form molecular orbitals.

Nonbonding Orbitals and other Factors

Nonbonding molecular orbitals have energies essentially equal to that of atomic orbitals. These can form in larger molecules, for example when there are three atomic orbitals of the same symmetry and similar energies, a situation that requires the formation of three molecular orbitals. Most commonly, one molecular orbital formed is a low-energy bonding orbital, one is a high-energy antibonding orbital, and one is of intermediate energy and is a nonbonding orbital.

In addition to symmetry, the second major factor that must be considered in forming molecular orbitals is the relative energy of the atomic orbitals. When the interacting atomic orbitals have the same energy, the interaction is strong, and the resulting molecular orbitals have energies well below (bonding) and above (antibonding) that of the original atomic orbitals. When the two atomic orbitals have quite different energies, the interaction is weaker, and the resulting molecular orbitals have energies and shapes closer to the original atomic orbitals. For example, although they have the same symmetry, 1s orbitals do not combine significantly with 2s orbitals of the other atom in diatomic molecules such as N_2, because their energies are too far apart. The general rule is that the closer the energy match, the stronger the interaction.

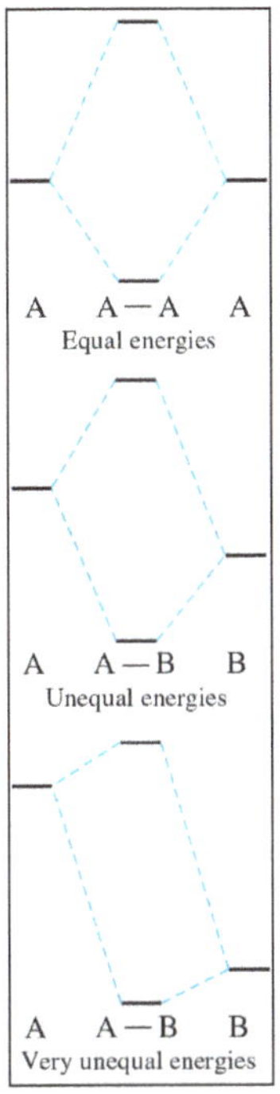

Energy Match and Molecular Orbital Formation.

Homonuclear Diatomic Molecules

Because of their simplicity, diatomic molecules provide convenient examples to illustrate how the orbitals of individual atoms interact to form orbitals in molecules.

Molecular Orbitals

Although apparently satisfactory Lewis electron-dot structures of N_2, O_2, and F_2 can be drawn, the same is not true with Li_2, Be_2, B_2, and C_2, which violate the octet rule. In addition, the Lewis structure of O_2 predicts a double-bonded, diamagnetic (all electrons paired) molecule ($\ddot{O}{=}\ddot{O}$) but experiment has shown O_2 to have two unpaired electrons, making it paramagnetic. The molecular orbital description predicts this paramagnetism, and is more in agreement with experiment. Figure shows the full set of molecular orbitals for the homonuclear diatomic molecules of the first 10 elements, based on the energies appropriate for O_2. The diagram shows the order of energy levels for the molecular orbitals, assuming significant interactions only between atomic orbitals of identical energy.

The energies of the molecular orbitals change in a periodic way with atomic number, since the energies of the interacting atomic orbitals decrease across a period, but the general order of the molecular orbitals remains similar (with some subtle changes) even for heavier atoms lower in the periodic table. Electrons fill the molecular orbitals according to the same rules that govern the filling of atomic orbitals, filling from lowest to highest energy (aufbau principle), maximum spin multiplicity consistent with the lowest net energy (Hund's rules), and no two electrons with identical quantum numbers (Pauli exclusion principle). The most stable configuration of electrons in the molecular orbitals is always the configuration with minimum energy, and the greatest net stabilization of the electrons.

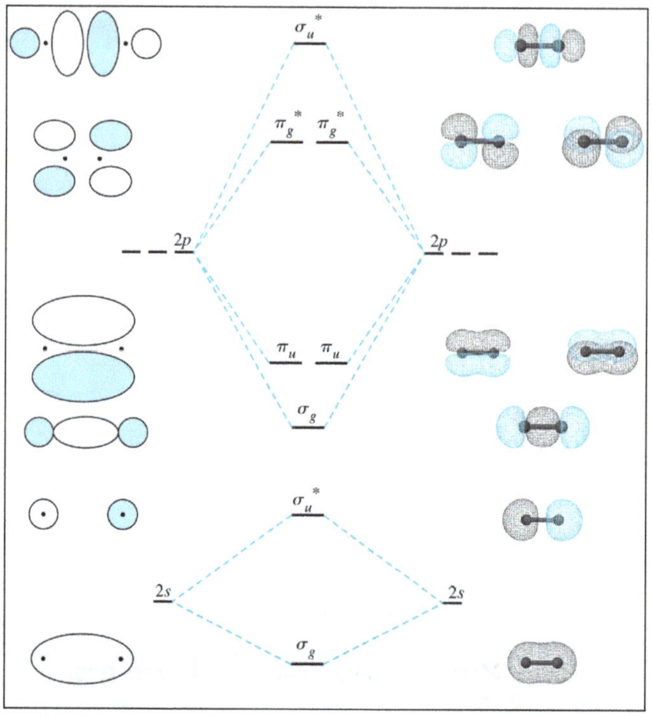

Molecular Orbitals for the First 10 Elements, Assuming Significant Interactions Only between the Valence Atomic Orbitals of Identical Energy.

The overall number of bonding and antibonding electrons determines the number of bonds (bond order):

$$\text{Bond order} = \frac{1}{2}\left[\begin{pmatrix}\text{number of electrons}\\ \text{in bonding orbitals}\end{pmatrix} - \begin{pmatrix}\text{number of electrons}\\ \text{in antibonding orbitals}\end{pmatrix}\right]$$

It is generally sufficient to consider only valence electrons. For example, O_2, with 10 electrons in bonding orbitals and 6 electrons in antibonding orbitals, has a bond order of 2, a double bond. Counting only valence electrons, 8 bonding and 4 antibonding, gives the same result. Because the molecular orbitals derived from the 1s orbitals have the same number of bonding and antibonding electrons, they have no net effect on the bond order. Generally electrons in atomic orbitals lower in energy than the valence orbitals are considered to reside primarily on the original atoms and to engage only weakly in bonding and antibonding interactions, as shown for the 1s orbitals in figure above, the difference in energy between the σ_g and σ_u* orbitals is slight. Because such interactions are so weak, we will not include them in other molecular orbital energy level diagrams.

Additional labels describe the orbitals. The subscripts g for gerade, orbitals symmetric to inversion, and u for ungerade, orbitals antisymmetric to inversion (those whose signs change on inversion), are commonly used. The g or u notation describes the symmetry of the orbitals without a judgment as to their relative energies. Figure above has examples of both bonding and antibonding orbitals with g and u designations.

Orbital Mixing

In the previous figure, we only considered interactions between atomic orbitals of identical energy. However, atomic orbitals with similar, but unequal, energies can interact if they have appropriate symmetries. We now outline two approaches to analyzing this phenomenon, one in which we first consider the atomic orbitals that contribute most to each molecular orbital before consideration of additional interactions and one in which we consider all atomic orbital interactions permitted by symmetry simultaneously.

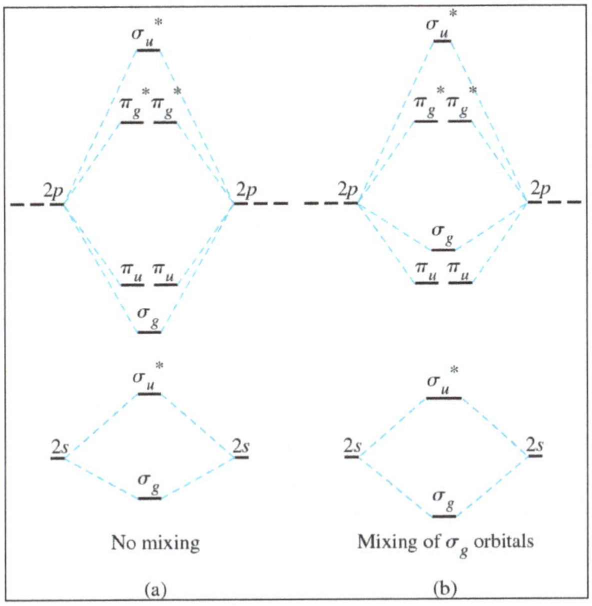

Interaction between Molecular Orbitals. Mixing molecular orbitals of the same symmetry results in a greater energy difference between the orbitals. The σ orbitals mix strongly; the σ* orbitals differ more in energy and mix weakly.

Figure (a) shows the familiar energy levels for a homonuclear diatomic molecule where only interactions between degenerate (having the same energy) atomic orbitals are considered. However, when two molecular orbitals of the same symmetry have similar energies, they interact to lower the energy of the lower orbital and raise the energy of the higher orbital. For example, in the homonuclear diatomics, the σg (2s) and σg (2p) orbitals both have σg symmetry (symmetric to infinite rotation and inversion); these orbitals interact to lower the energy of the σg (2s) and to raise the energy of the σg (2p) as shown in figure (b). Similarly, the $\sigma_u^*(2s)$ and $\sigma_u^*(2p)$ orbitals interact to lower the energy of the $\sigma_u^*(2s)$ and to raise the energy of the $\sigma_u^*(2p)$. This phenomenon is called mixing, which takes into account that molecular orbitals with similar energies interact if they have appropriate symmetry. When two molecular orbitals of the same symmetry mix, the one with higher energy moves still higher in energy, and the one with lower energy moves lower. Mixing results in additional electron stabilization, and enhances the bonding.

A perhaps more rigorous approach to explain mixing considers that the four s molecular orbitals (MOs) result from combining the four atomic orbitals (two 2s and two $2p_z$) that have similar energies. The resulting molecular orbitals have the following general form, where a and b identify the two atoms, with appropriate normalization constants for each atomic orbital:

$$\Psi = c_1\psi(2d_a) \pm c_2\psi(2s_b) \pm c_3\psi(2p_a) \pm c_4\psi(2p_b)$$

For homonuclear diatomic molecules, $c_1 = c_2$ and $c_3 = c_4$ in each of the four MOs. The lowest energy MO has larger values of c_1 and c_2, the highest has larger values of c_3 and c_4, and the two intermediate MOs have intermediate values for all four coefficients. The symmetry of these four orbitals is the same as those without mixing, but their shapes are changed somewhat by having significant contributions from both the s and p atomic orbitals. In addition, the energies are shifted relative to their placement if the upper two exhibited nearly exclusive contribution from $2p_z$ while the lower two exclusive contribution from 2s.

It is clear that s-p mixing often has a detectable influence on molecular orbital energies. For example, in early second period homonuclear diatomics (Li_2 to N_2), the σ_g orbital formed from $2p_z$ orbitals is higher in energy than the π_u orbitals formed from the $2p_x$ and $2p_y$ orbitals. This is an inverted order from that expected without s-p mixing. For B_2 and C_2, s-p mixing affects their magnetic properties. Mixing also changes the bonding–antibonding nature of some orbitals. The orbitals with intermediate energies may, on the basis of s-p mixing, gain either a slightly bonding or slightly antibonding character and contribute in minor ways to the bonding. Each orbital must be considered separately on the basis of its energy and electron distribution.

Diatomic Molecules of the First and Second Periods

Before proceeding with examples of homonuclear diatomic molecules, we must define two types of magnetic behavior, paramagnetic and diamagnetic. Paramagnetic compounds are attracted by an external magnetic field. This attraction is a consequence of one or more unpaired electrons behaving as tiny magnets. Diamagnetic compounds, on the other hand, have no unpaired electrons and are repelled slightly by magnetic fields.

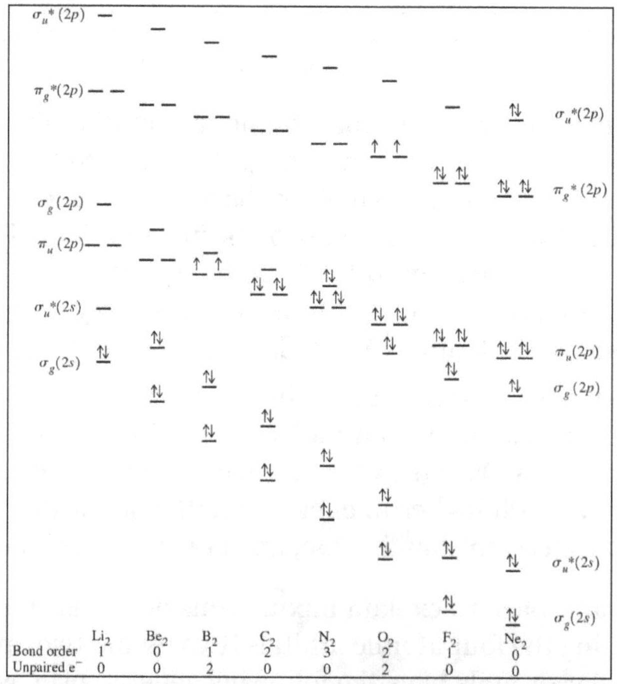

Energy Levels of the Homonuclear Diatomics of the Second Period.

H_2, He_2, and the homonuclear diatomic species are shown in above figure. Atomic orbital energies decrease across a row in the Periodic Table as the increasing effective nuclear charge attracts the electrons more strongly. The result is that the molecular orbital energies for the corresponding homonuclear diatomics also decrease across the row. This decrease in energy is larger for s orbitals than for π orbitals, due to the greater overlap of the atomic orbitals that participate in σ interactions.

$H_2[\sigma_g^2(1s)]$

This is the simplest diatomic molecule. The MO description shows a single σ orbital containing one electron pair; the bond order is 1, representing a single bond. The ionic species H_2^+, with a single electron in the a σ orbital and a bond order of $\frac{1}{2}$, has been detected in low-pressure gas-discharge systems. As expected, H_2^+ has a weaker bond than H_2 and therefore a considerably longer bond distance than H_2 (105.2 pm vs. 74.1 pm).

$He_2[\sigma_g^2\sigma_u^{*2}(1s)]$

The molecular orbital description of He_2 predicts two electrons in a bonding orbital and two in an antibonding orbital, with a bond order of zero—in other words, no bond. This is what is observed experimentally. The noble gas He has no significant tendency to form diatomic molecules and, like the other noble gases, exists in the form of free atoms. He_2 has been detected only in very low-pressure and low-temperature molecular beams. It has an extremely low binding energy, approximately 0.01 J/mol; for comparison, H_2 has a bond energy of 436 kJ/mol.

$Li_2[\sigma_g^2(2s)]$

The MO model predicts a single Li–Li bond in Li_2, in agreement with gas-phase observations of the molecule.

$Be_2[\sigma_g^2\sigma_u^{*2}(2p)]$

Be_2 has the same number of antibonding and bonding electrons and consequently a bond order of zero. Hence, like He_2, Be_2 is an unstable species.

$B_2[\pi_u^1\pi_u^1(2p)]$

Here is an example in which the MO model has a distinct advantage over the Lewis dot model. B_2 is a gas-phase species; solid boron exists in several forms with complex bonding, primarily involving B_{12} icosahedra. B_2 is paramagnetic. This behavior can be explained if its two highest energy electrons occupy separate π orbitals. The Lewis dot model cannot account for the paramagnetic behavior of this molecule.

The energy-level shift caused by s-p mixing is vital to understand the bonding in B_2. In the absence of mixing, the $\sigma_g(2p)$ orbital would be expected to be lower in energy than the $\pi_u(2p)$ orbitals, and the molecule would likely be diamagnetic. However, mixing of the $\sigma_g(2s)$ orbital with the $\sigma_g(2p)$ orbital lowers the energy of the $\sigma_g(2s)$ orbital and increases the energy of the $\sigma_g(2p)$ orbital to a higher level than the p orbitals, giving the order of energies. As a result, the last two electrons are unpaired in the degenerate π orbitals, as required by Hund's rule of maximum multiplicity, and the molecule is paramagnetic. Overall, the bond order is one, even though the two π electrons are in different orbitals.

$C_2[\pi_u^2\pi_u^2(2p)]$

The MO model of C_2 predicts a doubly bonded molecule, with all electrons paired, but with both highest occupied molecular orbitals (HOMOs) having π symmetry. C_2 is unusual because it has two π bonds and no σ bond. Although C_2 is a rarely encountered allotrope of carbon (carbon is significantly more stable as diamond, graphite, fullerenes and other polyatomic forms), the acetylide ion, C_2^{2-}, is well known, particularly in compounds with alkali metals, alkaline earths, and lanthanides. According to the molecular orbital model, C_2^{2-} should have a bond order of 3 (configuration $\pi_u^2\pi_u^2\sigma_g^2$). This is supported by the similar C–C distances in acetylene and calcium carbide (acetylide).

C–C Distance	(pm)
C=C(gas phase)	124.2
H–C≡C–H	120.5
CaC_2	119.1

$N_2[\pi_u^2\pi_u^2\sigma_g^2(2p)]$

N_2 has a triple bond according to both the Lewis and the molecular orbital models. This agrees with its very short N–N distance (109.8 pm) and extremely high bond dissociation energy (942 kJ/mol). Atomic orbitals decrease in energy with increasing nuclear charge Z; as the effective nuclear charge increases, the energies of all orbitals are reduced. The varying shielding abilities of electrons in different orbitals and electron–electron interactions cause the difference between the 2s and 2p energies to increase as Z increases, from 5.7 eV for boron to 8.8 eV for carbon and 12.4 eV for nitrogen. The radial probability functions indicate that 2s electrons have a higher probability of being close to the nucleus relative to 2p electrons, rendering the 2s electrons more susceptible to the increasing nuclear charge as Z increases. As a result, the $\sigma_g(2s)$ and $\sigma_g(2p)$ levels of N_2 interact (mix) less than the corresponding B_2 and C_2 levels, and the N_2 $\sigma_g(2p)$ and $\pi_u(2p)$ are very close in energy.

$O_2[\sigma_g^2\pi_u^2\pi_u^2\pi_g^{*1}\pi_g^{*1}(2p)]$

O_2 is paramagnetic. As for B_2, this property cannot be explained by the Lewis dot structure :Ö=Ö:, but it is evident from the MO picture, which assigns two electrons to the degenerate πg^* orbitals. The paramagnetism can be demonstrated by pouring liquid O_2 between the poles of a strong magnet; O_2 will be held between the pole faces until it evaporates. Several charged forms of diatomic oxygen are known, including O_2^+, O_2^-, and O_2^{2-}. The internuclear O–O distance can be conveniently correlated with the bond order predicted by the molecular orbital model, as shown in the following table:

	Bond Order	Internuclear Distance (pm)
O_2^+ (dioxygenyl)	2.5	111.6
O_2 (dioxygen)	2.0	120.8
O_2^- (superoxide)	1.5	135
O_2^{2-} (peroxide)	1.0	150.4

Oxygen–oxygen distances in O_2^- and O_2^{2-} are influenced by the cation. This influence is especially strong in the case of O_2^{2-} and is one factor in its unusually long bond distance, which should be considered approximate. The disproportionation of KO_2 to O_2 and O_2^{2-} in the presence of hexacarboximide crypt and results in encapsulation of O_2^{2-} in the crypt and via hydrogen-bonding interactions. This O–O peroxide distance was determined as 150.4(2) pm. The extent of mixing is not sufficient in O_2 to push the $\sigma_g(2p)$ orbital to higher energy than the $\pi_u(2p)$ orbitals. The order of molecular orbitals shown is consistent with the photoelectron spectrum.

$F_2[\sigma_g^2\pi_u^2\pi_u^2\pi_g^{*2}\pi_g^{*2}(2p)]$

The MO model of F_2 shows a diamagnetic molecule having a single fluorine–fluorine bond, in agreement with experimental data. The bond order in N_2, O_2, and F_2 is the same whether or not mixing is taken into account, but the order of the $\sigma_g(2p)$ and $\pi_u(2p)$ orbitals is different in N_2 than in O_2 and F_2. The energy difference between the 2s and 2p orbitals of the second row main group elements increases with increasing Z, from 5.7 eV in boron to 21.5 eV in fluorine. As this difference increases, the s-p interaction (mixing) decreases, and the "normal" order of molecular orbitals returns in O_2 and F_2. The higher $\sigma_g(2p)$ orbital (relative to $\pi_u(2p)$) occurs in many heteronuclear diatomic molecules, such as CO.

Ne_2

All the molecular orbitals are filled, there are equal numbers of bonding and antibonding electrons, and the bond order is therefore zero. The Ne_2 molecule is a transient species, if it exists at all. One triumph of molecular orbital theory is its prediction of two unpaired electrons for O_2. Oxygen had long been known to be paramagnetic, but early explanations for this phenomenon were unsatisfactory. For example, a special "three-electron bond" was proposed. The molecular orbital description directly explains why two unpaired electrons are required. In other cases, experimental observations (paramagnetic B_2, diamagnetic C_2) require a shift of orbital energies, raising σ_g above π_u, but they do not require major modifications of the model.

Bond Lengths in Homonuclear Diatomic Molecules

Figure shows the variation of bond distance with the number of valence electrons in second-period p-block homonuclear diatomic molecules having 6 to 14 valence electrons. Beginning at the left, as the number of electrons increases the number in bonding orbitals also increases; the bond strength becomes greater, and the bond length becomes shorter.

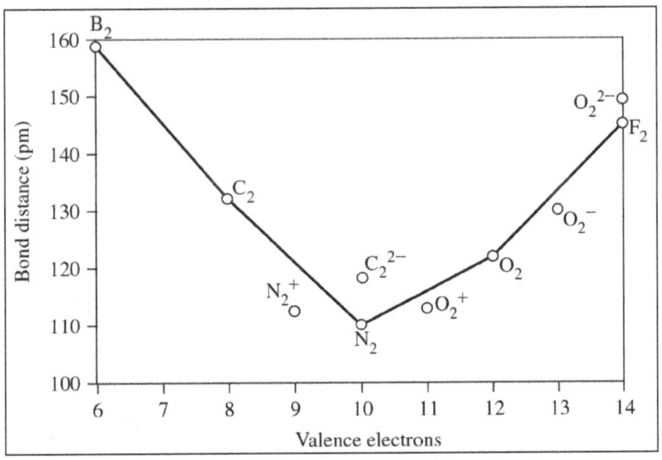

Bond Distances of Homonuclear Diatomic Molecules and Ions.

This continues up to 10 valence electrons in N_2, where the trend reverses, because the additional electrons occupy antibonding orbitals. The ions N_2^+, O_2^+, O_2^-, and O_2^{2-} are also shown in the figure and follow a similar trend. The minimum in figure above occurs even though the radii of the free atoms decrease steadily from B to F. Figure below shows the change in covalent radius for these atoms(defined for single bonds), decreasing as the number of valence electrons increases, primarily because the increasing nuclear charge pulls the electrons closer to the nucleus. For the elements boron through nitrogen, the trends shown in figures are similar: as the covalent radius of the atom decreases, the bond distance of the matching diatomic molecule also decreases. However, beyond nitrogen these trends diverge.

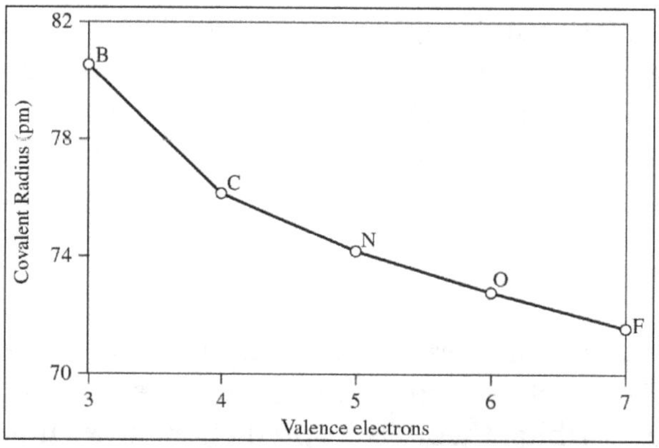

Covalent Radii of Second-Period Atoms.

Even though the covalent radii of the free atoms continue to decrease (N > O > F), the bond distances in their diatomic molecules increase ($N_2 < O_2 < F_2$) with the increasing population of antibonding orbitals. In general the bond order is the more important factor, overriding the covalent radii of the component atoms. Bond lengths of homonuclear and heteronuclear diatomic species are given in Table.

Table: Bond Distances in Diatomic Species.

Formula	Valence Electrons	Internuclear Distance (pm)
H_2^+	1	105.2
H_2	2	74.1
B_2	6	159.0
C_2	8	124.2
C_2^{2-}	10	119.1
N_2^+	9	111.6
N_2	10	109.8
O_2^+	11	111.6
O_2	12	120.8
O_2^-	13	135
O_2^{2-}	14	150.4
F_2	14	141.2

CN	9	117.2
CN^-	10	115.9
CO	10	112.8
NO^+	10	106.3
NO	11	115.1
NO^-	12	126.7

Photoelectron Spectroscopy

In addition to data on bond distances and energies, specific information about the energies of electrons in orbitals can be determined from photoelectron spectroscopy. In this technique, ultraviolet (UV) light or X-rays eject electrons from molecules:

$$O_2 + hv\text{ (photons)} \rightarrow O_2^+ + e^-$$

The kinetic energy of the expelled electrons can be measured; the difference between the energy of the incident photons and this kinetic energy equals the ionization energy (binding energy) of the electron:

$$\text{Ionization energy} = hv(\text{energy of photons}) - \text{kinetic energy of the expelled electron}$$

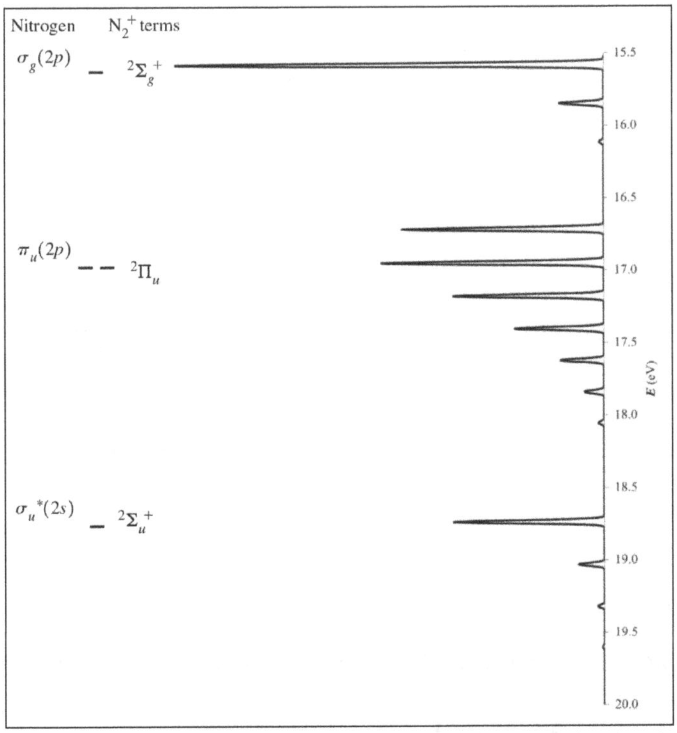

Photoelectron Spectrum and Molecular Orbital Energy Levels of N_2.

UV light removes outer electrons; X-rays are more energetic and can remove inner electrons. Figures show photoelectron spectra for N_2 and O_2, respectively, and the relative energies of the highest occupied orbitals of the ions. The lower energy peaks (at the top in the figure) are for the higher energy orbitals (less energy required to remove electrons). If the energy levels of the ionized molecule are assumed to be essentially the same as those of the uncharged molecule, the observed energies can be directly correlated with the molecular orbital energies. The levels in the N_2 spectrum are more closely spaced than those in the O_2 spectrum, and some theoretical calculations have disagreed about the order of the highest occupied orbitalsin N_2. Stowasser and Hoffmann have compared different calculation methods and showed that the different order of energy levels wassimply a function ofthe calculation method; the methods favored by the authors agree with the experimental results, with σ_g above π_u.

The photoelectron spectrum shows the π_u lower than σ_g in N_2. In addition to the ionization energies of the orbitals, the spectrum provides evidence of the quantized electronic and vibrational energy levels of the molecule. Because vibrational energy levels are much more closely spaced than electronic levels, any collection of molecules will include molecules with different vibrational energies even when the molecules are in the ground electronic state. Therefore, transitions from electronic levels can originate from different vibrational levels, resulting in multiple peaks for a single electronic transition. Orbitals that are strongly involved in bonding have vibrational fine structure (multiple peaks); orbitals that are less involved in bonding have only a few peaks at each energy level. The N_2 spectrum indicates that the π_u orbitals are more involved in the bonding than either of the s orbitals. The CO photoelectron spectrum has a similar pattern. The O_2 photoelectron spectrum has much more vibrational fine structure for all the energy levels, with the π_u levels again more involved in bonding than the other orbitals. The photoelectron spectra of O_2 and of CO show the expected order of energy levels for these molecules.

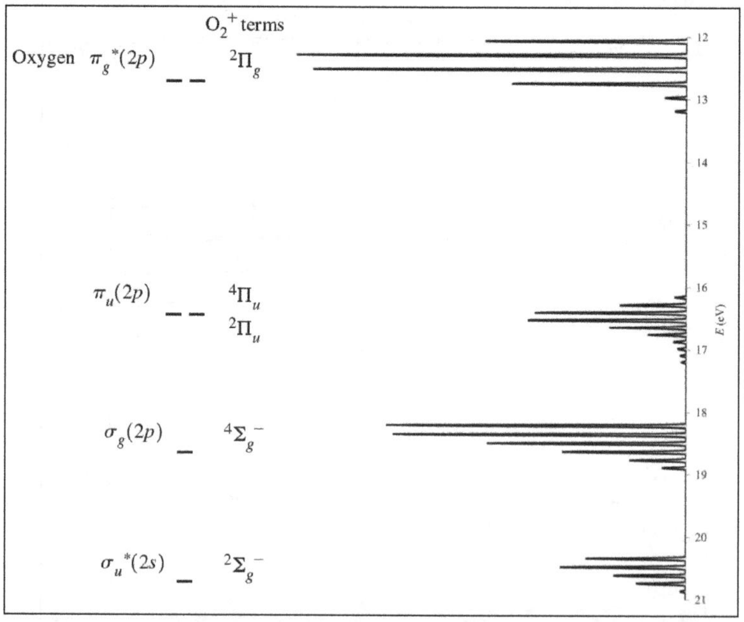

Photoelectron Spectrum and Molecular Orbital Energy Levels of O_2.

Heteronuclear Diatomic Molecules

The electron density within the occupied molecular orbitals is evenly distributed over each atom. A discussion of heteronuclear diatomic molecules provides an introduction into how molecular orbital theory treats molecules that are polar, with an unequal distribution of the electron density in the occupied orbitals.

Polar Bonds

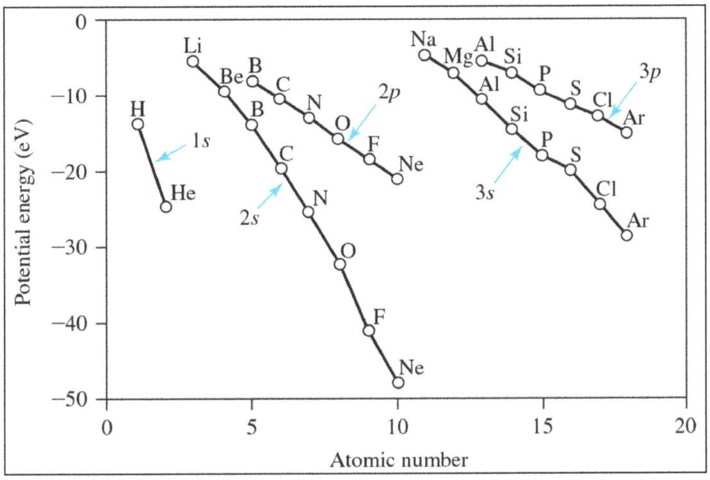

Orbital Potential Energies.

The application of molecular orbital theory to heteronuclear diatomic molecules is similar to its application to homonuclear diatomics, but the different nuclear charges of the atoms require that interactions occur between orbitals of unequal energies and shifts the resulting molecular orbital energies. In dealing with these heteronuclear molecules, it is necessary to estimate the energies of the atomic orbitals that may interact. These potential energies are negative, because they represent attraction between valence electrons and atomic nuclei. The values are the average energies for all electrons in the same level (for example, all 3p electrons), and they are weighted averages of all the energy states that arise due to electron–electron interactions. For this reason, the values do not show the variations of the ionization energies but steadily become more negative from left to right within a period, as the increasing nuclear charge attracts all the electrons more strongly.

Table: Orbital Potential Energies.

Orbital Potential Energy (eV)								
Atomic Number	Element	1s	2s	2p	3s	3p	4s	4p
1	H	-13.61						
2	He	-24.59						
3	Li		-5.39					
4	Be		-9.32					
5	B		-14.05	-8.30				
6	C		-19.43	-10.66				
7	N		-25.56	-13.18				
8	O		-32.38	-15.85				
9	F		-40.17	-18.65				
10	Ne		-48.47	-21.59				
11	Na				-5.14			
12	Mg				-7.65			

13	Al				-11.32	-5.98			
14	Si				-15.89	-7.78			
15	P				-18.84	-9.65			
16	S				-22.71	-11.62			
17	Cl				-25.23	-13.67			
18	Ar				-29.24	-15.82			
19	K							-4.34	
20	Ca							-6.11	
30	Zn							-9.39	
31	Ga							-12.61	-5.93
32	Ge							-16.05	-7.54
33	As							-18.94	-9.17
34	Se							-21.37	-10.82
35	Br							-24.37	-12.49
36	Kr							-27.51	-14.22

The atomic orbitals of the atoms that form homonuclear diatomic molecules have identical energies, and both atoms contribute equally to a given MO. Therefore, in the molecular orbital equations, the coefficients associated with the same atomic orbitals of each atom (such as the $2p_z$) are identical. In heteronuclear diatomic molecules, such as CO and HF, the atomic orbitals have different energies, and a given MO receives unequal contributions from these atomic orbitals; the MO equation has a different coefficient for each of the atomic orbitals that contribute to it. As the energies of the atomic orbitals get farther apart, the magnitude of the interaction decreases. The atomic orbital closer in energy to an MO contributes more to the MO, and its coefficient is larger in the wave equation.

Carbon Monoxide

The most efficient approach to bonding in heteronuclear diatomic molecules employs the same strategy as for homonuclear diatomics with one exception: the more electronegative element has atomic orbitals at lower potential energies than the less electronegative element. Carbon monoxide, shows this effect, with oxygen having lower energies for its 2s and 2p orbitals than the matching orbitals of carbon. The result is that the orbital interaction diagram for CO resembles that for a homonuclear diatomic, with the right (more electronegative) side pulled down in comparison with the left. In CO, the lowest set of π orbitals is lower in energy than the lowest s orbital with significant contribution from the 2p subshells the same order occurs in N_2. This is the consequence of significant interactions between the $2p_z$ orbital of oxygen and both the 2s and $2p_z$ orbitals of carbon. Oxygen's $2p_z$ orbital (-15.85 eV) is intermediate in energy between carbon's 2s (-19.43 eV) and $2p_z$ (-10.66 eV), so the energy match for both interactions is favorable.

The 2σ orbital has more contribution from (and is closer in energy to) the lower energy oxygen 2s atomic orbital; the $2\sigma^*$ orbital has more contribution from (and is closer in energy to) the higher energy carbon 2s atomic orbital. In the simplest case, the bonding orbital is similar in energy and shape to the lower energy atomic orbital, and the antibonding orbital is similar in energy and shape to the higher energy atomic orbital. In more complicated cases, such as the 2σ* orbital of CO, other orbitals (the oxygen $2p_z$ orbital) also contribute, and the molecular orbital shapes and energies are not as easily predicted. As a practical matter, atomic orbitals with energy differences greater than about 10 eV to 14 eV usually do not interact significantly.

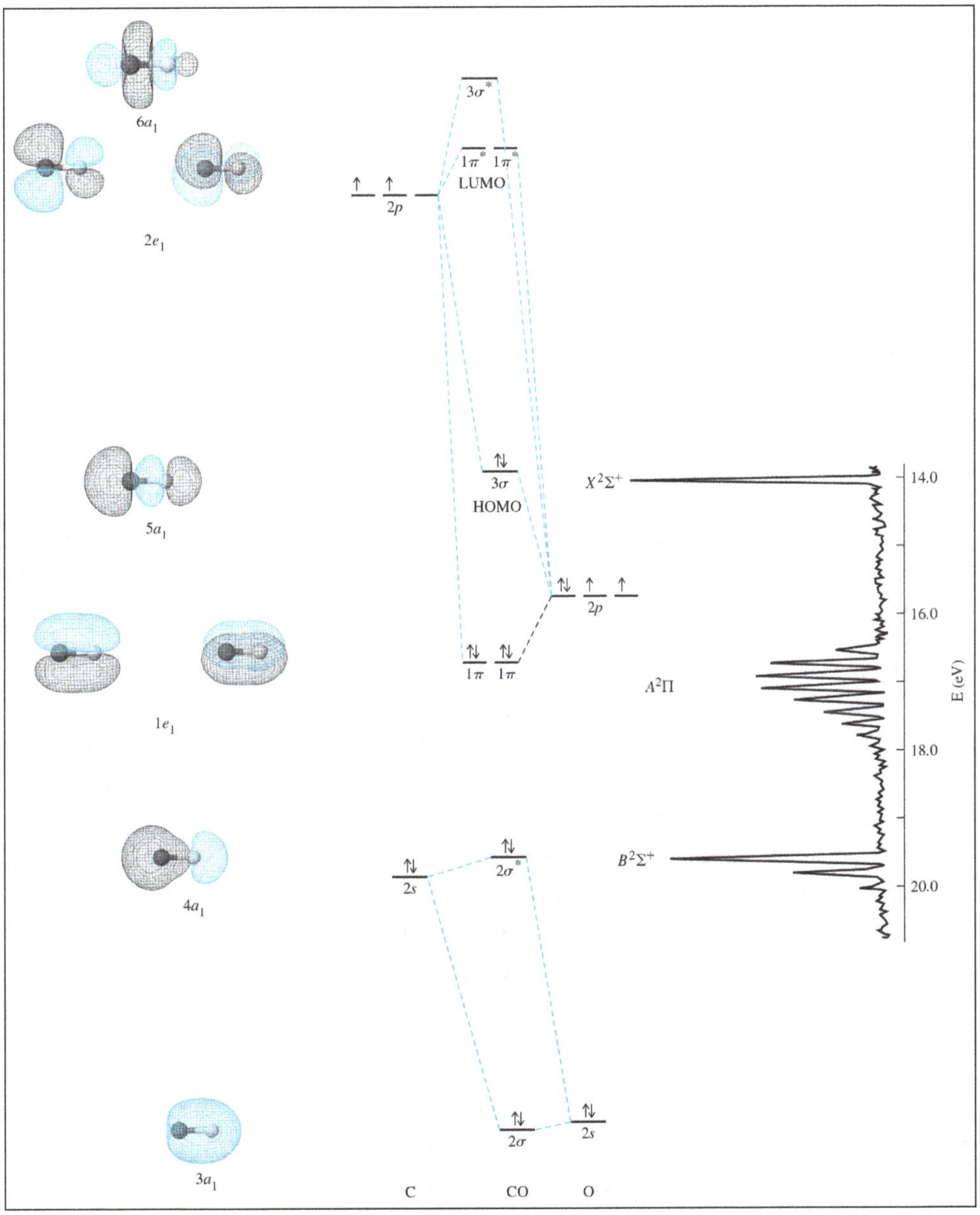

Molecular Orbitals and Photoelectron Spectrum of CO. Molecular orbitals 1σ and 1σ* are from the 1s orbitals and are not shown.

Mixing of the σ and σ* levels, like that seen in the homonuclear σ_g and σ_u orbitals, causes a larger split in energy between the 2σ* and 3σ, and the 3σ is higher than the 1π levels. The shape of the 3σ orbital is interesting, with a very large lobe on the carbon end. This is a consequence of the ability of both the 2s and $2p_z$ orbitals of carbon to interact with the $2p_z$ orbital of oxygen (because of the favorable energy match in both cases); the orbital has significant contributions from two orbitals on carbon but only one on oxygen, leading to a larger lobe on the carbon end. The pair of electrons in the 3σ orbital most closely approximates the carbon-based lone pair in the Lewis structure of CO, but the electron density is still delocalized over both atoms.

The p_x and p_y orbitals also form four molecular orbitals, two bonding (1π) and two antibonding (1π*). In the bonding orbitals the larger lobes are concentrated on the side of the more electronegative oxygen, reflecting the better energy match between these orbitals and the $2p_x$ and $2p_y$ orbitals of oxygen. In contrast, the larger lobes of the π* orbitals are on carbon, a consequence of the better energy match of these antibonding orbitals with the $2p_x$ and $2p_y$ orbitals of carbon. The distribution of electron density in the 3σ and 1π* orbitals is vital to understand how CO binds to transition metals. When the electrons are filled in, the valence orbitals form four bonding pairs and one antibonding pair for a net bond order of 3.

The molecular orbitals that are typically of greatest interest for reactions are the highest occupied molecular orbital (HOMO) and the lowest unoccupied molecular orbital (LUMO), collectively known asfrontier orbitals because they lie at the occupied–unoccupied frontier. The MO diagram of CO helps explain its reaction chemistry with transition metals, which is different than that predicted by electronegativity considerations that would suggest more electron density on the oxygen. On the sole basis of the carbon–oxygen electronegativity difference (and without considering the MO diagram), compounds in which CO is bonded to metals, called carbonyl complexes, would be expected to bond as M–O–C with the more electronegative oxygen attached to the electropositive metal. One impact of this electronegativity difference within the MO model is that the 2σ and 1π molecular orbitals in CO feature greater electron density on the more electronegative.

However, the structure of the vast majority of metal carbonyl complexes, such as $Ni(CO)_4$ and $Mo(-CO)_6$, has atoms in the order M–C–O. The HOMO of CO is 3σ, with a larger lobe, and therefore higher electron density, on the carbon (as explained above on the basis of s-p mixing). The electron pair in this orbital is more concentrated on the carbon atom, and can form a bond with a vacant orbital on the metal. The electrons of the HOMO are of highest energy (and least stabilized) in the molecule; these are generally the most energetically accessible for reactions with empty orbitals of other reactants. The LUMOs are the 1π* orbitals; like the HOMO, these are concentrated on the less electronegative carbon, a feature that also predisposes CO to coordinate to metals via the carbon atom. Indeed, the frontier orbitals can both donate electrons (HOMO) and accept electrons (LUMO) in reactions.

Ionic Compounds and Molecular Orbitals

Ionic compounds can be considered the limiting form of polarity in heteronuclear diatomic molecules. As the atoms forming bonds differ more in electronegativity, the energy gap between the interacting atomic orbitals also increases, and the concentration of electron density is increasingly biased toward the more electronegative atom in the bonding molecular orbitals. At the limit, the electron is transferred completely to the more electronegative atom to form a negative ion, leaving a positive ion with a high-energy vacant orbital. When two elements with a large difference in their electronegativities (such as Li and F) combine, the result is an ionic bond.

However, in molecular orbital terms, we can treat an ion pair like we do a covalent compound. In figure, the atomic orbitals and an approximate indication of molecular orbitals for such a diatomic molecule, LiF, are given. On formation of the diatomic molecule LiF, the electron from the Li 2s orbital is transferred to the bonding orbital formed from interaction between the Li 2s orbital and the F $2p_z$ orbital.

Indeed, the s orbital deriving from the $2s/2p_z$ interaction has a significantly higher contribution from $2p_z$ relative to 2s on the basis of the large energy gap. Both electrons, the one originating from Li and the one originating from the F $2p_z$ orbital, are stabilized. Note that the level of theory used to calculate the orbital surfaces in Figure below suggests essentially no covalency in diatomic LiF.

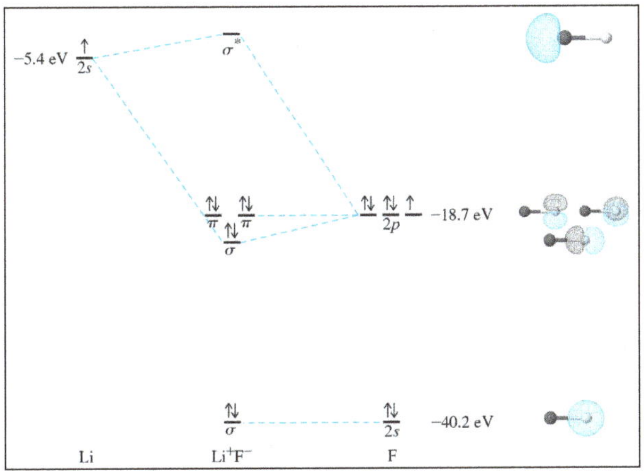

Approximate LiF Molecular Orbitals.

Lithium fluoride exists as a crystalline solid; this form of LiF has significantly lower energy than diatomic LiF. In a three-dimensional crystalline lattice containing many formula units of a salt, the ions are held together by a combination of electrostatic (ionic) attraction and covalent bonding. There is a small covalent bonding contribution in all ionic compounds, but salts do not exhibit directional bonds, in contrast to molecules with highly covalent bonds that adopt geometries predicted by the VSEPR model. In the highly ionic LiF, each Li^+ ion is surrounded by six F^- ions, each of which in turn is surrounded by six Li^+ ions.

Addition of these elementary steps, beginning with solid Li and gaseous F_2, results in formation of the corresponding gas-phase ions, and provides the net enthalpy change for this chemical change:

Elementary Step	Chemical/Physical Change	ΔH°(kJ/mol)
$Li(s) \rightarrow Li(g)$	Sublimation	161
$Li(g) \rightarrow Li^+(g) + e^-$	Ionization	520
$\frac{1}{2}F_2(g) \rightarrow F(g)$	Bond dissociation	79
$F(g) + e^- \rightarrow F^-(g)$	−Electron affinity	-328
$Li(s) + \frac{1}{2}F_2(g) \rightarrow Li^+(g) + F^-(g)$	Formation of gas-phase ions from the elements in their standard states	432

The free energy change ($\Delta G = \Delta H - T\Delta S$) must be negative for a reaction to proceed spontaneously. The very large and positive ΔH (432 kJ/mol) associated with the formation of these gas-phase ions renders the ΔG positive for this change despite its positive ΔS. However, if we combine these isolated ions, the large coulombic attraction between them results in a dramatic decrease in electronic energy, releasing 755 kJ/mol on formation of gaseous Li^+F^- ion pairs and 1050 kJ/mol on formation of a LiF crystal containing 1 mol of each ion.

Elementary Step	Chemical Change	ΔH°(kJ/mol)
$Li^+(g)+F^-(g) \rightarrow LiF(g)$	Formation of gaseous ion pairs	−755
$Li^+(g)+F^-(g) \rightarrow LiF(s)$	Formation of crystalline solid	−1050

The lattice enthalpy for crystal formation is sufficiently large and negative to render ΔG for $Li(s)+\frac{1}{2}F_2(g) \rightarrow LiF(s)$ negative. Consequently, the reaction is spontaneous, despite the net endothermic contribution for generating gaseous ions from the parent elements and the negative entropy change associated with gas-phase ions coalescing into a crystalline solid.

Symmetry Properties of Orbitals

Of particular importance in this book are the symmetry properties of a collection of orbitals and the electronic states that arise by populating them with electrons. For a given basis set of valence atomic orbitals for a molecule $\{\chi_i\}$, it is of interest to ask what set of irreducible representations they fall into (in the mathematical terms, in the foregoing discussion). There are two types of functions to consider: those associated with atoms, which lie at the point through which all of the symmetry elements pass and those that do not. In the molecules the orbitals of the central atom fall into the first class and those of the hydrogen atoms fall into the second. They may both be treated analogously. Consider as an example the set of hydrogen 1s orbitals in the NH_3 molecule below, which we use as a representation.

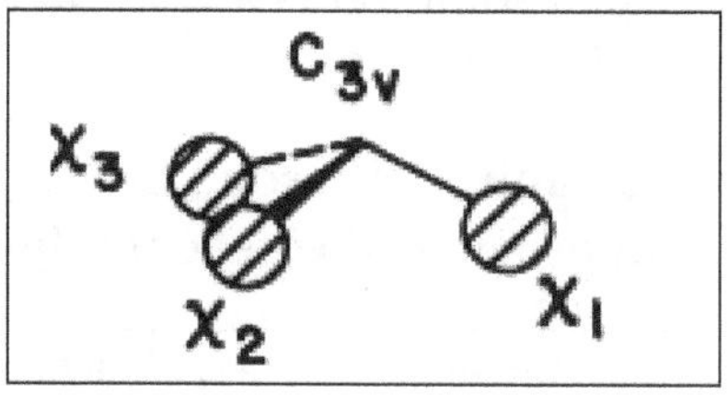

The first task is to generate the character of this representation, by examining the trace of the relevant transformation matrices, D(P) or D(K). This is easier than it seems at first sight. If a particular basis function, χ_j, is left unchanged during a symmetry operation, then a + 1 will reside on the diagonal (the jj entry) of the matrix. If a basis function is transformed to another function then, a 0 will remain on the diagonal. If the function is converted into minus itself then a - 1 will appear on the diagonal. With this set of rules, it is easy to generate the set of characters appropriate for the representation using the three hydrogen 1s orbitals of the ammonia molecule as a basis (Γ_{sH}). It is given by,

C_{3v}	E	$2C_3$	$3\sigma_v$
Γ_{sH}	3	0	1

Obviously, since the E operation is the null operation, the entry γ(E) is equal to the number of basis orbitals (3 in this case). The C_3 operation leads to all basis functions changing positions; the same is true for the C_3^2 operation. This is a general result; $\gamma_i(\mathcal{K}) = \gamma_i(\mathcal{R})$. In this instance, $\gamma(2C_3) = 0$. Using any one of the mirror planes changes positions of two members of the basis, and the third (which resides on the mirror plane) is left unshifted, so $\gamma(3\sigma) = 1$. Certainly, by inspection, this representation is reducible since its dimension is greater than the largest dimension permissible in this point group. The process of reducing a reducible representation of this type to a set of irreducible representations is an important step in the construction of molecular orbitals. The way we do this makes use of the properties of the irreducible representations.

Let us assume that some reducible representation may be written as the sum of a set of irreducible representations as in equation,

$$\Gamma_{red} = m_1\Gamma_1 + m_2\Gamma_2 + \cdots + m_h\Gamma_h$$

that is,

$$\gamma^{red}(\mathcal{R}) = \sum_{i=1}^{h} m_i\gamma_i(\mathcal{R}) \quad \text{or} \quad \gamma^{red}(\mathcal{K}) = \sum_{i=1}^{h} m_i\gamma_i(\mathcal{K})$$

Then,

$$\Gamma_i\Gamma_{red} = m_1\Gamma_i\Gamma_1 + m_2\Gamma_i\Gamma_2 + \cdots + m_i\Gamma_i\Gamma_i + \cdots + m_h\Gamma_i\Gamma_h$$

Using the shorthand notation introduced in equations $g_k\Gamma_i(\mathcal{K})\Gamma_j(\mathcal{K}) = g\delta_{ij}$ and $\Gamma_i(\mathcal{R})\Gamma_j(\mathcal{R}) = g\delta_{ij}$ all such products will be zero except that for j = i.

$$\Gamma_i\Gamma_{red} = {}_i\Gamma_i\Gamma_i$$

and so

$$m_i = \frac{\Gamma_i \cdot \Gamma_{red}}{\Gamma_i \cdot \Gamma_i}$$

which in terms of the characters is just,

$$m_i = \frac{1}{g}\sum_{\mathcal{K}} g_k\gamma^{red}(\mathcal{K})\gamma_i(\mathcal{K})$$

or

$$m_i = \frac{1}{g}\sum_{\mathcal{R}} \gamma^{red}(\mathcal{R})\gamma_i(\mathcal{R})$$

Using this result for the hydrogen 1s orbitals of NH_3 (equation, $E^{-1} = E$, $C_m^{-1} = C_3^{\,2}$, $(C_3^{\,2})^{-1} = C_3$, $\sigma i^{-1} = \sigma_i (i = a, b, c)$) leads to a reduction of the reducible representation as,

$$m_{a_1} = \frac{1}{6}\left[(1)(3)(1) + (2)(0)(1) + (3)(1)(1)\right] = 1$$

$$m_{a_2} = \frac{1}{6}\left[(1)(3)(1) + (2)(0)(1) + (3)(1)(-1)\right] = 0$$

$$m_e = \frac{1}{6}\left[(1)(3)(2) + (2)(0)(-1) + (3)(1)(0)\right] = 1$$

or $\Gamma_{sH} = a_1 + e$. We say then that the 1s orbitals of hydrogen span, or belong to the $a_1 + e$ representations or transform as the $a_1 + e$ symmetry species of this point group or are just of $a_1 + e$ symmetry in this point group. Notice we have used lowercase letters here for the Mulliken labels. Conventionally, uppercase characters are used for electronic states, whereas lowercase labels apply to atomic and molecular orbitals and also for describing molecular vibrations.

A similar process may be employed for the hydrogen 1s orbitals of water,

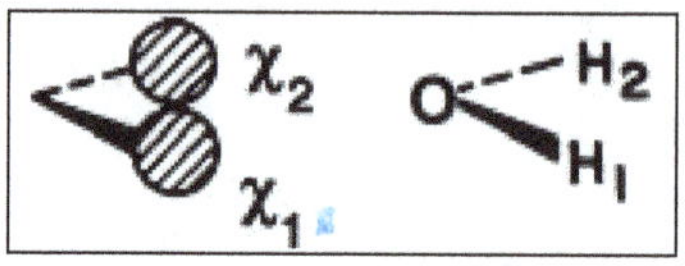

The reducible representation is simply generated as in equation,

C_{2v}	E	C_2	$\sigma_v(xz)$	$\sigma_v(yz)$
Γ_{sH}	2	0	0	2

and applying equation $m_i = \frac{1}{g}\sum_{\mathcal{R}} \gamma^{red}(\mathcal{R})\gamma_i(\mathcal{R})$ leads to,

$$m_{a_1} = \frac{1}{4}[(2)(1)+(0)(1)+(0)(1)+(2)(1)] = 1$$

$$m_{a_2} = \frac{1}{4}[(2)(1)+(0)(1)+(0)(-1)+(2)(-1)] = 0$$

$$m_{b_1} = \frac{1}{4}[(2)(1)+(0)(-1)+(0)(1)+(2)(-1)] = 0$$

$$m_{b_2} = \frac{1}{4}[(2)(1)+(0)(-1)+(0)(-1)+(2)(1)] = 1$$

So the pair of hydrogen ls orbitals in water transform as $a_1 + b_2$.

Another example is shown in below:

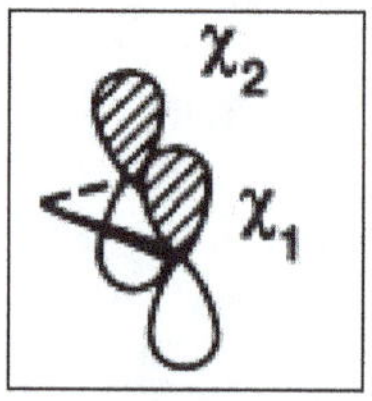

Suppose that instead of OH_2, the molecule of, interest was OF_2 and we wished to know how the two p_x orbitals on the fluorine atoms transformed. Then the set of characters for the two orbitals is given by,

C_{2v}	E	C_2	$\sigma_v(xz)$	$\sigma_v(yz)$
Γ_{xF}	2	0	0	-2

The entry under σv(yz) is of different sign from the corresponding one, since this symmetry operation sends each F px orbital to -px. Reduction of equation above either by inspection or using equation $m_i = \frac{1}{g}\sum_{\mathcal{R}} \gamma^{red}(\mathcal{R})\gamma_i(\mathcal{R})$ gives us the result that these two px orbitals transform as a2 + bl. Although equation $m_i = \frac{1}{g}\sum_{\mathcal{R}} \gamma^{red}(\mathcal{R})\gamma_i(\mathcal{R})$ allows access to the results of interest, with a little practice, the reducible representation may be reduced by inspection of the character table.

It is important at this point to comment on one consequence of these results. All of the molecular orbitals of the water molecule which contain hydrogen ls character must be either of a_1 or b_2 symmetry. The wavefunction of al symmetry must be symmetric with respect to all symmetry operations of the molecule, while that of b_2 symmetry is symmetric with respect to some and antisymmetric with respect to other symmetry operations. If of a_1 symmetry, then χ_1 and χ_2 will have the same sign and magnitude; if of b_2 symmetry, then χ_1 and χ_2 will have equal weight but their orbital coefficients will be of opposite sign.

Evaluation of the symmetry species of orbitals on the central atoms proceeds analogously. For the oxygen 2s orbital of the water molecule this is particularly simple. The characters of the representation given by the oxygen 2s orbital (Γ_{sO}) are written as,

C_{2v}	E	C_2	$\sigma_v(xz)$	$\sigma_v(yz)$
Γ_{sO}	1	1	1	1

By inspection, $\Gamma_{sO} = a_1$. This is a general result; s orbitals located on atoms, through which all the symmetry elements of the point group pass, always transform as the totally symmetric representation. The characters of the representation given by the oxygen 2p orbital (Γ_{pO}) are generated in the same way. Here, of course, a contribution of 1 is found whenever a given R sends one of these functions to minus itself.

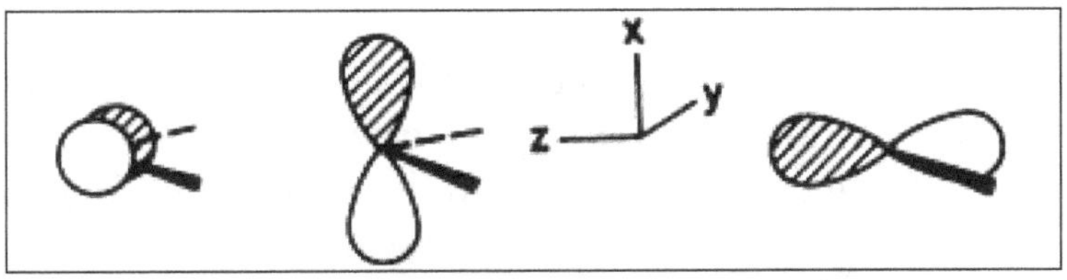

C_{2v}	E	C_2	$\sigma_v(xz)$	$\sigma_v(yz)$
Γ_{pO}	3	-1	1	1

Reduction of this representation leads to $\Gamma_{pO} = a_1 + b_1 + b_2$. These p orbitals change sign on going from +x, +y, or +z to -x, -y, or -z, respectively. In sets of character tables, the last columns often show the transformation properties of useful functions. In general, all that needs to be done to identify the symmetry species of such atomic p orbitals located at the point in space, which contains all of the point symmetry elements, is to identify those of x, y, and z. Notice that in the C_{3v} point group x, y, and z transform as $\Gamma_{xyz} = e + a_1$. Thus, the p_x and p_y orbitals on the central atom are degenerate (e) in this point group. Similar considerations apply to d orbitals; d_{xy} and d_{xz} transform as the functions xy and xz, for example. Inspection of the character table for the D_{4h} point group then immediately tells us the symmetry species of the 3d orbitals needed to construct a molecular orbital diagram for PtH_4^{2-}. They are $\Gamma_{dPt} = a_{1g} + b_{1g} + b_{2g} + e_g$.

Symmetry-Adapted Wavefunctions

After having identified the symmetry properties of basis sets of atomic orbitals, we need to explore how to build wavefunctions with given symmetry properties from a particular basis. First we need to introduce the concept of symmetry-equivalent atoms or orbitals. Clearly, the two hydrogen ls orbitals of water are equivalent, specifically because any symmetry operation of the C_{2v} point group will either send χ_1 to itself or to χ_2 and likewise send χ_2 to itself or to χ_1. Thus, $\{\chi_1, \chi_2\}$ form a completely equivalent set. The SF_4 molecule of,

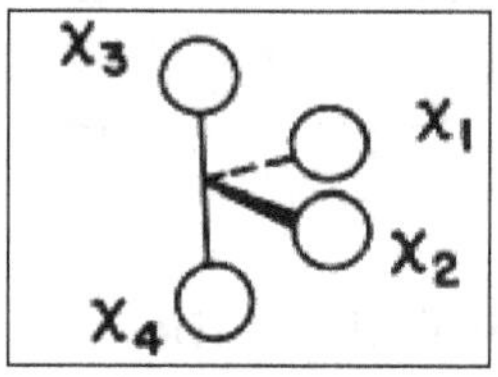

also belongs to the C_{2v} point group. Here, however, no symmetry operation sends χ_1 or χ_2 to χ_3 or χ_4, or vice versa. The orbitals χ_1 and χ_2 are permuted amongst themselves, as are the orbitals χ_3 and χ_4, by the symmetry operations of the group. So $\{\chi_1, \chi_2\}$ and $\{\chi_3, \chi_4\}$ form two different sets of symmetry-equivalent orbitals. We now generate the orbitals of a_1 and b_2 symmetry for the hydrogen ls orbitals of water. Since the symmetry properties of the resulting function must satisfy each entry $\gamma_i(\mathcal{R})$ of the relevant row of the character table, starting with the full complement of symmetry-equivalent orbitals (χ_1 and χ_2 in this case), we force the resulting wavefunction to have particular symmetry properties by construction using the character table itself. We write,

$$\phi(i)\alpha\sum_{\mathcal{R}}\gamma_i(\mathcal{R})\mathcal{R}\xi_1$$

Here, ξ_1 is any member of a complete symmetry-equivalent set. It does not matter whether we start off with χ_1 or χ_2 in this equation. Since they are symmetry equivalent, the same result will be found. For the C_{2v} point group, the results of the operations $\mathcal{R}(\chi_1)$ and $\mathcal{R}(\chi_2)$ are given by,

$$\begin{array}{ll} E\cdot\chi_1=\chi_1; & E\cdot\chi_2=\chi_2 \\ C_2\cdot\chi_1=\chi_2; & C_2\cdot\chi_2=\chi_1 \\ \sigma_v(yz)\cdot\chi_1=\chi_1; & \sigma_v(yz)\cdot\chi_2=\chi_2 \\ \sigma_v(xz)\cdot\chi_1=\chi_2; & \sigma_v(xz)\cdot\chi_2=\chi_1 \end{array}$$

The characters $\gamma_i(\mathcal{R})$ are given in table below For i = a_1,

$$\begin{aligned}\phi(a_1)\,&\alpha\,\chi_1+\chi_2+\chi_2+\chi_1\\ &\alpha\,\chi_1+\chi_2\end{aligned}$$

Table: The Group Table for C_{2v}.

C_{2v}	E	C_2	$\sigma_v(xz)$	$\sigma_v(yz)$	
A_1	1	1	1	1	z

A_2	1	1	-1	-1	
B_1	1	-1	1	-1	x
B_2	1	-1	-1	1	y

This is shown pictorially in molecule below and may be normalized if desired.

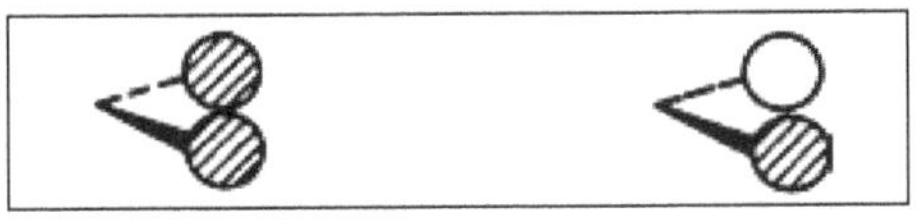

For i = b_2 then,

$$\phi(b_2)\,\alpha\,\chi_1 - \chi_2 - \chi_2 + \chi_1$$
$$\alpha\,\chi_1 - \chi_2$$

This is shown pictorially in molecule above. This technique using equation $\phi(i)\alpha\sum_{R}\gamma_i(R)R\xi_1$ is a very general and powerful way to construct functions of this type. The operator in this equation is called a projection operator.

Group theory from equation,

$$m_{a_1} = \frac{1}{4}\left[(2)(1)+(0)(1)+(0)(1)+(2)(1)\right]=1$$
$$m_{a_2} = \frac{1}{4}\left[(2)(1)+(0)(1)+(0)(-1)+(2)(-1)\right]=0$$
$$m_{b_1} = \frac{1}{4}\left[(2)(1)+(0)(-1)+(0)(1)+(2)(-1)\right]=0$$
$$m_{b_2} = \frac{1}{4}\left[(2)(1)+(0)(-1)+(0)(-1)+(2)(1)\right]=1$$

describe that only functions of a_1 and b_2 symmetry were permitted. What happens if we try to construct a function of a_2 symmetry?

$$\phi(a_2)\,\alpha\,\chi_1 + \chi_2 - \chi_2 - \chi_1 = 0$$

The result is meaningless.

The case of the water molecule involved a low symmetry point group where all of the possible representations are nondegenerate. In general, this appears in the character table with a + 1 entry under the identity operation E. Higher symmetry point groups contain degenerate representations. Any point group that contains a threefold (or higher) rotation axis will contain at least one degenerate representation and as we have noted, the entry under E reveals the order of the degeneracy. Everything we have said so far holds for degenerate representations, except we need to be a little careful in using equation $\phi(i)\alpha\sum_{R}\gamma_i(R)R\xi_1$ for generating the wavefunctions. An example shows the problem encountered and the way it is overcome.

The three ls orbitals on the hydrogen atoms in ammonia, which belongs to the C3v point group. Using the approach described earlier, we showed that these three orbitals $\chi_1 - \chi_3$ transform as a1 + e, that is, one nondegenerate plus one doubly degenerate representation. Generation of the a1 wavefunction is easy. Using equation $m_i = \frac{1}{g}\sum_{\mathcal{R}} \gamma^{red}(\mathcal{R})\gamma_i(\mathcal{R})$

$$\phi(a_1)\ \alpha\ \chi_1 + \chi_2 + \chi_3 + \chi_1 + \chi_2 + \chi_3$$
$$\alpha\ \chi_1 + \chi_2 + \chi_3$$

The same answer is obtained from equation $m_i = \frac{1}{g}\sum_{\mathcal{R}} \gamma^{red}(\mathcal{R})\gamma_i(\mathcal{R})$, irrespective of which of χ_1, χ_2, or χ_3 is used as ξ_1 in equation $m_i = \frac{1}{g}\sum_{\mathcal{R}} \gamma^{red}(\mathcal{R})\gamma_i(\mathcal{R})$. For the doubly degenerate wavefunction, however, using equation $m_i = \frac{1}{g}\sum_{\mathcal{R}} \gamma^{red}(\mathcal{R})\gamma_i(\mathcal{R})$ with χ_1 gives,

$$\phi_a(e)\ \alpha\ 2\chi_1 - \chi_2 - \chi_3$$

Now if instead of using χ_1 in equation,

$$E = \begin{pmatrix} 1 & 0 & 0 \\ 0 & 1 & 0 \\ 0 & 0 & 1 \end{pmatrix} \quad C_3 = \begin{pmatrix} -\frac{1}{2} & -\frac{\sqrt{3}}{2} & 0 \\ \frac{\sqrt{3}}{2} & -\frac{1}{2} & 0 \\ 0 & 0 & 1 \end{pmatrix} \quad C_3^2 = \begin{pmatrix} -\frac{1}{2} & \frac{\sqrt{3}}{2} & 0 \\ -\frac{\sqrt{3}}{2} & -\frac{1}{2} & 0 \\ 0 & 0 & 1 \end{pmatrix}$$

$$\sigma_a = \begin{pmatrix} -1 & 0 & 0 \\ 0 & 1 & 0 \\ 0 & 0 & 1 \end{pmatrix} \quad \sigma_b = \begin{pmatrix} \frac{1}{2} & -\frac{\sqrt{3}}{2} & 0 \\ -\frac{\sqrt{3}}{2} & -\frac{1}{2} & 0 \\ 0 & 0 & 1 \end{pmatrix} \quad \sigma_c = \begin{pmatrix} \frac{1}{2} & \frac{\sqrt{3}}{2} & 0 \\ \frac{\sqrt{3}}{2} & -\frac{1}{2} & 0 \\ 0 & 0 & 1 \end{pmatrix}$$

we use χ_2, the result is,

$$\phi'(e)\ \alpha\ 2\chi_2 - \chi_1 - \chi_3$$

And finally, if χ_3 is used, we find,

$$\phi''(e)\ \alpha\ 2\chi_3 - \chi_1 - \chi_2$$

The result is three (not two!) different wavefunctions, which apparently describe a doubly degenerate situation. The problem here is that φa(e), φ'(e) and φ"(e) are not orthogonal. The solution is to use a technique known as the Schmidt orthogonalization process. First we accept equation $\phi_a(e) \, \alpha \, 2\chi_1 - \chi_2 - \chi_3$ as one component of the doubly degenerate pair.

The atomic orbitals x_1, x_2, and x_3 are normalized, and in equation $\phi_a(e) \, \alpha \, 2\chi_1 - \chi_2 - \chi_3$ the overlap between any two of them are the same. Thus,

$$\langle \chi_1 | \chi_1 \rangle = \langle \chi_2 | \chi_2 \rangle = \langle \chi_3 | \chi_3 \rangle = 1$$

$$\langle \chi_1 | \chi_2 \rangle = \langle \chi_2 | \chi_3 \rangle = \langle \chi_3 | \chi_1 \rangle = S$$

Consequently, we obtain,

$$\langle 2\chi_1 - \chi_2 - \chi_3 | 2\chi_1 - \chi_2 - \chi_3 \rangle = 6(1-S)$$

so that $\phi_a(e)$ is normalized as,

$$\phi_a(e) = \frac{1}{\sqrt{6(1-S)}}(2\chi_1 - \chi_2 - \chi_3)$$

Similarly, $\phi'(e)$ and $\phi''(e)$ are normalized as follows:

$$\phi'(e) = \frac{1}{\sqrt{6(1-S)}}(2\chi_2 - \chi_3 - \chi_1)$$

$$\phi''(e) = \frac{1}{\sqrt{6(1-S)}}(2\chi_3 - \chi_1 - \chi_2)$$

The orbital $\phi'(e)$ is not orthogonal to $\phi_a(e)$ since,

$$\langle \phi'(e) | \phi_a(e) \rangle = \frac{1}{6(1-S)} \langle 2\chi_1 - \chi_2 - \chi_3 | 2\chi_2 - \chi_3 - \chi_1 \rangle$$

$$= \frac{1}{6(1-S)}[3S-3] = -\frac{1}{2}$$

However, the orbital $\phi_b(e)$ defined in the special way of equation below is orthogonal to $\phi_a(e)$,

$$\phi_b(e) \, \alpha \, \phi'(e) - \langle \phi'(e) | \phi_a(e) \rangle \phi_a(e)$$

$$\alpha \, \chi_2 - \chi_3$$

Using above equation then "projects out" that portion of $\phi'(e)$ which is not orthogonal to $\phi_a(e)$. Since,

$$\langle \chi_2 - \chi_3 | \chi_2 - \chi_3 \rangle = 2(1-S)$$

$\phi_b(e)$ is normalized as,

$$\phi_b(e) = \frac{1}{\sqrt{2(1-S)}}(\chi_2 - \chi_3)$$

We note that use of the orbital $\phi''(e)$ in place of $\phi'(e)$ in equation $\phi_b(e) \alpha \phi'(e) - \langle \phi'(e) | \phi_a(e) \rangle \phi_a(e) \alpha \chi_2 - \chi_3$ does not produce a new orbital that is orthogonal to both $\phi a(e)$ and $\phi b(e)$ since,

$$\phi''(e) - \langle \phi''(e) | \phi_a(e) \rangle \phi_a(e) - \langle \phi''(e) | \phi_b(e) \rangle \phi_b(e) = 0$$

Therefore, $\psi_a(e)$ and $\psi_b(e)$ are a set of two orthonormal orbitals belonging to the e representation. In fact, there is considerable choice as to the form of the degenerate wavefunctions. Any linear combination of $\psi a(e)$ and $\psi b(e)$ are also valid wavefunctions. Figure below shows the form of the functions of equations $\phi_a(e) \alpha 2\chi_1 - \chi_2 - \chi_3$ and $\phi_b(e) \alpha \phi'(e) - \langle \phi'(e) | \phi_a(e) \rangle \phi_a(e) \alpha \chi_2 - \chi_3$.

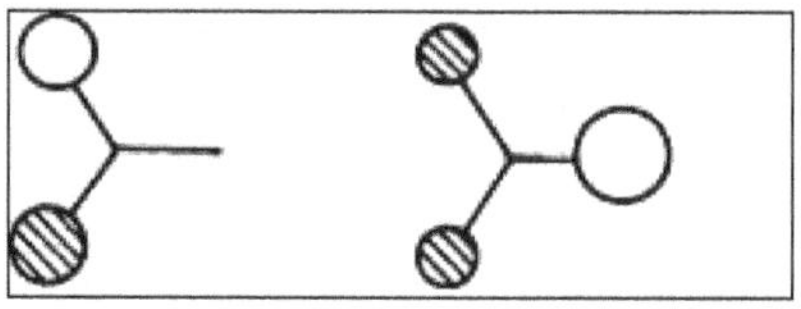

Direct Products

Very often we need to be able to write down the symmetry species of a function which is a simple product of two functions whose symmetry we know. For example, we may wish to know the symmetry species of the electronic state which arises from the orbital occupancy.

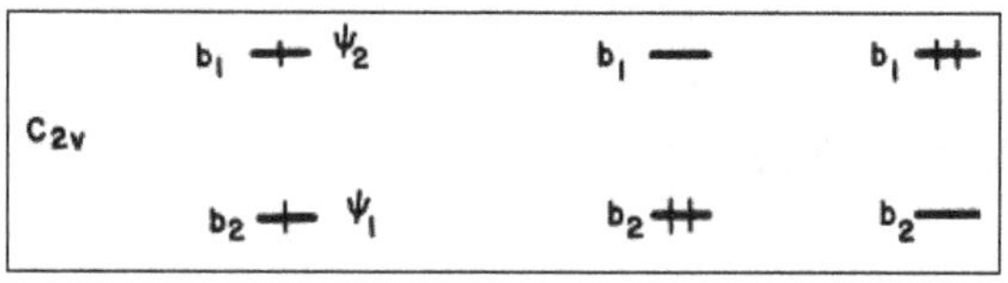

The symmetry labels that describe the wavefunctions ψ_1 and ψ_2 are written alongside the molecular orbital levels, using lowercase letters. The simplest function that might be expected to describe the behavior of this electronic state is just $\psi_1 \cdot \psi_2$. The symmetry species of such a product function is obtained by multiplying the respective characters $\gamma b_1(\mathcal{R}) \times \gamma b_2(\mathcal{R})$ as in equation below:

C_{2v}	E	C_2	$\sigma_v(xz)$	$\sigma_v(yz)$
B_1	1	-1	1	-1
B_2	1	-1	-1	1
$B_1 \otimes B_2$	1	1	-1	-1

The result is clearly the set of characters corresponding to the a_2 irreducible representation, that is, $b_1 \otimes b_2 = a_2$. An A_2 state, where uppercase letters are used to distinguish state symmetry from orbital symmetry. Depending upon whether the electron spins are oriented parallel or antiparallel, respectively, a triplet ($^3A_2$) or a singlet ($^1A_2$) state results. Using the same approach, it may readily be shown that the orbital occupation situations of equation $m_1^2 + m_2^2 + \cdots + m_k^2 = g$ give rise to $^1A_1$ electronic states, that is, $b_2 \otimes b_2 = a_1$ and $b_1 \otimes b_1 = a_1$. This result is found in all point groups. Any product of the two functions ψ_a and ψ_b which have the same symmetry species will always belong to the totally symmetric representation of the point group. This observation leads to the simplifying result that all closed shells of electrons give rise to electronic states that are totally symmetric.

In determining the symmetry species corresponding to a given orbital occupation, all we need to do, as a result, is to look at those electrons outside of completely filled orbitals (i.e., closed shells). Thus equation shown in above figure may just represent the valence orbitals of a system, which are partially filled. The symmetry species of the electronic state is not influenced by the core of completely filled levels.

Two examples involving degenerate levels are shown:

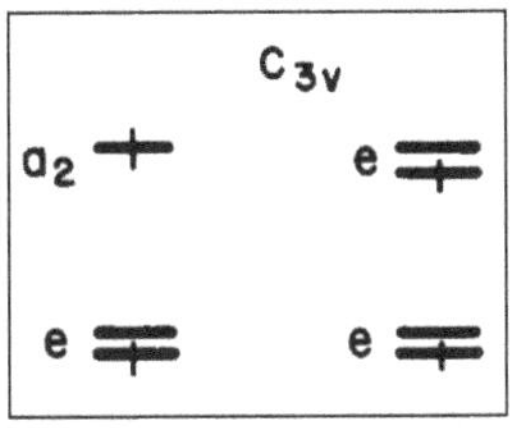

It is readily seen from table below that $a_2 \otimes e = e$, which leads to ^1E and ^3E electronic states, depending upon the orientation of the electron spins. In above molecule, the product of the characters leads to a reducible representation,

C_{3v}	E	$2C_3$	$3\sigma_v$
E	2	-1	0
E	2	-1	0
$E \otimes E$	4	1	0

and we need to use equation $m_i = \frac{1}{g}\sum_{\mathcal{R}} \gamma^{red}(\mathcal{R})\gamma_i(\mathcal{R})$ to break it down into a sum of irreducible representations. In this way, or by inspection, we find $e \otimes e = a_1 + a_2 + e$. Depending, then, upon the electron spin orientation, the electronic states that result are $^1A_1$, $^1A_2$, ^1E and $^3A_1$, $^3A_2$, ^3E. Note that in each of these products, the total number of singlets or total number of triplets corresponded to the simple product of the dimensions of the representations involved. So $b_1 \otimes b_2$ gave a $(1 \times 1 = 1)$ one-dimensional product (a_2), $e \otimes a_2$ gave a $(2 \times 1 = 2)$ two-dimensional product (e), and $e \otimes e$ gave a $(2 \times 2 = 4)$ four-dimensional product $(a_1 + a_2 + e)$.

Problems involving degenerate levels are not always as simple. Molecule shows double occupancy of an e orbital. There can only be three singlets and one triplet and not four of each.

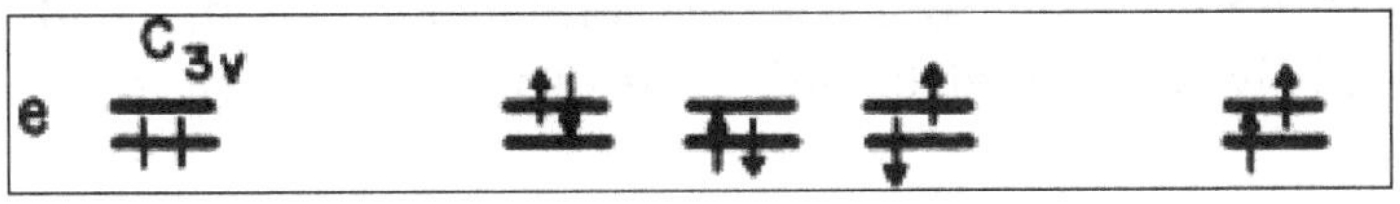

In this case, we need to define a different sort of direct product. The usual simple product leads to the characters defined in equation:

$$\gamma_i \otimes_j = \gamma_i(\mathcal{R})\,\gamma_j(\mathcal{R})$$

But the symmetric direct product which we use to generate the singlet levels are defined by,

$$\chi^{+}_{i^2}(\mathcal{R}) = 1/2\left[\gamma_i(\mathcal{R})\gamma_i(\mathcal{R}) + \gamma_i(\mathcal{R}^2)\right]$$

The antisymmetric direct product, which we use to generate the triplet levels of previous molecule, is defined by,

$$\chi^{-}_{i^2}(\mathcal{R}) = 1/2\left[\gamma_i(\mathcal{R})\gamma_i(\mathcal{R}) - \gamma_i(\mathcal{R}^2)\right]$$

First, we need to identify the operations corresponding to $\mathcal{R}^2$ in above two equations. For the C_{3v} point group,

$$E \times E = E;\ C_3^{\,2} \times C_3^{\,2} = C_3$$

$$C_3 \times C_3 = C_3^{\,2};\ \sigma_v = \sigma_v = E$$

and we may readily evaluate the relevant characters as,

C_{3v}	E	C_3	$C_3^{\,2}$	σ_a	σ_b	σ_c
$\gamma_e(\mathcal{R})$	2	-1	-1	0	0	0
$\gamma_e(\mathcal{R}^2)$	2	-1	-1	2	2	2
$\gamma_e(\mathcal{R}^2)$	4	1	1	0	0	0
$\chi^{+}_{e^2}(\mathcal{R})$	3	0	0	1	1	1
$\chi^{-}_{e^2}(\mathcal{R})$	1	1	1	-1	-1	-1

The set of characters for $\chi^{-}_{e^2}(\mathcal{R})$ are seen to give rise to an a_1 + e representation with the aid of Table below.

Table: The Group Table for C_{3v}.

C_{3v}	E	$2C_3$	$\sigma_v(xz)$	$\sigma_v(yz)$
A_1	1	1	1	Z
A_2	1	1	-1	
E	2	-1	0	(x, y)

The set for $\chi^{-}_{e^2}(\mathcal{R})$ corresponds to a_2. So in the C_3v point group, a configuration with two electrons in an e symmetry orbital gives rise to electronic states with symmetry $^1A_1$, ^1E, and $^3A_2$.

Orbital Symmetry Rules and Reactions

Figure shows the usual adiabatic plot of potential energy versus reaction coordinate. The points marked A, B and C will be used to derive the symmetry rules since they represent characteristic features of such a plot. Any point on the diagram corresponds to some arrangement of the nuclei of the reactants. This arrangement will automatically generate a certain point group (T_d, C_{3v}, C_s, etc.). All of the symmetry properties are now contained in the irreducible representations or symmetry species of that point group.

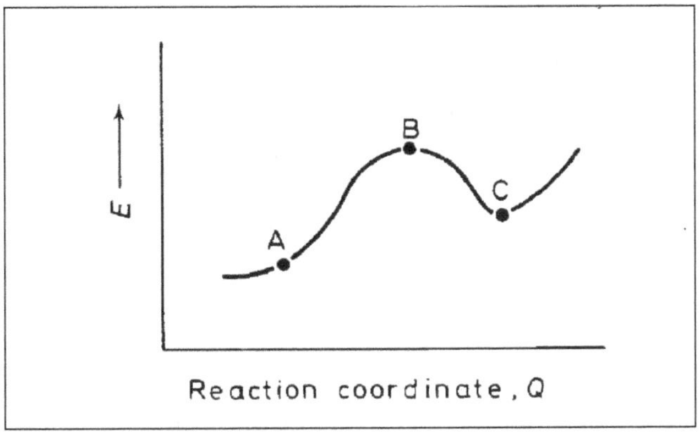

The wave equation for the system is now assumed to be solved exactly. This gives rise to a number of eigen states $\psi_0, \psi_1 \ldots \psi_k$, and corresponding eigen values $E_0, E_1 \ldots E_k$, where ψ_0 and E_0 refer to the ground electronic state. Now all the wave functions must belong to one of the symmetry species A, B, E etc. of the point group. Indirectly then, each energy value has a symmetry label tied to it.

Any arbitrary small motion of the nuclei away from the original configuration can be analysed as a sum of displacements corresponding to the normal modes of the pseudo-molecule representing the reactants. Each of these normal modes (of vibration) belongs to one of the symmetry species of the point group. Hence the reaction coordinate can also be classified as having a symmetry label depending on which nuclei are moved, and in what direction.

We now use quantum-mechanics in the form of perturbation theory to relate potential energy, E, to the reaction coordinate. For the ground electronic state, the energy becomes,

$$E = E_0 + Q\langle\psi_0|\partial U/\partial Q|\psi_0\rangle + \frac{1}{2}Q^2\langle\psi_0|\partial^2 U/\partial Q^2|\psi_0\rangle$$
$$+ Q^2\sum_k\left[\langle\psi_0|\partial U/\partial Q|\psi_k\rangle\right]^2/(E_0 - E_k)$$

where Q represents the reaction coordinate and also the magnitude of the small displacement from the original point on the diagram, Q_0, for which $E = E_0$. While above equation is valid only for Q very small, we can select Q_0 anywhere on Figure above. Hence, above equation is general for the purpose of displaying symmetry properties. Q and $(\partial U/\partial Q)$ have the same symmetry, where U is the nuclear—nuclear and electron—nuclear potential energy.

The bracket symbol, ⟨...⟩, represents integration over the electron coordinates, covering all space. We can now use a group theory rule to decide whether the integrals in above equation are exactly zero or not. The rule is that the direct product of three functions must contain the totally symmetrical species, or the integral over all space is zero.

Let us consider the term in above equation which is linear in Q. At any maximum or minimum in the potential energy curve the integral must be identically zero, independent of symmetry. At all other points this term must be the dominant one, since Q is small. If ψ_0 belongs to a degenerate symmetry species (E or T), the term usually leads to the first order Jahn—Teller effect, which removes the degeneracy. Since this is not important in the present context, we will assume that ψ_0 is non-degenerate.

Since the direct product of a non-degenerate species with itself is always totally symmetric, we derive our first symmetry rule: all reaction coordinates belong to the totally symmetrical representation. That is, $(\partial U/\partial Q)$ and also Q, must be totally symmetrical, otherwise its product with ψ_o^2 will not be symmetric and the integral will be zero. But it must be non-zero for all of the rising and falling parts of figure.

This means that once a reaction embarks on a particular reaction path it must stay within the same point group until it reaches energy maximum or minimum. A totally symmetrical set of nuclear motions can change bond angles and distances, but it cannot change the point group. This restriction on the point group is not as absolute as it sounds since an energy maximum may also be encountered in a normal mode orthogonal to the reaction coordinate. This then allows a non-symmetrical nuclear motion to change the point group.

We now consider point A on figure above. The integral $\langle \psi_o | \partial U / \partial Q | \psi_o \rangle$ has a positive value since the reaction has a positive activation energy. Instead of trying to evaluate the integral we accept that its value is the slope of figure at the point A. The terms in Q^2 in previous equation now become important. Their sum determines the curvature of the potential energy plot. For a reaction with small activation energy, the curvature should be as small as possible (or negative).

The integral $\langle \psi_o | \partial^2 U / \partial Q^2 | \psi_o \rangle$ has a non-zero value by symmetry since $(\partial^2 U/\partial Q^2)$ is totally symmetrical. Furthermore it will be positive for all molecules. It represents the force constant which resists moving any set of nuclei away from an original configuration for which ψ_o^2 is the electron density distribution. The last term in previous equation represents the change in energy that results from changing the electron distribution to one more suited to the new nuclear positions determined by Q. Its value is always negative since $E_o - E_k$ is a negative number.

This can be seen more easily if the equation for the wave function is written down from perturbation theory.

$$\psi = \psi_o + Q \sum \langle \psi_o | \partial U / \partial Q | \psi_k \rangle \psi_k / (E_o - E_k)$$

The summations in equations $E = E_o + Q\langle \psi_o | \partial U / \partial Q | \psi_o \rangle + \frac{1}{2} Q^2 \langle \psi_o | \partial^2 U / \partial Q^2 | \psi_o \rangle + Q^2 \sum_k [\langle \psi_o | \partial U / \partial Q | \psi_k \rangle]^2 / (E_o - E_k)$ and above equation are over all excited states. Each excited state wave function is mixed into the ground state wave function by an amount shown in above equation. The wave function is changed because the resulting electron distribution, ψ_2, is better suited to the new nuclear positions.

Now we can use group theory to show that only excited state wave functions, ψ_k, which have the same symmetry as ψ_o can mix in and lower the potential energy barrier. This follows because we have already shown that $(\partial U/\partial Q)$ must be totally symmetric. Hence the direct product of ψ_o and ψ_k must be totally symmetric, but this requires that they have the same symmetry. We can conclude that for a chemical reaction to occur with a reasonable activation energy, there must be low-lying excited states for the reacting system of the same symmetry as the ground state. Such a reaction is said to be symmetry allowed. A symmetry forbidden reaction is simply one which has a very high activation energy because of the absence of suitable excited states.

Previous two equations are exact, as are the symmetry rules derived from them. For practical applications, some rather drastic assumption must now be made. One is that LCAO MO theory will be used in place of the exact wave functions, ψ_o and ψ_k. Since we are interested only in the symmetry properties, this creates no serious error, since MO theory has the great virtue of accurately showing the symmetries of the various electronic states.

The second assumption is more serious, since we will replace the infinite sum of excited states in previous two equations by only a few lowest lying states. This procedure will work because we are not trying to evaluate the sum but only to decide if it has a substantial value. It can be shown that the various states contributing to previous two equations fall off very rapidly as the difference $|E_o - E_k|$ becomes large. This is because the integral $\langle \psi_o | \partial U / \partial Q | \psi_k \rangle$ decreases very rapidly for two wave functions of quite different energy.

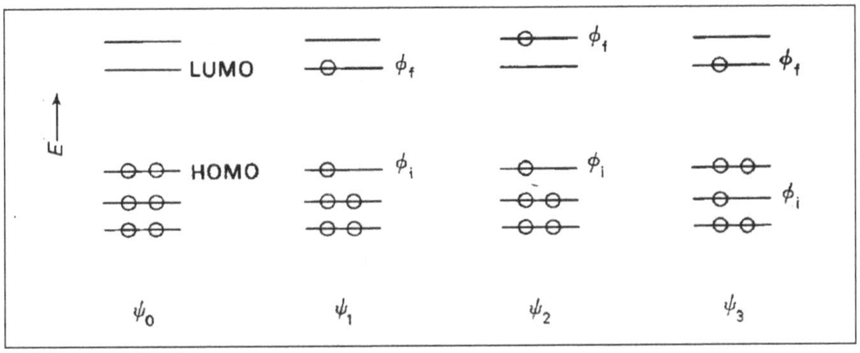

Molecular orbital description of ground state and first few excited states.

Figure above shows bow we use MO theory to represent the ground and excited states that are needed. The symmetry of $\psi_o \psi_k$ is replaced by $\phi_i \psi_f$, where ϕ_i is the occupied MO in the ground state and ϕ_f is the MO occupied in its place in the excited state. Positions of special importance are occupied by the highest occupied (HOMO) and lowest unoccupied (LUMO) molecular orbitals, since excitation of an electron from HOMO to LUMO defines the lowest excited state.

These two orbitals are called frontier orbitals by Fukui. At this time it is helpful to point out that the requirement that two orbitals, ϕ_i and ϕ_f, have the same symmetry is the same as saying that they must have a net positive overlap. Two molecular, or atomic, orbitals of different symmetry species have exactly zero overlap.

Bimolecular Reactions

Let us consider a bimolecular reaction which has reached point A in figure. Two molecules have approached each other with a definite orientation. They have started to interact with each other, but the interaction energy is still small. This means that the MOs of the two separate molecules are still a good starting point for considering the combined system. Those of the same symmetry (positive overlap) will interact more and more strongly as the reaction coordinate is traversed and at the transition state quite different MOs will be produced.

For the reaction to be allowed by symmetry, we must have transfer of electrons from high energy occupied MOs (ϕi) to low energy empty MOs (ϕf) which have positive overlap. This will lower the energy of the system via the last term in equation $E = E_o + Q\langle \psi_o | \partial U / \partial Q | \psi_o \rangle + \frac{1}{2} Q^2 \langle \psi_o | \partial^2 U / \partial Q^2 | \psi_o \rangle +$

$Q^2\sum_k\left[\langle\psi_o|\partial U/\partial Q|\psi_k\rangle\right]^2/(E_o - E_k)$ and prevent an excessive energy barrier. Now we can add an additional requirement on ϕi and ϕf using chemical knowledge rather than mathematical or quantum mechanical arguments.

All chemical reactions consist of the breaking of certain bonds and the making of new bonds. All MOs correspond to the bonding together of certain atoms, anti-bonding of other atoms, and non-bonding of the remaining atoms. It follows that ϕ_i must represent bonds that are broken and ϕ_f bonds that are made during the reaction, for their bonding parts. The reverse statement holds for their anti-bonding parts.

Also we know that some atoms are much more electronegative than other atoms. Therefore electrons will move more easily from ϕ_i to ϕ_f when they move in the direction of the more electronegative atoms. In such cases $|E_o - E_k|$ will be small and the stabilizing effect of electron movement will be large. Conversely electron movement from an occupied MO in a halogen molecule to an empty MO in an alkali metal molecule, for example, would correspond to a large value of $|E_o - E_k|$.

Reactions of hydrogen are particularly easy to describe. The only MOs of reasonable energy are the bonding σ_g, which is occupied and the anti-bonding σ_u^*, which is empty. One of the simplest of chemical reactions would be isotope exchange between hydrogen and deuterium.

$$H_2 + D_2 \rightleftharpoons 2HD$$

Let us assume that above reaction occurs by a bimolecular mechanism in which H_2 and D_2 collide broadside, giving rise to a four-centre transition state.

```
H—H
|  |
D—D
```

The point-group of this transition state is C_{2v}. Also, at point A in previous figure the point group is C_{2v}. The MOs of H_2 and D_2 should now be classified as A_1 for the bonding σ_g and B_1 for the anti-bonding.

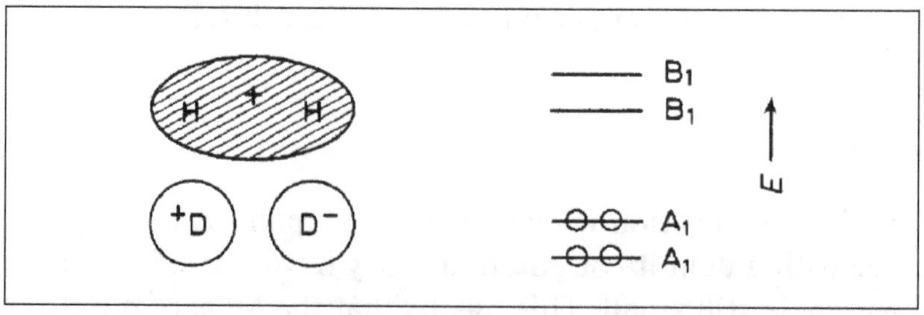

Molecular orbitals in $H_2 + D_2$ reaction. Occupied orbitals are shaded. Symmetry labels are for C_{2v} point group.

The filled MO of one hydrogen or deuterium molecule has zero overlap with the empty MO of the other. This is the same as showing that there is no low-flying excited state of the same symmetry as that of the ground state. Hence the exchange reaction $H_2 + D_2 \rightleftharpoons 2HD$ is forbidden by orbital symmetry. This simply means that the selected mechanism and transition state would have an excessively high energy barrier.

Indeed reaction $H_2 + D_2 \rightleftharpoons 2HD$ does not occur in a single elementary step. Instead a series of allowed steps occurs:

$$D_2 \rightleftharpoons 2D$$
$$D + H_2 \rightarrow HD + H$$
$$H + D_2 \rightarrow HD + D$$

Reactions of free radicals and atoms rarely have serious symmetry restrictions and are often found. Four-centre reactions of diatomic molecules, on the other hand, are almost always symmetry forbidden and either do not occur, or occur with high activation energies. An important example of a forbidden reaction is the decomposition of nitric oxide.

$$2NO \rightarrow N_2 + O_2$$

Since this reaction is highly exothermic (by 43 kcal), one would expect it to occur rapidly. In fact it is extremely slow, having an activation energy of 50 kcal. This is a symmetry imposed barrier. This can be seen most readily by looking at the reverse reaction, since a symmetry barrier which exists for a forward reaction must also exist for the reverse. Figure below shows the important MOs of N_2 and O_2. The π_g^* anti-bonding MO of O_2 is half-filled. It can act either as an electron acceptor or donor. By symmetry, electrons cannot flow from the filled π_u, MO of N_2 into the π_g^* of O_2, since the overlap is zero. Electrons could flow from O_2 to N_2 by symmetry, since the π_g^* MO of N_2 is empty. However, this is chemically unacceptable, since O_2 is more electronegative than N_2. In addition, the O—O bond would be strengthened rather than weakened by emptying the π_g^* anti-bonding MO of O_2.

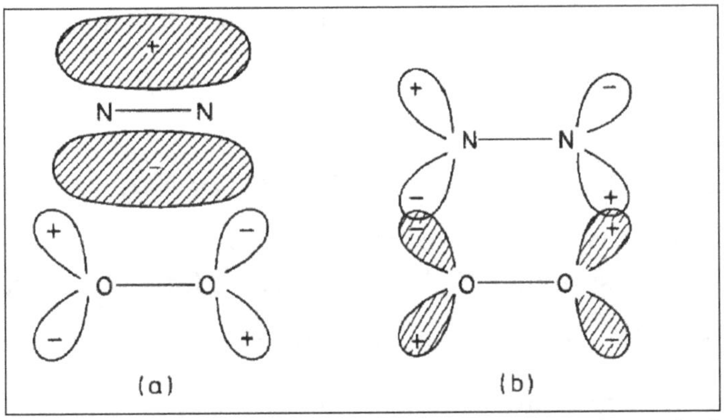

Molecular orbitals in $N_2 + O_2$ reaction. Occupied orbitals are shaded.

Orbital symmetry arguments can be used to decide between two proposed mechanisms for the oxidation of NO to NO_2.

$$2NO + O_2 \rightarrow 2NO_2$$

This reaction, while third-order in kinetic behaviour, is believed to occur by a series of bimolecular steps.

$$2NO \rightleftharpoons N_2O_2$$
$$N_2O_2 + O_2 \rightarrow 2NO_2$$

Or,

$$NO + O_2 \rightleftharpoons NO_3$$
$$NO_3 + NO \rightarrow 2NO_2$$

Figure (a) shows the relevant MOs for the reaction of the N_2O_2 dimer with O_2. Flow of electrons from the filled B_1 orbital of N_2O_2 to the half-empty B_1 orbital of O_2 results in the breaking of the O_2 bond by filling up an anti-bonding MO of O_2. The N—N bond is strengthened in N_2O_2 because an anti-bonding MO is emptied. Two new N—O bonds are formed by positive overlap between the B_1 orbitals.

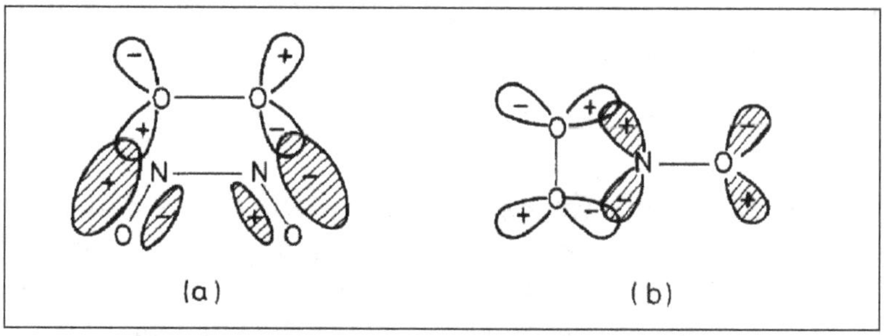

(a). Molecular orbitals in O_2 + N_2O_2 reaction. Product is N_2O_4.
(b). Molecular orbitals in NO + O_2 reaction.

All of this seems reasonable except that clearly the reaction is,

$$N_2O_2 + O_2 \rightarrow N_2O_4$$

which is not the same as reaction $N_2O_2 + O_2 \rightarrow 2NO_2$. In fact, it is forbidden by spin conservation rules if the oxygen molecule is in the normal $^3\Sigma_g^-$ ground state. It could only occur for the excited $^1\Delta_g$ state.

The alternative mechanism, shown in reactions $NO + O_2 \rightleftharpoons NO_3$ and $NO_3 + NO \rightarrow 2NO_2$ is allowed both by orbital symmetry and spin conservation. Figure (b) shows the MO situation for reaction $NO_3 + NO \rightarrow 2NO_2$ assuming a Y-shaped structure for NO_3. There is some evidence that a peroxy structure exists for NO_3, but its reaction with NO is also allowed.

The third-order reaction of nitric oxide with hydrogen,

$$2NO + 2H_2 \rightarrow N_2 + H_2O$$

is believed to go through a mechanism in which H_2O_2 is formed as an intermediate,

$$N_2O_2 + H_2 \rightarrow N_2 + H_2O_2$$

Followed by a rapid reaction of H_2O_2 with H_2 to give water. Figure below shows that above reaction is symmetry allowed and is chemically reasonable. The H—H bond is broken, two O—H bonds are formed, the weak N—N bond of N_2O_2 is strengthened and two strong N—O bonds are weakened. While O—O bonding also is increased, it still remains questionable whether H_2O_2 or two OH radicals would result from above reaction.

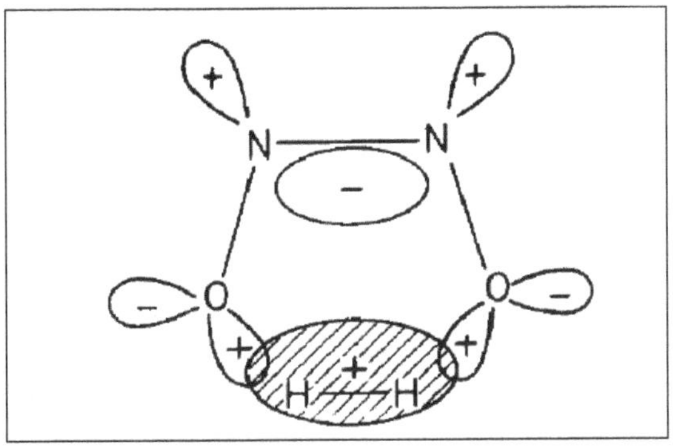

Molecular orbitals in $H_2 + N_2O_2$ reaction. Products are $N_2 + H_2O_2$ or $N_2 + 2OH$.

Nucleophilic Substitution Reactions

It is particularly instructive to look at old, well established mechanisms from the new viewpoint of orbital symmetry. A case in hand is the bimolecular nucleophilic substitution, or S_N2 reaction. Clearly it is the HOMO of the nucleophile and the LUMO of the substrate which are the critical orbitals. Figure shows these orbitals for a typical case of organic chemistry, e.g.

$$I^- + CH_3Cl \rightarrow CH_3I + Cl^-$$

The anti-bonding σ* orbital is a critical one in all molecules which can be regarded as X—Y, two groups joined together by a single bond. Filling this orbital will clearly break the bond between X and Y. Molecules of this type include H_2, Cl_2, HCI, CH_3CI, CH_3CH_3, CH_3H, etc. The symmetry of the σ* orbital will always be roughly that shown in Figure above.

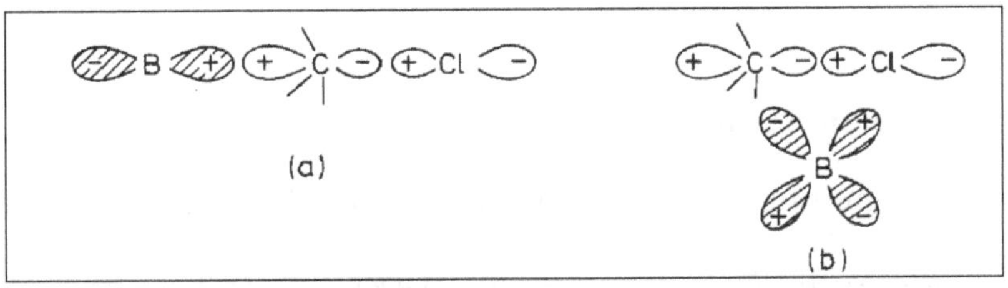

Nucleophilic displacement with (a) inversion of configuration, (b) retention of configuration at carbon atom.

We see in figure above the characteristic features of bond making between the nucleophile, B, and the carbon atom, as well as bond breaking between carbon and halogen. Also inversion of configuration at the carbon atom is clearly indicated. This results from the easy availability of the σ* orbital at this position. A similar situation would be found for many other four-coordinated central atoms of the non-transition elements.

If we consider tetrahedral complexes of the transition elements, a general conclusion emerges. Filling up the d orbitals has the effect of decreasing the availability of the anti-bonding σ* orbital. The most advantageous situation would be to have an empty e orbital. The next best would be to have an empty t_2^* orbital. Interestingly, using an e orbital corresponds to edge attack, and using the t_2^* orbital to face attack on the tetrahedron. Evaluating the relative rates of a series of dn complexes would require assessing a number of different effects, including changes in crystal field stabilization energies.

A square planar complex offers a different situation. The LUMO now is (a_{2u}), the p_z orbital perpendicular to the plane. It is a non-bonding, rather than an anti-bonding MO. This means that bond making will be considerably in excess of bond breaking. The reaction coordinate can lead directly to an intermediate of D_{3h} symmetry. Unlike the tetrahedral case, the entering and leaving groups will be equatorial, rather than axial.

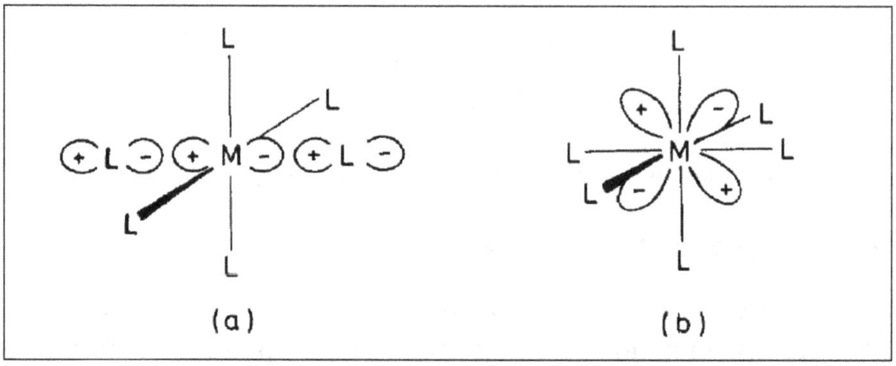

(a) Empty anti-bonding MO of T_{1u}, symmetry in octahedral complex.
(b) Non-bonding MO of T_{2g} symmetry in octahedral complex.

Figure (a) shows a typical LUMO in an octahedral complex. This particular one is of t^*_{1u} symmetry. Those of a^*_{1g} or a^*_g symmetry would have similar properties. The most significant feature is that the orbital is quite inaccessible to a nucleophilic reagent. The six ligands effectively block off access. Again, filling up the d orbitals would have an inhibiting effect. An empty t2g orbital could be used as the LUMO and, lithe ligands were not too bulky, would allow for an S_N2 mechanism. It is a familiar fact that S_N2 mechanisms are extremely rare for the substitution reactions of octahedral complexes. On the other hand, nucleophilic attack on a ligand, such as a CO group, can easily occur.

Oxidative-Addition Reactions

There has been much recent interest in the mechanism of oxidative-addition reactions such as,

$$\mathrm{Ir(CO)Cl}\left[\mathrm{P}\left(\mathrm{C_6H_5}\right)_3\right]_2 + \mathrm{CH_3I} \rightarrow \mathrm{IrCO}\left[\mathrm{P}\left(\mathrm{C_6H_5}\right)_3\right]_2\mathrm{CH_3I}$$

Recently several observations have been made which give detailed information on reactions such as this. These observations lead to the conclusions that (1) the reaction occurs in a one-step, concerted process without the formation of intermediates, and in the absence of solvents; (2) that trans addition to the square plane can occur even with these restrictions; and (3) retention of configuration results at the carbon atom, when a more complex alkyl group than methyl is used.

These results are completely in agreement with orbital symmetry considerations. Figures (a) and (b) show that the d orbitals of the metal which are of e_g symmetry constitute the HOMO. It is well known that soft metal atoms act as nucleophiles, or Lewis bases, in their reactions. It is particularly important that in oxidative-addition, both the groups X and Y finish up bound to the metal. It can be seen that a metal complex using a d orbital as the HOMO could also react with an alkyl halide to give inversion of configuration, figure (c).

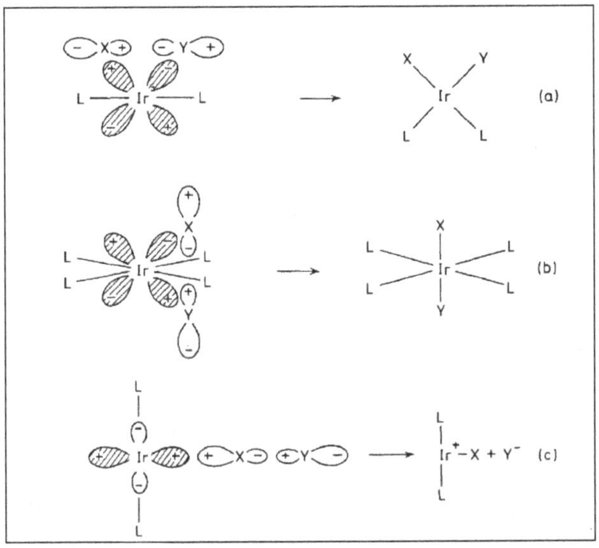

Mechanisms of oxidative addition: (a) one step cis addition of XY with retention at X. (b) one step trans addition with retention at X. (c) two step addition with inversion of configuration at X.

This mechanism is expected when only one group, X, is bound to the metal and Y is displaced. A number of examples of this type are known. For example, the reaction,

$$R^*Br + Mn(CO)_5^- \rightarrow R^*Mn(CO)_5 + Br^-$$

occurs with inversion of configuration at the carbon atom of R. It is also possible that the group Y in Figure (c) could later add to the five-coordinated product first formed. This would be a two-step mechanism for oxidative addition. It appears that such a mechanism would be favoured in a polar solvent.

It is increasingly being realized that many examples of both homogeneous and heterogeneous catalysis on transition metals involve oxidative-addition, and its reverse, as key steps. These two processes, together with the ligand migration reaction, are the steps involved in most detailed mechanisms. In agreement with this concept, it has been found that silicon hydrides add to both metal complexes and to metal surfaces with retention of configuration at silicon. Years ago it was shown that optically active hydrocarbons are adsorbed on metal surfaces with retention of configuration.

Oxidative-addition reactions may occur more commonly than supposed with transition metal compounds. Many reactions with electrophilic reagents may proceed by prior addition of the reagent, followed rapidly by the reverse process. In such cases the product of the addition step need not be a stable species. As an example, bromine cleavage of the optically active product formed in above reaction proceeds with retention of configuration.

$$Br_2 + R^*Mn(CO)_5 \xrightarrow{fast} Br-Mn(CO)_5 + R^*Br$$

It is tempting to postulate that Br_2 adds first to a filled d orbital of the manganese, and then that R*Br splits out. In the same way cleavage by acid could follow the course,

$$HCl + CH_3Mn(CO)_5 \xrightarrow{slow} CH_3MnHCl(CO)_5$$

$$CH_3MnHCl(CO)_5 \xrightarrow{fast} CH_4 + ClMn(CO)_5$$

Such an oxidative-addition, reductive-elimination mechanism could account for the fact that molecular HCI is the cleaving agent. Thus $HC1O_4$ is ineffective and H^+ plus CI^- in a polar solvent is not effective. A similar mechanism would also account for the well-known retention of configuration at carbon those results in the acid cleavage of organomercury compounds. Other electrophilic reagents also give retention with the organomercurials.

In at least one case of a d^8 planar complex, the addition-elimination mechanism for alkyl—metal bond cleavage seems well established.

$$\text{trans}-\text{Pt}\left[\text{P}\left(\text{C}_2\text{H}_5\right)_3\right]_2\text{CH}_3\text{Cl}+\text{HCl}\rightarrow\text{trans}-\text{Pt}\left[\text{P}\left(\text{C}_2\text{H}_5\right)_3\right]_2\text{Cl}_2+\text{CH}_4$$

The intermediate would be a six-coordinated platinum(Iv) complex. While not stable itself, certainly it would have many stable analogues. In methanol, the solvent where above reaction was studied, the solvated H^+ can also cause cleavage. Apparently a one-step addition of HCl is not required.

Unimolecular Reactions

We now go to a consideration of points B and C in Figure. B refers to an activated complex and C to a single molecular species, which is unstable with respect to isomerization, or breakdown to other products. In either case, the theory is changed somewhat from that of the bimolecular reactions discussed earlier. For unimolecular processes, the theory is also called the second-order, or pseudo, Jahn—Teller effect.

The term linear in Q in equation $E=E_o+Q\langle\psi_o|\partial U/\partial Q|\psi_o\rangle+\frac{1}{2}Q^2\langle\psi_o|\partial^2U/\partial Q^2|\psi_o\rangle+Q^2\sum_k\left[\langle\psi_o|\partial U/\partial Q|\psi_k\rangle\right]^2/(E_o-E_k)$ now vanishes, since we are at an extremum in the potential energy plot. As before, the first quadratic term is positive, and the second one is negative. Clearly at a maximum, point B, the second term is larger than the first. At a minimum, point C, the first term dominates, but the magnitude of the second term determines whether we lie in a deep potential well or a shallow one.

Again, the existence of low lying states, ψ_k, of the correct symmetry to match with ψ_o is critical. Now there is no restriction on the symmetry of the reaction coordinates. It need no longer be totally symmetrical. However, ψ_o, $(\partial U/\partial Q)$ and ψ_k are still bound by the symmetry requirement that their direct product must contain the totally symmetrical representation.

II we consider rather symmetrical molecules to begin with, it will usually be found that the reaction coordinate and $(\partial U/\partial Q)$ are asymmetric. The reason for this is that maximum and minimum potential energies are usually found for nuclear arrangements with a high degree of symmetry. Any disturbance of the nuclear positions will now reduce the symmetry. But this corresponds to a change in the point group, which can only come about by an asymmetric vibrational mode.

Conversely, it may be pointed out that a number of point groups depend upon a unique value of Q_o. For example, a tetrahedral molecule has uniquely determined bond angles. All such cases must correspond to either maxima or minima in Figure, if the reaction coordinate is taken either as the bond angles or relative bond distances.

In molecular orbital theory the product $\psi_o\psi_k$ is again replaced by $\phi_i\phi_f$, where both the occupied and empty MOs must be in the same molecule. Electron transfer from ϕ_i to ϕ_f results in a shift in charge density in the molecule. Electron density increases in the regions where ϕ_i and ϕ_f have the same sign (positive overlap), and decreases where they have opposite signs (negative overlap). The positively charged nuclei then move in the direction of increased electron density. The motion of the nuclei defines a reaction coordinate. The symmetry of Q is the same as that of the product $\phi_i \times \phi_f$.

The size of the energy gap between ϕ_i and ϕ_f is critical. A small gap means an unstable structure, unless no vibrational mode of the right symmetry exists for the molecule capable of changing its structure. A large energy gap between the HOMO and the LUMO means a stable molecular structure. Reactions can occur, but only with a high activation energy.

For an activated complex (point B) there must necessarily be at least one excited state of low energy. The symmetry of this state and the ground state then determines the mode of decomposition of the activated complex. By the principle of microscopic reversibility, this can then be used to decide on the mechanism of formation of the activated complex.

As an example of a molecule in a shallow potential well (point C), we will consider the molecule NO_3. Assuming a planar structure of D_{3h} symmetry, the MO sequence is,

$$\ldots(2e')^4(1e'')^4(3e')^4(1a_2')^1(2a_2'')^0$$

The transition $(3e') \rightarrow (1a_2')$ requires very little energy, according to the calculations. The direct product $E' \times A_2' = E'$, a normal mode which distorts the molecule into a Y-shaped structure, with two oxygen atoms close together. The predicted mode of dissociation of NO_3 is therefore into NO and O_2, in agreement with the facts.

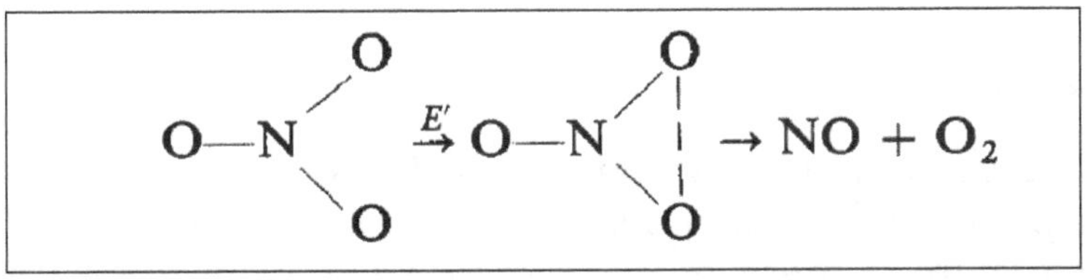

If we were to add one more electron, as in NO_3^- or SO_3, the easily available excited state would be blocked. Thus these molecules would be much more stable towards dissociation. The NO_3 molecule absorbs visible light, with a maximum at 660 nm. The NO_3^- ion, however, does not absorb until 300 nm. Coloured molecules are less stable than colourless molecules, providing the symmetry rules can be obeyed. Predicting the mode of decomposition of a molecule lying in a deep potential well is more difficult. Since the reaction requires much energy, quite high-lying excited states may be important. Nevertheless, the symmetry rules can be very useful.

Suppose we know that a unimolecular reaction occurs in which certain bonds are broken and some other bonds are made. These bonds can then select ϕ_i and ϕ_f the relevant molecular orbitals. These MOs, in turn, can determine the symmetry of the reaction coordinate, Q. We need to know only the symmetries of ϕ_i and ϕ_f. Considerations of this kind have been of great importance in elucidating complex organic reactions. Applications to inorganic chemistry are still unknown.

Alternatively, we can pick the symmetry of the reaction coordinate and then seek suitable molecular orbitals. For example, the unimolecular (S_N1) dissociation of a tetrahedral complex requires a vibration of T_2 symmetry. Hence we seek a HOMO and LUMO whose direct product is of this symmetry type. The MO sequence for CH_4 is,

$$(a_1)^2(t_2)^6(a_1^*)^0(t_2^*)^0$$

Excitation of an electron from a bonding (t_2) to an anti-bonding (a_1^*) orbital will promote dissociation into CH_3 and H.

A transition metal complex inserts the d manifold between the above bonding and anti-bonding MOs.

$$\left|(e)^m(t_2^*)^n\right|$$

Since $E \times T_2 = T_1 + T_2$, it is possible that a hole in the d shell will promote unimolecular dissociation of tetrahedral complexes. At the moment there is little evidence one way or another on this point.

In an octahedral complex it is a T_{1u} vibration which corresponds to dissociation of one ligand. We see that the d manifold,

$$\left|(t_{2g})^m(e_g^*)^n\right|$$

cannot contribute excitations that will promote dissociation, since $T_{2g} \times E_g = T_{1g} + T_{2g}$. This raises the suspicion that, because of the inherent gerade properties of d orbitals d—d transitions in general will not effectively promote the dissociation of even less symmetrical complexes.

Molecular Structure

One of the interesting applications that can be made of the symmetry rules is the prediction of the stable shapes of molecules. Molecules with formulas XY_n or X_2Y_n usually are found with rather regular structures. A decision can be made as to which of several alternative structures is the most stable using equation $E = E_o + Q\langle\psi_o|\partial U/\partial Q|\psi_o\rangle + \frac{1}{2}Q^2\langle\psi_o|\partial^2 U/\partial Q^2|\psi_o\rangle + Q^2\sum_k\left[\langle\psi_o|\partial U/\partial Q|\psi_k\rangle\right]^2/(E_o - E_k)$. It is necessary to assume that the term linear in Q has a coefficient of zero. That is, we can only decide if a certain point group is stable for the molecule. We cannot find the best values for the bond angles and bond distances within the point group.

The procedure is to test a given molecule in two (or more) possible structures, say, square planar and tetrahedral. One structure usually corresponds to a maximum in Figure and the other to a minimum. The reaction coordinate is the normal mode which interconverts the two structures. The stable structure will have a large energy gap between the HOMO and LUMO that matches up with this transition. The unstable structure, conversely, must have a small energy gap. Occasionally both structures are unstable, indicating an intermediate structure, say of D_{2d} symmetry (squashed tetrahedron).

A complex molecule with many atoms will have many possible structures. If an accurate MO sequence is available for such a molecule in an unstable structure, it is possible to predict which normal mode is favoured. This is actually a prediction of the stable structure into which the original configuration will distort.

The geometric isomers of XY_n, which can be interconverted without breaking bonds, have been called polytopal isomers. More complex molecules, such as X_2Y_n, have many more structures possible, some of which can be interconverted only by the breaking of bonds.

Consider the possible isomerization of four-coordinated complexes of the transition metal ions between tetrahedral (T_d) and square-planar (D_{4h}) structures. The normal mode which takes a tetrahedral structure into a planar one is of E symmetry. The molecular orbital scheme (needed to express the symmetries) is generally agreed to be,

$$(t_1) \text{ or } (t_2) \Big| (e)(t_2^*) \Big| (a_1)$$

with the d manifold separated by dashed lines.

Since T_1 (or T_2) × E = $T_1 + T_2$, there is no low-lying transition which causes a tetrahedral complex to rearrange to a planar complex. This is true for all systems from d^o to d^{10}. However, first-order Jahn—Teller effects are important for many of these complexes. Only d^o, d^2, high-spin d^5, d^7 and d^{10} complexes have A_1 or A_2 ground states and are stable. All others are E, T_1 or T_2 states. Since E × E = $A_1 + A_2 + E$, and $T_1 \times T_1 = T_2 \times T_2 = A_1 + E + T_1 + T_2$, all of these allow a distortion in the direction leading toward a planar structure. Since a D_{2d}, or distorted tetrahedral structure, is sufficient to lift the orbital degeneracy in every case except low-spin d^3, we cannot tell whether the initial distortions will continue on to a planar structure. To get further information we must now look at the stability of the possible planar forms. The generally agreed upon MO scheme for D_{4h} symmetry is,

$$(a_{2g})(b_{2u}) \Big| (a_{1g})(e_g)(b_{2g})(b_{1g}) \Big| (a_{2u})$$

The vibration which takes a planar complex into a tetrahedral one is of B_{2u} symmetry. The transitions which can give the correct symmetry are $(b_{2u}) \rightarrow (a_{1g})$ and (b_{1g}) - (a_{2u}). It seems likely that the energy gap in both cases is only 2—3 eV.

This leads to the conclusion that low-spin d^2, d^4, d^5, d^6, d^7 and d^8 complexes are stable as square-planar structures. Low-spin d^3 is orbitally degenerate. In all other cases an excitation is possible which favours the B_{2u} vibration. This means a low, or zero, activation energy for conversion into either D_{2d} or Td symmetry. In combination with the conclusions from the first-order Jahn—Teller effects, we can assign a most stable structure for each four-coordinate system according to the number of d electrons and the spin state. Table below summarizes the results for all systems.

Table: Stable structures for MX_4 systems.

High spin		Low spin	High spin		Low spin
d^o	T_d	T_d	d^6	D_{2d}	D_{4h}
d^1	D_{2d}	D_{2d}	d^7	T_d	D_{4h}
d^2	T_d	D_{4h}	d^8	D_{2d}	D_{4h}

d^3	D_{2d}		d^9	D_{2d}	D_{2d}
d^4	D_{2d}	D_{4h}	d^{10}	T_d	T_d
d^5	T_d	D_{4h}			

The predictions are in accord with experimental facts where known. The conclusions for D_{4h} are based on no interaction with other groups above and below the plane. Such interactions could raise the (a_{2u}) level markedly and stabilize the planar form. Four simple ligands such as halide or cyanide lead to D_{2d} structures in solution for copper(u) complexes, so that axial solvent perturbations are not sufficient in these cases to create planarity. It appears that a combination of first-order and second-order Jabn—Teller effects can be used to predict the stable structures of molecules. In practice an MO scheme is required which need be only qualitatively correct. It also follows that an incorrect MO scheme may predict structures wrongly. It seems reasonable to apply a test for second-order Jahn—Teller distortions to MO calculations in general.

References

- Molecular-Orbitals: pearsonhighered.com , Retrieved 19, August 2020
- Orbital-interactions-in-chemistry: fzu.cz, Retrieved 03, February 2020
- Orbital-symmetry-rules-and-the-mechanism-of-inorgamc-reactions: iupac.org, Retrieved 27, July 2020

3

Representation of Point Groups

The different symmetry operations which can be performed on a molecule which result in a conformation that is identical to the original are described by a point group. The topics elaborated in this chapter will help in gaining a better perspective about the representation of point groups as well as matrix representation of groups and irreducible representations.

Matrix Representation of Groups

The symmetry operations in a group may be represented by a set of transformation matrices Γ(g), one for each symmetry element g. Each individual matrix is called a representative of the corresponding symmetry operation, and the complete set of matrices is called a matrix representation of the group. The matrix representatives' act on some chosen basis set of functions, and the actual matrices making up a given representation will depend on the basis that has been chosen. The representation is then said to span the chosen basis. Previously we were looking at the effect of some simple transformation matrices on an arbitrary vector (x,y). The basis was therefore a pair of unit vectors pointing in the x and y directions.

Before proceeding any further, we must check that a matrix representation of a group obeys all of the rules set out in the formal mathematical definition of a group:

1. The first rule is that the group must include the identity operation E (the 'do nothing' operation). We showed above that the matrix representative of the identity operation is simply the identity matrix. As a consequence, every matrix representation includes the appropriate identity matrix.

2. The second rule is that the combination of any pair of elements must also be an element of the group (the group property). If we multiply together any two matrix representatives, we should get a new matrix which is a representative of another symmetry operation of the group. In fact, matrix representatives multiply together to give new representatives in exactly the same way as symmetry operations combine according to the group multiplication table. For example, in the C_{3v} point group, we showed that the combined symmetry operation $C_{3\sigma v}$ is equivalent to σv". In a matrix representation of the group, if the matrix representatives of C_3 and σ_v are multiplied together, the result will be the representative of σ_v".

3. The third rule states that every operation must have an inverse, which is also a member of the group. The combined effect of carrying out an operation and its inverse is the same as the identity operation. It is fairly easy to show that matrix representatives satisfy this criterion. For example, the inverse of a reflection is another reflection, identical to the first. In matrix terms we would therefore expect that a reflection matrix was its own inverse, and that two

identical reflection matrices multiplied together would give the identity matrix. This turns out to be true, and can be verified using any of the reflection matrices in the examples above. The inverse of a rotation matrix is another rotation matrix corresponding to a rotation of the opposite sense to the first.

4. The final rule states that the rule of combination of symmetry elements in a group must be associative. This is automatically satisfied by the rules of matrix multiplication.

Example: A matrix representation of the C_{3v} point group (the ammonia molecule). The first thing we need to do before we can construct a matrix representation is to choose a basis. For NH_3, we will select a basis (s_N,s_1,s_2,s_3) that consists of the valence s orbitals on the nitrogen and the three hydrogen atoms. We need to consider what happens to this basis when it is acted on by each of the symmetry operations in the C_{3v} point group, and determine the matrices that would be required to produce the same effect. The basis set and the symmetry operations in the C_{3v} point group are shown in the figure below:

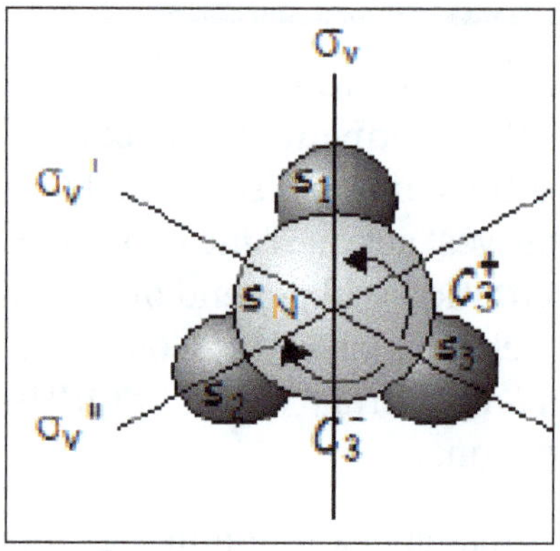

The effects of the symmetry operations on our chosen basis are as follows:

- E $(s_N,s_1,s_2,s_3)\rightarrow(s_N,s_1,s_2,s_3)$
- C_3^+ $(s_N,s_1,s_2,s_3)\rightarrow(s_N,s_2,s_3,s_1)$
- C_3^- $(s_N,s_1,s_2,s_3)\rightarrow(s_N,s_3,s_1,s_2)$
- σ_v $(s_N,s_1,s_2,s_3)\rightarrow(s_N,s_1,s_3,s_2)$
- σ_v' $(s_N,s_1,s_2,s_3)\rightarrow(s_N,s_2,s_1,s_3)$
- σ_v'' $(s_N,s_1,s_2,s_3)\rightarrow(s_N,s_3,s_2,s_1)$

By inspection, the matrices that carry out the same transformations are:

$$\Gamma(E) \qquad (s_N,s_1,s_2,s_3)\begin{pmatrix}1&0&0&0\\0&1&0&0\\0&0&1&0\\0&0&0&1\end{pmatrix}=(s_N,s_1,s_2,s_3)$$

$$\Gamma(C_3^+) \quad (s_N,s_1,s_2,s_3)\begin{pmatrix}1&0&0&0\\0&0&0&1\\0&1&0&0\\0&0&1&0\end{pmatrix}=(s_N,s_2,s_3,s_1)$$

$$\Gamma(C_3^-) \quad (s_N,s_1,s_2,s_3)\begin{pmatrix}1&0&0&0\\0&0&1&0\\0&0&0&1\\0&1&0&0\end{pmatrix}=(s_N,s_3,s_1,s_2)$$

$$\Gamma(\sigma_v) \quad (s_N,s_1,s_2,s_3)\begin{pmatrix}1&0&0&0\\0&1&0&0\\0&0&0&1\\0&0&1&0\end{pmatrix}=(s_N,s_1,s_3,s_2)$$

$$\Gamma(\sigma_v') \quad (s_N,s_1,s_2,s_3)\begin{pmatrix}1&0&0&0\\0&0&1&0\\0&1&0&0\\0&0&0&1\end{pmatrix}=(s_N,s_2,s_1,s_3)$$

$$\Gamma(\sigma_v'') \quad (s_N,s_1,s_2,s_3)\begin{pmatrix}1&0&0&0\\0&0&0&1\\0&0&1&0\\0&1&0&0\end{pmatrix}=(s_N,s_3,s_2,s_1)$$

These six matrices therefore form a representation for the C_{3v} point group in the (s_N,s_1,s_2,s_3) basis. They multiply together according to the group multiplication table and satisfy all the requirements for a mathematical group. We have written the vectors representing our basis as row vectors. This is important. If we had written them as column vectors, the corresponding transformation matrices would be the transposes of the matrices above, and would not reproduce the group multiplication table.

Example: A matrix representation of the C_{2v} point group (the allyl radical). In this example, we'll take as our basis a p orbital on each carbon atom (p_1,p_2,p_3).

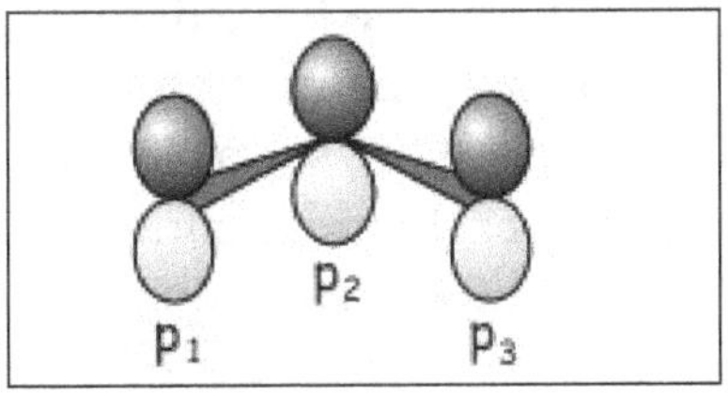

The p orbitals are perpendicular to the plane of the carbon atoms (this may seem obvious, but if you're visualising the basis incorrectly it will shortly cause you a not inconsiderable amount of confusion). The symmetry operations in the C_{2v} point group, and their effect on the three p orbitals, are as follows:

- E $(p_1,p_2,p_3) \rightarrow (p_1,p_2,p_3)$
- C_2 $(p_1,p_2,p_3) \rightarrow (-p_3,-p_2,-p_1)$
- σ_v $(p_1,p_2,p_3) \rightarrow (-p_1,-p_2,-p_3)$
- σ_v' $(p_1,p_2,p_3) \rightarrow (p_3,p_2,p_1)$

The matrices that carry out the transformation are:

$$\Gamma(E) \quad (p_1,p_2,p_3)\begin{pmatrix} 1 & 0 & 0 \\ 0 & 1 & 0 \\ 0 & 0 & 1 \end{pmatrix} = (p_1,p_2,p_3)$$

$$\Gamma(C_2) \quad (p_1,p_2,p_3)\begin{pmatrix} 0 & 0 & -1 \\ 0 & -1 & 0 \\ -1 & 0 & 0 \end{pmatrix} = (-p_3,-p_2,-p_1)$$

$$\Gamma(\sigma_v) \quad (p_1,p_2,p_3)\begin{pmatrix} -1 & 0 & 0 \\ 0 & -1 & 0 \\ 0 & 0 & -1 \end{pmatrix} = (-p_1,-p_2,-p_3)$$

$$\Gamma(\sigma_v') \quad (p_1,p_2,p_3)\begin{pmatrix} 0 & 0 & 1 \\ 0 & 1 & 0 \\ 1 & 0 & 0 \end{pmatrix} = (p_3,p_1,p_2)$$

Properties of Matrix Representations

Now that we've learnt how to create a matrix representation of a point group within a given basis, we will move on to look at some of the properties that make these representations so powerful in the treatment of molecular symmetry.

Similarity Transforms

Suppose we have a basis set $(x_1,x_2,x_3,...x_n)$, and we have determined the matrix representatives for the basis in a given point group. There is nothing particularly special about the basis set we have chosen, and we could equally well have used any set of linear combinations of the original functions (provided the combinations were linearly independent). The matrix representatives for the two basis sets will certainly be different, but we would expect them to be related to each other in some way. As we shall show shortly, they are in fact related by a similarity transform. It will be far from obvious at this point why we would want to carry out such a transformation, but similarity transforms will become important later on when we use group theory to choose an optimal basis set with which to generate molecular orbitals.

Consider a basis set $(x_1', x_2', x_3', \ldots x_n')$, in which each basis function x_i' is a linear combination of our original basis $(x_1, x_2, x_3, \ldots x_n)$.

$$x_j' = \sum_i x_i c_{ji} = x_1 c_{j1} + x_2 c_{j2} + \ldots$$

The c_{ji} appearing in the sum are coefficients; c_{ji} is the coefficient multiplying the original basis function x_i in the new linear combination basis function x_j'. We could also represent this transformation in terms of a matrix equation x' = xC:

$$\left(x_1', x_2', \ldots x_n'\right) = \left(x_1, x_2, \ldots x_n\right)\begin{pmatrix} c_{11} & c_{12} & \cdots & c_{1n} \\ c_{21} & c_{22} & \cdots & c_{2n} \\ \cdots & \cdots & \cdots & \cdots \\ c_{n1} & c_{n2} & \cdots & c_{nn} \end{pmatrix}$$

Now we look at what happens when we apply a symmetry operation g to our two basis sets. If Γ(g) and Γ'(g) are matrix representatives of the symmetry operation in the x and x' bases, then we have:

$$\begin{aligned} gx' &= x'\Gamma'(g) \\ gxC &= xC\Gamma'(g) \\ gx &= xC\Gamma'(g)C^{-1} \\ &= x\Gamma(g) \end{aligned}$$

since x' = xC multiplying on the right by C^{-1} and using $CC^{-1} = I$.

We can therefore identify the similarity transform relating Γ(g), the matrix representative in our original basis, to Γ'(g), the representative in the transformed basis. The transform depends only on the matrix of coefficients used to transform the basis functions:

$$\Gamma(g) = C\Gamma'(g)C^{-1}$$

Also, $\Gamma'(g) = C^{-1}\Gamma(g)C$

Characters of Representations

The trace of a matrix representative Γ(g) is usually referred to as the character of the representation under the symmetry operation g. We will soon come to see that the characters of a matrix representation are often more useful than the matrix representatives themselves. Characters have several important properties:

- The character of a symmetry operation is invariant under a similarity transform.
- Symmetry operations belonging to the same class have the same character in a given representation. Note that the character for a given class may be different in different representations, and that more than one class may have the same character.

Reduction of Representations I

If we look at the matrices carefully we see that they all take the same block diagonal form (a square matrix is said to be block diagonal if all the elements are zero except for a set of submatrices lying along the diagonal).

$$\Gamma(E) = \begin{pmatrix} 1&0&0&0\\ 0&1&0&0\\ 0&0&1&0\\ 0&0&0&1 \end{pmatrix} \quad \Gamma(C_3^+) = \begin{pmatrix} 1&0&0&0\\ 0&0&1&0\\ 0&0&0&1\\ 0&1&0&0 \end{pmatrix} \quad \Gamma(C_3^-) = \begin{pmatrix} 1&0&0&0\\ 0&0&0&1\\ 0&1&0&0\\ 0&0&1&0 \end{pmatrix}$$

$$\Gamma(\sigma_v) = \begin{pmatrix} 1&0&0&0\\ 0&1&0&0\\ 0&0&0&1\\ 0&0&1&0 \end{pmatrix} \quad \Gamma(\sigma_v') = \begin{pmatrix} 1&0&0&0\\ 0&0&1&0\\ 0&1&0&0\\ 0&0&0&1 \end{pmatrix} \quad \Gamma(\sigma_v'') = \begin{pmatrix} 1&0&0&0\\ 0&0&0&1\\ 0&0&1&0\\ 0&1&0&0 \end{pmatrix}$$

$$\chi(E) = 4 \quad \chi(C_3^+) = 1 \quad \chi(C_3^-) = 1 \quad \chi(\sigma_v) = 2 \quad \chi(\sigma_v') = 2 \quad \chi(\sigma_v'') = 2$$

A block diagonal matrix can be written as the direct sum of the matrices that lie along the diagonal. In the case of the C_{3v} matrix representation, each of the matrix representatives may be written as the direct sum of a 1×1 matrix and a 3×3 matrix, in which the bracketed superscripts denote the dimensionality of the matrices.

$$\Gamma^{(4)}(g) = \Gamma^{(1)}(g) \oplus \Gamma^{(3)}(g)$$

In which the bracketed superscripts denote the dimensionality of the matrices. A direct sum is very different from ordinary matrix addition since it produces a matrix of higher dimensionality. A direct sum of two matrices of orders n and m is performed by placing the matrices to be summed along the diagonal of a matrix of order n+m and filling in the remaining elements with zeroes.

The reason why this result is useful in group theory is that the two sets of matrices $\Gamma^{(1)}(g)$ and $\Gamma^{(3)}(g)$ also satisfy all of the requirements for a matrix representation. Each set contains the identity and an inverse for each member, and the members multiply together associatively according to the group multiplication table. Recall that the basis for the original four-dimensional representation had the s orbitals (s_N,s_1,s_2,s_3) of ammonia as its basis. The first set of reduced matrices, $\Gamma^{(1)}(g)$, forms a one-dimensional representation with (s_N) as its basis. The second set, $\Gamma^{(3)}(g)$ forms a three-dimensional representation with the basis (s_1,s_2,s_3). Separation of the original representation into representations of lower dimensionality is called reduction of the representation. The two reduced representations are shown below:

g	E	C_3^+	C_3^-	σ_v	σ_v'	σ_v''	
$\Gamma^{(1)}(g)$	(1)	(1)	(1)	(1)	(1)	(1)	1D representation spanned by (sN)
$\Gamma^{(3)}(g)$	$\begin{pmatrix} 1&0&0\\ 0&1&0\\ 0&0&1 \end{pmatrix}$	$\begin{pmatrix} 0&1&0\\ 0&0&1\\ 1&0&0 \end{pmatrix}$	$\begin{pmatrix} 0&0&1\\ 1&0&0\\ 0&1&0 \end{pmatrix}$	$\begin{pmatrix} 1&0&0\\ 0&0&1\\ 0&1&0 \end{pmatrix}$	$\begin{pmatrix} 0&1&0\\ 1&0&0\\ 0&0&1 \end{pmatrix}$	$\begin{pmatrix} 0&0&1\\ 0&1&0\\ 1&0&0 \end{pmatrix}$	3D representation spanned by (s_1,s_2,s_3)

The logical next step is to investigate whether or not the three dimensional representation $\Gamma^{(3)}(g)$ can be reduced any further. As it stands, the matrices making up this representation are not in block diagonal form (some of you may have noted that the matrices representing E and σ_v are block diagonal, but in order for a representation to be reducible all of the matrix representatives must be in the same block diagonal form) so the representation is not reducible. However, we can carry out a similarity transformation to a new representation spanned by a new set of basis functions (made up of linear combinations of (s_1,s_2,s_3)), which is reducible. In this case, the appropriate (normalised) linear combinations to use as our new basis functions are:

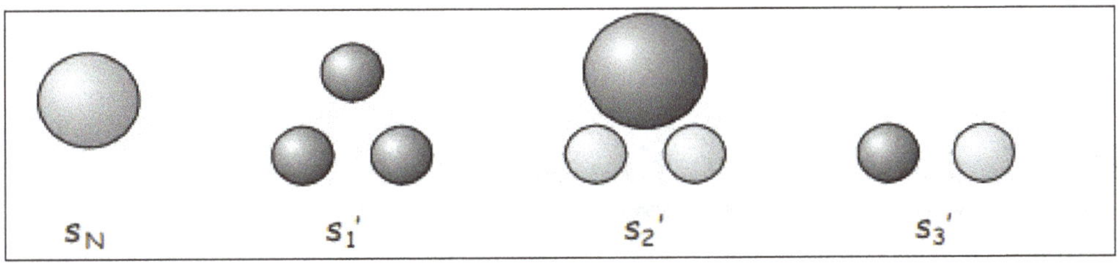

$$s_1' = \frac{1}{\sqrt{3}}(s_1 + s_2 + s_3)$$

$$s_2' = \frac{1}{\sqrt{6}}(2s_1 - s_2 - s_3)$$

$$s_3' = \frac{1}{\sqrt{2}}(s_2 - s_3)$$

or in matrix form,

$$(s_1', s_2', s_3') = (s_1, s_2, s_3)\begin{pmatrix} 1/\sqrt{3} & 2/\sqrt{6} & 0 \\ 1/\sqrt{3} & -1/\sqrt{6} & 1/\sqrt{2} \\ 1/\sqrt{3} & -1/\sqrt{6} & -1/\sqrt{2} \end{pmatrix}$$

$$x' \quad = \quad x \qquad C$$

The matrices in the new representation are found from $\Gamma'(g) = C^{-1}\Gamma(g)C$ to be,

	E	C_3^+	C_3^-	σ_v	σ_v'	σ_v''
$\Gamma^{(3)}(g)$	$\begin{pmatrix} 1 & 0 & 0 \\ 0 & 1 & 0 \\ 0 & 0 & 1 \end{pmatrix}$	$\begin{pmatrix} 1 & 0 & 0 \\ 0 & -1/2 & \sqrt{3}/2 \\ 0 & -\sqrt{3}/2 & -1/2 \end{pmatrix}$	$\begin{pmatrix} 1 & 0 & 0 \\ 0 & -1/2 & -\sqrt{3}/2 \\ 0 & \sqrt{3}/2 & -1/2 \end{pmatrix}$	$\begin{pmatrix} 1 & 0 & 0 \\ 0 & 1 & 0 \\ 0 & 0 & -1 \end{pmatrix}$	$\begin{pmatrix} 1 & 0 & 0 \\ 0 & -1/2 & \sqrt{3}/2 \\ 0 & \sqrt{3}/2 & 1/2 \end{pmatrix}$	$\begin{pmatrix} 1 & 0 & 0 \\ 0 & -1/2 & -\sqrt{3}/2 \\ 0 & -\sqrt{3}/2 & 1/2 \end{pmatrix}$

We see that each matrix is now in block diagonal form, and the representation may be reduced into the direct sum of a 1 × 1 representation spanned by (s_1') and a 2 × 2 representation spanned by (s_2',s_3'). The complete set of reduced representations obtained from the original 4D representation is:

E	C_3^+	C_3^-	σ_v	σ_v'	σ_v''	
(1)	(1)	(1)	(1)	(1)	(1)	1D representation spanned by (s_N)
(1)	(1)	(1)	(1)	(1)	(1)	1D representation spanned by (s_1')
$\begin{pmatrix} 1 & 0 \\ 0 & 1 \end{pmatrix}$	$\begin{pmatrix} -1/2 & \sqrt{3}/2 \\ -\sqrt{3}/2 & -1/2 \end{pmatrix}$	$\begin{pmatrix} -1/2 & -\sqrt{3}/2 \\ \sqrt{3}/2 & -1/2 \end{pmatrix}$	$\begin{pmatrix} 1 & 0 \\ 0 & -1 \end{pmatrix}$	$\begin{pmatrix} -1/2 & \sqrt{3}/2 \\ \sqrt{3}/2 & 1/2 \end{pmatrix}$	$\begin{pmatrix} -1/2 & -\sqrt{3}/2 \\ -\sqrt{3}/2 & 1/2 \end{pmatrix}$	2D representation spanned by (s_2',s_3')

This is as far as we can go in reducing this representation. None of the three representations above can be reduced any further, and they are therefore called irreducible representations, or 'irreps', of the point group. Formally, a representation is an irrep if there is no similarity transform that can simultaneously convert all of the representatives into block diagonal form. The linear combination of basis functions that converts a matrix representation into block diagonal form, allowing reduction of the representation, is called a symmetry adapted linear combination.

Irreducible Representations

The dimensionality of a representation is equal to the dimensionality of each of its matrices, which is in turn equal to the number of rows or columns of the matrix. These representations are not unique. For example, by performing a similarity (or equivalence, or canonical) transformation $UD(A)U^{-1}$ we generate a new set of matrices which provides an equally good representation. A simple physical example for this transformation is the rotation of reference axes, such as (x, y, z) to (x', y', z'). We can also generate another representation by taking one or more representations and combining them according to,

$$\begin{pmatrix} D(A) & \mathcal{O} \\ \mathcal{O} & D'(A) \end{pmatrix}$$

where $\mathcal{O}$ = (m × n) matrix of zeros, not necessarily a square zero matrix. The matrices D(A) and D'(A) can be either two distinct representations or they can be identical representations. To overcome the difficulty of non-uniqueness of a representation with regard to a similarity transformation, we often just deal with the traces of the matrices which are invariant under similarity transformations. The trace of a matrix is defined as the sum of the diagonal matrix elements. To overcome the difficulty of the ambiguity of representations in general, we introduce the concept of irreducible representations.

Consider the representation made up of two distinct or identical representations for every element in the group,

$$\begin{pmatrix} D(A) & \mathcal{O} \\ \mathcal{O} & D'(A) \end{pmatrix}$$

This is a reducible representation because the matrix corresponding to each and every element of the group is in the same block form. We could now carry out a similarity transformation which would mix up all the elements so that the matrices are no longer in block form. But still the representation is reducible.

If by one and the same equivalence transformation, all the matrices in the representation of a group can be made to acquire the same block form, then the representation is said to be reducible; otherwise it is irreducible. Thus, an irreducible representation cannot be expressed in terms of representations of lower dimensionality. We will now consider three irreducible representations for the permutation group P(3):

	E	A	B
Γ_1:	(1)	(1)	(1)
$\Gamma_{1'}$:	(1)	(-1)	(-1)
Γ_2:	$\begin{pmatrix} 1 & 0 \\ 0 & 1 \end{pmatrix}$	$\begin{pmatrix} -1 & 0 \\ 0 & 1 \end{pmatrix}$	$\begin{pmatrix} \frac{1}{2} & -\frac{\sqrt{3}}{2} \\ -\frac{\sqrt{3}}{2} & -\frac{1}{2} \end{pmatrix}$

	C	D	F
Γ_1:	(1)	(1)	(1)
$\Gamma_{1'}$:	(-1)	(1)	(1)
Γ_2:	$\begin{pmatrix} \frac{1}{2} & \frac{\sqrt{3}}{2} \\ \frac{\sqrt{3}}{2} & -\frac{1}{2} \end{pmatrix}$	$\begin{pmatrix} -\frac{1}{2} & \frac{\sqrt{3}}{2} \\ -\frac{\sqrt{3}}{2} & -\frac{1}{2} \end{pmatrix}$	$\begin{pmatrix} -\frac{1}{2} & -\frac{\sqrt{3}}{2} \\ \frac{\sqrt{3}}{2} & -\frac{1}{2} \end{pmatrix}$

A reducible representation containing these three irreducible representations is,

$$\Gamma_R: \quad \overset{E}{\begin{pmatrix} 1 & 0 & 0 & 0 \\ 0 & 1 & 0 & 0 \\ 0 & 0 & 1 & 0 \\ 0 & 0 & 0 & 1 \end{pmatrix}} \overset{A}{\begin{pmatrix} 1 & 0 & 0 & 0 \\ 0 & -1 & 0 & 0 \\ 0 & 0 & -1 & 0 \\ 0 & 0 & 0 & 1 \end{pmatrix}} \overset{B}{\begin{pmatrix} 1 & 0 & 0 & 0 \\ 0 & -1 & 0 & 0 \\ 0 & 0 & \frac{1}{2} & -\frac{\sqrt{3}}{2} \\ 0 & 0 & -\frac{\sqrt{3}}{2} & -\frac{1}{2} \end{pmatrix}} \ldots,$$

where Γ_R is of the form,

$$\left(\begin{array}{c|c|c} \Gamma_1 & 0 & \mathcal{O} \\ \hline 0 & \Gamma_{1'} & \mathcal{O} \\ \hline \mathcal{O} & \mathcal{O} & \Gamma_2 \end{array}\right)$$

It is customary to list the irreducible representations contained in a reducible representation Γ_R as,

$$\Gamma_R = \Gamma_1 + \Gamma_{1'} + \Gamma_2.$$

In working out problems of physical interest, each irreducible representation describes the transformation properties of a set of eigenfunctions and corresponds to a distinct energy eigenvalue. Assume Γ_R is a reducible representation for some group G but an irreducible representation for some other group G'. If Γ_R contains the irreducible representations $\Gamma_1 + \Gamma_{1'} + \Gamma_2$ as illustrated earlier for the group P(3), this indicates that some interaction is breaking up a fourfold degenerate level in group G' into three energy levels in group G: two nondegenerate ones and a doubly degenerate one. Group theory does not tell us what these energies are, nor their ordering. Group theory only specifies the symmetries and degeneracies of the energy levels. In general, the higher the symmetry, meaning the larger the number of symmetry operations in the group, the higher the degeneracy of the energy levels. Thus when a perturbation is applied to lower the symmetry, the degeneracy of the energy levels tends to be reduced. Group theory provides a systematic method for determining exactly how the degeneracy is lowered.

Representation theory is useful for the treatment of physical problems because of certain orthogonality theorems. To prove the orthogonality theorems we need first to prove some other theorems.

The Unitarity of Representations

The following theorem shows that in most physical cases, the elements of a group can be represented by unitary matrices, which have the property of preserving length scales. This theorem is then used to prove lemmas leading to the proof of the "Wonderful Orthogonality Theorem".

Theorem: Every representation with matrices having nonvanishing determinants can be brought into unitary form by an equivalence (similarity) transformation.

Proof: By unitary form we mean that the matrix elements obey the relation $\left(A^{-1}\right)_{ij} = A^{\dagger}_{ij} = A^{*}_{ij}$, where A is an arbitrary matrix of the representation. The proof is carried out by actually finding the corresponding unitary matrices if the Aij matrices are not already unitary matrices.

Let $A_1, A_2, \cdots, A_h$ denote matrices of the representation. We start by forming the matrix sum,

$$H = \sum_{x=1}^{h} A_x A_x^{\dagger},$$

where the sum is over all the elements in the group and where the adjoint of a matrix is the transposed complex conjugate matrix $\left(A_x^{\dagger}\right)_{ij} = \left(A_x\right)_{ji}^{*}$. The matrix H is Hermitian because,

$$H^{\dagger} = \sum_{x}\left(A_x A_x^{\dagger}\right)^{\dagger} = \sum_{x} A_x A_x^{\dagger}.$$

Any Hermitian matrix can be diagonalized by a suitable unitary transformation. Let U be a unitary matrix made up of the orthonormal eigenvectors which diagonalize H to give the diagonal matrix d:

$$\begin{aligned} d &= U^{-1}HU \\ &= \sum_{x} U^{-1} A_x A_x^{\dagger} U \\ &= \sum_{x} U^{-1} A_x U U^{-1} A_x^{\dagger} U \\ &= \sum_{x} \hat{A}_x \hat{A}_x^{\dagger}, \end{aligned}$$

where we define $\hat{A}_x = U^{-1}A_xU$ for all x. The diagonal matrix d is a special kind of matrix and contains only real, positive diagonal elements since,

$$\begin{aligned} d_{kk} &= \sum_x \sum_j \left(\hat{A}_x\right)_{kj} \left(\hat{A}_x^\dagger\right)_{jk} \\ &= \sum_x \sum_j \left(\hat{A}_x\right)_{kj} \left(\hat{A}_x\right)^*_{kj} \\ &= \sum_x \sum_j \left|\left(\hat{A}_x\right)_{kj}\right|^2 . \end{aligned}$$

Out of the diagonal matrix d, one can form two matrices ($d^{1/2}$ and $d^{-1/2}$) such that,

$$d^{1/2} \equiv \begin{pmatrix} \sqrt{d_{11}} & & \mathcal{O} \\ & \sqrt{d_{22}} & \\ \mathcal{O} & & \ddots \end{pmatrix}$$

and $$d^{-1/2} \equiv \begin{pmatrix} \frac{1}{\sqrt{d_{11}}} & & \mathcal{O} \\ & \frac{1}{\sqrt{d_{22}}} & \\ \mathcal{O} & & \ddots \end{pmatrix}$$

where $d^{1/2}$ and $d^{-1/2}$ are real, diagonal matrices. We note that the generation of $d^{-1/2}$ from $d^{1/2}$ requires that none of the d_{kk} vanish. These matrices clearly obey the relations,

$$\begin{aligned} \left(d^{1/2}\right)^\dagger &= d^{1/2} \\ \left(d^{-1/2}\right)^\dagger &= d^{-1/2} \\ \left(d^{1/2}\right)\left(d^{1/2}\right) &= d \end{aligned}$$

so that,

$$d^{1/2}d^{-1/2} = d^{-1/2}d^{1/2} = \hat{1} = \text{unit matrix} .$$

From equation,

$$\begin{aligned} d &= U^{-1}HU \\ &= \sum_x U^{-1}A_xA_x^\dagger U \\ &= \sum_x U^{-1}A_xUU^{-1}A_x^\dagger U \\ &= \sum_x \hat{A}_x\hat{A}_x^\dagger , \end{aligned}$$

we can also write,

$$d = d^{1/2}d^{1/2} = \sum_{x} \hat{A}_x \hat{A}_x^{\dagger}.$$

We now define a new set of matrices,

$$\hat{\hat{A}}_x \equiv d^{-1/2}\hat{A}_x d^{1/2}$$

and

$$\hat{A}_x^{\dagger} = \left(U^{-1}A_xU\right)^{\dagger} = U^{-1}A_x^{\dagger}U$$

$$\hat{\hat{A}}_x^{\dagger} = \left(d^{-1/2}\hat{A}_x d^{1/2}\right)^{\dagger} = d^{1/2}\hat{A}_x^{\dagger}d^{-1/2}$$

We now show that the matrices $\hat{\hat{A}}_x$ are unitary:

$$\begin{aligned}\hat{\hat{A}}_x\hat{\hat{A}}_x^{\dagger} &= \left(d^{-1/2}\hat{A}_x d^{1/2}\right)\left(d^{1/2}\hat{A}_x^{\dagger}d^{-1/2}\right)\\ &= d^{-1/2}\hat{A}_x d\hat{A}_x^{\dagger}d^{-1/2}\\ &= d^{-1/2}\sum_{y}\hat{A}_x\hat{A}_y\hat{A}_y^{\dagger}\hat{A}_x^{\dagger}d^{-1/2}\\ &= d^{-1/2}\sum_{y}\left(\hat{A}_x\hat{A}_y\right)\left(\hat{A}_x\hat{A}_y\right)^{\dagger}d^{-1/2}\\ &= d^{-1/2}\sum_{z}\hat{A}_z\hat{A}_z^{\dagger}d^{-1/2}\end{aligned}$$

by the rearrangement theorem. But from the relation,

$$d = \sum_{z}\hat{A}_z\hat{A}_z^{\dagger}$$

it follows that $\hat{\hat{A}}_x\hat{\hat{A}}_x^{\dagger} = \hat{1}$, so that $\hat{\hat{A}}_x$ is unitary.

Therefore we have demonstrated how we can always construct a unitary representation by the transformation:

$$\hat{\hat{A}}_x = d^{-1/2}U^{-1}A_xUd^{1/2}$$

Where

$$H = \sum_{x=1}^{h}\hat{A}_x\hat{A}_x^{\dagger}$$

$$d = \sum_{x=1}^{h}\hat{A}_x\hat{A}_x^{\dagger}$$

and where U is the unitary matrix that diagonalizes the Hermitian matrix H and $\hat{A}_x = U^{-1}A_xU$.

Schur's Lemma (Part 1)

Schur's lemmas (Parts 1 and 2) on irreducible representations are proved in order to prove the "Wonderful Orthogonality Theorem".

Lemma: A matrix which commutes with all matrices of an irreducible representation is a constant matrix, i.e., a constant times the unit matrix. Therefore, if a non-constant commuting matrix exists, the representation is reducible; if none exists, the representation is irreducible.

Proof: Let M be a matrix which commutes with all the matrices of the representation $A_1, A_2, \ldots, A_h$,

$$MA_x = A_x M.$$

Take the adjoint of both sides of above equation to obtain,

$$A_x^\dagger M^\dagger = M^\dagger A_x^\dagger.$$

Since A_x can in all generality be taken to be unitary, multiply on the right and left of above equation by A_x to yield,

$$M^\dagger A_x = A_x M^\dagger,$$

so that if M commutes with A_x so does $M^\dagger$, and so do the Hermitian matrices H_1 and H_2 defined by,

$$H_1 = M + M^\dagger$$
$$H_2 = i\left(M - M^\dagger\right),$$

$$H_j A_x = A_x H_j, \text{ where } j = 1, 2.$$

We will now show that a commuting Hermitian matrix is a constant matrix from which it follows that $M = H_1 - iH_2$ is also a constant matrix.

Since H_j (j = 1, 2) is a Hermitian matrix, it can be diagonalized. Let U be the matrix that diagonalizes H_j (for example H_1) to give the diagonal matrix d,

$$d = U^{-1} H_j U.$$

We now perform the unitary transformation on the matrices A_x of the representation $\hat{A} = U^{-1} A_x U$. From the commutation relations equations, $MA_x = A_x M$, $A_x^\dagger M^\dagger = M^\dagger A_x^\dagger$ and $H_j A_x = A_x H_j$, where $j = 1, 2$ a unitary transformation on all matrices $H_j A_x = A_x H_j$ yields,

$$\underbrace{\left(U^{-1} H_j U\right)}_{d}\underbrace{\left(U^{-1} A_x U\right)}_{\hat{A}_x} = \underbrace{\left(U^{-1} A_x U\right)}_{\hat{A}_x}\underbrace{\left(U^{-1} H_j U\right)}_{d}.$$

So now we have a diagonal matrix d which commutes with all the matrices of the representation. We now show that this diagonal matrix d is a constant matrix, if all the $\hat{A}_x$ matrices (and thus also the A_x matrices) form an irreducible representation. Thus, starting with above equation,

$$d\hat{A}_x = \hat{A}_x d$$

we take the ij element of both sides of above equation,

$$d_{ii}\left(\hat{A}_x\right)_{ij} = \left(\hat{A}_x\right)_{ij} d_{jj},$$

so that,

$$\left(\hat{A}_x\right)_{ij}\left(d_{ii} - d_{jj}\right) = 0$$

for all the matrices A_x.

If $d_{ii} \neq d_{jj}$, so that the matrix d is not a constant diagonal matrix, then $\left(\hat{A}_x\right)_{ij}$ must be 0 for all the $\hat{A}$. This means that the similarity or unitary transformation $U^{-1}A_xU$ has brought all the matrices of the representation $\hat{A}_x$ into the same block form, since any time $d_{ii} \neq d_{jj}$ all the matrices $\left(\hat{A}_x\right)_{ij}$ are null matrices. Thus by definition the representation A_x is reducible. But we have assumed the A_x to be an irreducible representation. Therefore $\left(\hat{A}_x\right)_{ij} \neq 0$ for all $\hat{A}_x$ so that it is necessary that $d_{ii} = d_{jj}$, and Schur's lemma Part 1 is proved.

Schur's Lemma (Part 2)

Lemma: If the matrix representations $D^{(1)}(A_1)$, $D^{(1)}(A_2)$,...,$D^{(1)}(A_h)$ and $D^{(2)}(A_1)$, $D^{(2)}(A_2)$,...,$D^{(2)}(A_h)$ are two irreducible representations of a given group of dimensionality ℓ_1 and ℓ_2, respectively, then, if there is a matrix of ℓ_1 columns and ℓ_2 rows M such that,

$$MD^{(1)}(A_x) = D^{(2)}(A_x)M$$

for all A_x, then M must be the null matrix ($M = \mathcal{O}$) if $\ell_1 \neq \ell_2$. If $\ell_1 = \ell_2$, then either $M = \mathcal{O}$ or the representations $D^{(1)}(A_x)$ and $D^{(2)}(A_x)$ differ from each other by an equivalence (or similarity) transformation.

Proof: Since the matrices which form the representation can always be transformed into unitary form, we can in all generality assume that the matrices of both representations $D^{(1)}(A_x)$ and $D^{(2)}(A_x)$ have already been brought into unitary form.

Assume $\ell_1 \leq \ell_2$, and take the adjoint of above equation,

$$\left[D^{(1)}(A_x)\right]^{\dagger} M^{\dagger} = M^{\dagger}\left[D^{(2)}(A_x)\right]^{\dagger}.$$

The unitary property of the representation implies $\left[D(A_x)\right]^{\dagger} = \left[D(A_x)\right]^{-1} = D\left(A_x^{-1}\right)$, since the matrices form a substitution group for the elements Ax of the group. Therefore we can write above equation as,

$$D^{(1)}\left(A_x^{-1}\right)M^{\dagger} = M^{\dagger}D^{(2)}\left(A_x^{-1}\right).$$

Then multiplying above equation on the left by M yields,

$$MD^{(1)}\left(A_x^{-1}\right)M^{\dagger} = MM^{\dagger}D^{(2)}\left(A_x^{-1}\right) = D^{(2)}\left(A_x^{-1}\right)MM^{\dagger},$$

which follows from applying equation $MD^{(1)}\left(A_x\right) = D^{(2)}\left(A_x\right)M$ to the element A_x^{-1} which is also an element of the group,

$$MD^{(1)}\left(A_x^{-1}\right) = D^{(2)}\left(A_x^{-1}\right)M.$$

We have now shown that if $MD^{(1)}\left(A_x\right) = D^{(2)}\left(A_x\right)M$ then $MM^{\dagger}$ commutes with all the matrices of representation (2) and $M^{\dagger}M$ commutes with all matrices of representation (1). But if $MM^{\dagger}$ commutes with all matrices of a representation, then by Schur's lemma (Part 1), $MM^{\dagger}$ is a constant matrix of dimensionality ($\ell 2 \times \ell 2$):

$$MM^{\dagger} = c\hat{1},$$

where $\hat{1}$ is the unit matrix.

First we consider the case $\ell_1 = \ell_2$. Then M is a square matrix, with an inverse,

$$M^{-1} = \frac{M^{\dagger}}{c}, \quad c \neq 0.$$

Then if $M-1 \neq \mathcal{O}$, multiplying equation $MD^{(1)}\left(A_x\right) = D^{(2)}\left(A_x\right)M$ by $M-1$ on the left yields,

$$D^{(1)}\left(A_x\right) = M^{-1}D^{(2)}\left(A_x\right)M$$

and the two representations differ by an equivalence transformation.

However, if $c = 0$ then we cannot write equation $M^{-1} = \frac{M^{\dagger}}{c}, \ c \neq 0$ but instead we have to consider $MM^{\dagger} = 0$,

$$\sum_k M_{ik}M_{kj}^{\dagger} = 0 = \sum_k M_{ik}M_{jk}^{*}$$

for all ij elements. In particular, for $i = j$ we can write,

$$\sum_k M_{ik}M_{ik}^{*} = \sum_k \left|M_{ik}\right|^2 = 0.$$

Therefore each element $M_{ik} = 0$ so that M is a null matrix. This completes proof of the case $\ell_1 = \ell_2$ and $M = \mathcal{O}$.

Finally we prove that for $\ell_1 \neq \ell_2$, then $M = \mathcal{O}$. Suppose that $\ell_1 \neq \ell_2$, then we can arbitrarily take

$\ell_1 < \ell_2$. Then M has ℓ_1 columns and ℓ_2 rows. We can make a square $(\ell_2 \times \ell_2)$ matrix out of M by adding $(\ell_2 - \ell_1)$ columns of zeros,

$$\ell_2 \text{ rows} \overbrace{\begin{pmatrix} & 0\,0 & \cdots & 0 \\ & 0\,0 & \cdots & 0 \\ M & 0\,0 & \cdots & 0 \\ & \vdots\,\vdots & & \vdots \\ & 0\,0 & \cdots & 0 \end{pmatrix}}^{\ell_1 \text{ columns}} = N = \text{square } (\ell_2 \times \ell_2) \text{ matrix.}$$

The adjoint of above equation is then written as,

$$\begin{pmatrix} & M^\dagger & & \\ 0\,0 & 0 & \cdots & 0 \\ 0\,0 & 0 & \cdots & 0 \\ \vdots\,\vdots & \vdots & & \vdots \\ 0\,0 & 0 & \cdots & 0 \end{pmatrix} = N^\dagger$$

so that,

$$NN^\dagger = MM^\dagger = c\hat{1} \quad \text{dimension } (\ell_2 \times \ell_2).$$

$$\sum_k N_{ik} N_{ki}^\dagger = \sum_k N_{ik} N_{ik}^* = c\,\hat{1}$$

$$\sum_{ik} N_{ik} N_{ik}^* = c\ell_2.$$

But if we carry out the sum over i we see by direct computation that some of the diagonal terms of $\sum_{k,i} N_{ik} N_{ik}^*$ are o, so that c must be zero. But this implies that for every element we have $N_{ik} = 0$ and therefore also $M_{ik} = 0$, so that M is a null matrix.

Wonderful Orthogonality Theorem

The orthogonality theorem which we now prove is so central to the application of group theory to quantum mechanical problems that it was named the "Wonderful Orthogonality Theorem" by Van Vleck, and is widely known by this name. The theorem is in actuality an orthonormality theorem.

Theorem: The orthonormality relation,

$$\sum_R D_{\mu\nu}^{(\Gamma_j)}(R) D_{\nu'\mu'}^{(\Gamma_{j'})}(R^{-1}) = \frac{h}{\ell_j} \delta_{\Gamma_j,\Gamma_{j'}} \delta_{\mu\mu'} \delta_{\nu\nu'},$$

is obeyed for all the inequivalent, irreducible representations of a group, where the summation is over all h group elements $A_1, A_2, \ldots, A_h$ and ℓ_j and $\ell_{j'}$ are, respectively, the dimensionalities of representations Γ_j and $\Gamma_{j'}$. If the representations are unitary, the orthonormality relation becomes,

$$\sum_R D_{\mu\nu}^{(\Gamma_j)}(R)\left[D_{\mu'\nu'}^{(\Gamma_j')}(R)\right]^* = \frac{h}{\ell_j}\delta_{\Gamma_j,\Gamma_{j'}}\delta_{\mu\mu'}\delta_{\nu\nu'}.$$

Example: To illustrate the meaning of the mathematical symbols of this theorem, consider the orthogonality between the Γ_1 and Γ_1 irreducible representations for the P(3) group using the statements of the theorem:

$$\sum_R D_{\mu\nu}^{(\Gamma_1)}(R)D_{\mu'\nu'}^{(\Gamma_{1'})^*}(R) = [(1)\cdot(1)]+[(1)\cdot(1)]+[(1)\cdot(1)]$$
$$+[(1)\cdot(-1)]+[(1)\cdot(-1)]+[(1)\cdot(-1)] = 0.$$

Proof: Consider the $\ell_{j'} \times \ell_j$ matrix,

$$M = \sum_R D^{(\Gamma_{j'})}(R)XD^{(\Gamma_j)}(R^{-1}),$$

where X is an arbitrary matrix with $\ell_{j'}$ rows and ℓ_j columns so that M is a rectangular matrix of dimensionality ($\ell_{j'} \times \ell_j$). Multiply M by $D^{(\Gamma_{j'})}(S)$ for some element S in the group:

$$\underbrace{D^{(\Gamma_{j'})}(S)M}_{\ell_{j'}\times\ell_j} = \sum_R D^{(\Gamma_{j'})}(S)D^{(\Gamma_{j'})}(R)\,X\,D^{(\Gamma_j)}(R^{-1}).$$

We then carry out the multiplication of two elements in a group,

$$\underbrace{D^{(\Gamma_{j'})}(S)M}_{\ell_{j'}\times\ell_j} = \sum_R D^{(\Gamma_{j'})}(SR)X\,D^{(\Gamma_j)}(R^{-1}S^{-1})D^{(\Gamma_j)}(S),$$

where we have used the group properties of the representations Γ_j and $\Gamma_{j'}$. By the rearrangement theorem, above equation can be rewritten,

$$D^{(\Gamma_{j'})}(S)M = \underbrace{\sum_R D^{(\Gamma_{j'})}(R)X\,D^{(\Gamma_j)}(R^{-1})}_{M}D^{(\Gamma_j)}(S) = M\,D^{(\Gamma_j)}(S).$$

Now apply Schur's lemma Part 2 for the various cases.

Case: $\ell_j \neq \ell_{j'}$, or if $\ell_j = \ell_{j'}$, and the representations are not equivalent. Since $D^{(\Gamma_{j'})}(S)M = MD^{(\Gamma_j)}(S)$, then by Schur's lemma Part 2, M must be a null matrix. From the definition of M we have,

$$0 = M_{\mu\mu'} = \sum_R\sum_{\gamma,\lambda} D_{\mu\gamma}^{(\Gamma_{j'})}(R)X_{\gamma\lambda}D_{\lambda\mu'}^{(\Gamma_j)}(R^{-1}).$$

But X is an arbitrary matrix. By choosing X to have an entry 1 in the vv' position and 0 everywhere else, we write,

$$X=\begin{pmatrix} 0\,0\,0\,0\,0\,0\cdots \\ 0\,0\,0\,1\,0\,0\cdots \\ 0\,0\,0\,0\,0\,0\cdots \\ 0\,0\,0\,0\,0\,0\cdots \\ \vdots\;\vdots\;\vdots\;\vdots\;\vdots\;\vdots \end{pmatrix},\; X_{\gamma\lambda}=\delta_{\gamma v}\delta_{\lambda v'}.$$

It then follows by substituting above equation into equation $0=M_{\mu\mu'}=\sum_R\sum_{\gamma,\lambda}D_{\mu\gamma}^{(\Gamma_{j'})}(R)X_{\gamma\lambda}D_{\lambda\mu'}^{(\Gamma_j)}(R^{-1})$ that,

$$0=\sum_R D_{\mu v}^{(\Gamma_{j'})}(R)D_{v'\mu'}^{(\Gamma_j)}(R^{-1}).$$

Case: $\ell_j=\ell_{j'}$ and the representations Γj and Γj' are equivalent. If the representations Γ_j and Γ_j' are equivalent, then $\ell_j=\ell_{j'}$ and Schur's lemma part 1 tells us that $M=c\hat{1}$. The definition for M in equation $M=\sum_R D^{(\Gamma_{j'})}(R)XD^{(\Gamma_j)}(R^{-1})$, gives,

$$M_{\mu v'}=c\delta_{\mu\mu'}=\sum_R\sum_{\gamma,\lambda}D_{\mu\gamma}^{(\Gamma_{j'})}(R)X_{\gamma\lambda}D_{\gamma\mu'}^{(\Gamma_{j'})}(R^{-1}).$$

Choose X in equation,

$$X=\begin{pmatrix} 0\,0\,0\,0\,0\,0\cdots \\ 0\,0\,0\,1\,0\,0\cdots \\ 0\,0\,0\,0\,0\,0\cdots \\ 0\,0\,0\,0\,0\,0\cdots \\ \vdots\;\vdots\;\vdots\;\vdots\;\vdots\;\vdots \end{pmatrix},\; X_{\gamma\lambda}=\delta_{\gamma v}\delta_{\lambda v'}$$

as above to have a nonzero entry at vv' and 0 everywhere else. Then $X_{\gamma\lambda}=c'_{vv'}\delta_{\gamma v}\delta_{\lambda v'}$, so that,

$$c''_{vv'}\delta_{\mu\mu'}=\sum_R D_{\mu v}^{(\Gamma_{j'})}(R)D_{v'\mu'}^{(\Gamma_{j'})}(R^{-1}),$$

where $c''_{vv'}=c/c'_{vv'}$. To evaluate $c''_{vv'}$ choose μ = μ' in above equation and sum on μ:

$$\underbrace{c''_{vv'}\sum_\mu\delta_{\mu\mu}}_{\ell_{j'}}=\sum_R\sum_\mu D_{\mu v}^{(\Gamma_{j'})}(R)D_{v'\mu}^{(\Gamma_{j'})}(R^{-1})=\sum_R D_{v'v}^{(\Gamma_{j'})}(R^{-1}R).$$

Since $D(\Gamma_{j'})(R)$ is a representation of the group and follows the multiplication table for the group. Therefore we can write,

$$c''_{vv'}\ell_{j'}=\sum_R D_{v'v}^{(\Gamma_{j'})}(R^{-1}R)=\sum_R D_{v'v}^{(\Gamma_{j'})}(E)=D_{v'v}^{(\Gamma_{j'})}(E)\sum_R 1.$$

But $D^{(\Gamma_{j'})}_{v'v}(E)$ is a unit $(\ell_{j'} \times \ell_{j'})$ matrix and the v'v matrix element is $\delta_{v'v}$. The sum of unity over all the group elements is h. Therefore we obtain,

$$c''_{vv} = \frac{h}{\ell_{j'}}\delta_{vv'}.$$

Substituting above equation into $c''_{vv}\delta_{\mu\mu'} = \sum_R D^{(\Gamma_{j'})}_{\mu v}(R) D^{(\Gamma_{j'})}_{v'\mu'}(R^{-1})$, gives:

$$\frac{h}{\ell_{j'}}\delta_{\mu\mu'}\delta_{vv'} = \sum_R D^{(\Gamma_{j'})}_{\mu v}(R) D^{(\Gamma_{j'})}_{v'\mu'}(R^{-1}).$$

We can write the results of Cases in compact form,

$$\sum_R D^{(\Gamma_j)}_{\mu v}(R) D^{(\Gamma_{j'})}_{v'\mu'}(R^{-1}) = \frac{h}{\ell_j}\delta_{\Gamma_j\Gamma_{j'}}\delta_{\mu\mu'}\delta_{vv'}.$$

For a unitary representation above equation can also be written as,

$$\sum_R D^{(\Gamma_j)}_{\mu v}(R) D^{(\Gamma_{j'})*}_{\mu' v'}(R) = \frac{h}{\ell_j}\delta_{\Gamma_j\Gamma_{j'}}\delta_{\mu\mu'}\delta_{vv'}.$$

This completes the proof of the wonderful orthogonality theorem, and we see explicitly that this theorem is an orthonormality theorem.

Representations and Vector Spaces

Let us spend a moment and consider what the representations in above equation mean as an orthonormality relation in a vector space of dimensionality *h*. Here *h* is the order of the group which equals the number of group elements. In this space, the representations $D^{(\Gamma_j)}_{\mu v}(R)$ can be considered as elements in this *h*-dimensional space:

$$V^{(\Gamma_j)}_{\mu,v} = \left[D^{(\Gamma_j)}_{\mu v}(A_1), D^{(\Gamma_j)}_{\mu v}(A_2), \ldots, D^{(\Gamma_j)}_{\mu v}(A_h) \right].$$

The three indices Γ_j, μ, v label a particular vector. All distinct vectors in this space are orthogonal. Thus two representations are orthogonal if any one of their three indices is different. But in an h-dimensional vector space, the maximum number of orthogonal vectors is h. We now ask how many vectors $V^{(\Gamma_j)}_{\mu,v}$ can we make? For each representation, we have lj choices for μ and v so that the total number of vectors we can have is $\sum_j \ell_j^2$ where we are now summing over representations Γj. This argument yields the important result,

$$\sum_j \ell_j^2 \le h.$$

Character Tables

C_{3v}	E	$2C_3$	$3\sigma_v$		
A_1	1	1	1	z	$z^3, x(x^2-3y^2)$
A_2	1	1	−1	R_z	$y(3x^2-y^2)$
E	2	−1	0	$(x,y),(R_x,R_y)$	$(xz^2,yz^2),[xyz, z(x^2-y^2)]$
$\Gamma_{x,y,z}$	3	0	1		

A general and rigorous method for deriving character tables is based on five theorems which in turn are based on something called The Great Orthogonality Theorem.

The five theorems are:

1. The number of irreducible representations is equal to the number of classes in the group. E.g. NH_3, point group C_{3v}:

Which C_{3v} symmetry operations are the inverse of which and which are together in one class?

$$E^{-1}=E \qquad \sigma_v^{-1}=\sigma_v \qquad C_3^{-1}=C^2_3$$

$$\sigma'^{-1}_v=\sigma'_v \qquad (C^2_3)^{-1}=C_3 \qquad \sigma''^{-1}_v=\sigma''_v$$

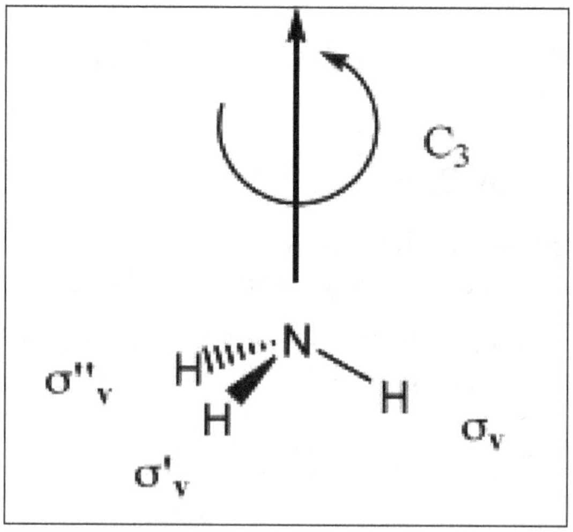

Using the above relationships we can set up the following similarity transformations:

$$\sigma_v \times C_3 \times \sigma_v = C^2_3 \qquad C^2_3 \times \sigma_v \times C_3 = \sigma''_v \qquad C^2_3 \times E \times C_3 = E$$

$$\sigma''_v \times C_3 \times \sigma''_v = C^2_3 \qquad \sigma_v \times \sigma_v \times \sigma_v = \sigma_v \qquad \sigma_v \times E \times \sigma_v = E$$

$$C^2_3 \times C^2_3 \times C_3 = C^2_3 \qquad C_3 \times \sigma_v \times C^2_3 = \sigma'_v \qquad \sigma''_v \times E \times \sigma''_v = E$$

$$\sigma'_v \times C^2_3 \times \sigma'_v = C_3 \qquad \sigma''_v \times \sigma'_v \times \sigma''_v = \sigma_v$$

$$\sigma'_v \times \sigma''_v \times \sigma'_v = \sigma_v$$

- Apparently $\{C_3, C^2_3\}$, $\{\sigma_v, \sigma'_v, \sigma''_v\}$, and $\{E\}$ are each in a class.
- In C_{3v} there are three classes and hence three irreducible representations.

2. The characters of all operations in the same class are equal in each given irreducible (or reducible) representation. In above example, all rotations C_3, C^2_3 will have the same character; all mirror planes σ_v, σ'_v, σ''_v will have the same character, etc. The character of a matrix is the sum of all its diagonal elements (also called the trace of a matrix).

Example: Consider the 3×3 matrix that represents the symmetry operation E as performed on a vector (x,y,z) in 3D space:

$$E = \begin{bmatrix} 1 & 0 & 0 \\ 0 & 1 & 0 \\ 0 & 0 & 1 \end{bmatrix} \rightarrow \text{trace} = 3$$

Therefore, A reducible representation for a vector (x,y,z) in 3D space will have a character 3 for the symmetry element E.

Example: Consider the 3×3 matrix that represents the symmetry operation C_3 as performed on a vector (x,y,z) in 3D space:

$$C_3 = \begin{bmatrix} \cos\frac{2\pi}{3} & -\sin\frac{2\pi}{3} & 0 \\ \sin\frac{2\pi}{3} & \cos\frac{2\pi}{3} & 0 \\ 0 & 0 & 1 \end{bmatrix} = \begin{bmatrix} -0.5 & -\frac{\sqrt{3}}{2} & 0 \\ \frac{\sqrt{3}}{2} & -0.5 & 0 \\ 0 & 0 & 1 \end{bmatrix} \rightarrow \text{trace} = 0$$

Therefore, A reducible representation for a vector (x,y,z) in 3D space will have a character 0 for the symmetry element C_3.

The reducible representation for a vector (x,y,z) in 3D space is often shown at the bottom of a character table. For the C_{3v} character table it is:

C_{3v}	E	C_3	C_3^2	σ_v	σ_v'	σ_v''
$\Gamma_{x,y,z}$	3	0	0	1	1	1

3. The sum of the squares of all characters in any irreducible representation is equal to the order of the group.

Order of the group = number of symmetry operators

C_{3v}	E	C_3	C_3^2	σ_v	σ_v'	σ_v''
A_1	1	1	1	1	1	1
A_2	1	1	1	−1	−1	−1
E	2	−1	−1	0	0	0

$$\text{Check } A_1:\ 1^2+1^2+1^2+1^2+1^2+1^2=6$$
$$\text{Check } A_2:\ 1^2+1^2+1^2+(-1)^2+(-1)^2+(-1)^2=6$$
$$\text{Check } E:\ 2^2+(-1)^2+(-1)^2+0^2+0^2+0^2=6$$

4. The point product of the characters of any two irreducible representations is 0.

Let's check that with the C^{3v} character table above:

$$\Gamma_{A1} * \Gamma_{A2} = (1x1)+(1x1)+(1x1)+(1x-1)+(1x-1)+(1x-1)=0$$
$$\Gamma_{A2} * \Gamma_{E} = (1x2)+(1x-1)+(1x-1)+(-1x0)+(-1x0)+(-1x0)=0$$

This is equivalent to saying that all irreducible representations are orthogonal.

5. The sum of the squares of the dimensions of the irreducible representations is equal to the order of the group. The dimension of a representation is the trace of the matrix of the identity operator (E).

Example: A vector (x,y,z) in 3D space is (obviously) 3-dimensional:

$$E=\begin{bmatrix}1 & 0 & 0\\ 0 & 1 & 0\\ 0 & 0 & 1\end{bmatrix} \rightarrow \text{trace}=3$$

Since the characters in the character table are the traces of matrices: In the point group C_3v, the irreducible representation, A_1 is 1 –dimensional, A_2 is 1 –dimensional and E is 1 –dimensional.

The order of the group is 6: $1^2 + 1^2 + 2^2 = 6$. A fully worked out example: The derivation of the C_{4v} character table. The symmetry operations in this point group are: E, C_4, $C^2_4 = C_2$, C^3_4, σ_v, σ'_v, σ_d, σ'_d.

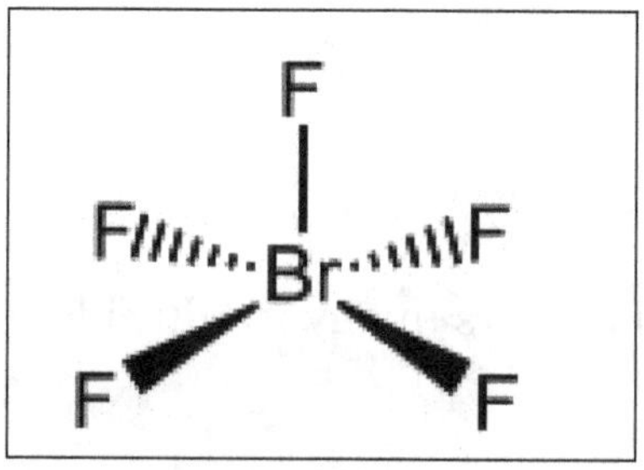

There are five classes of symmetry operations derived using a multiplication table: [E], [C_2], [C_4, C^3_4], [σ_v,σ'_v], and [σ_d,σ'_d], i.e. there will be five irreducible representations.

- There is always a totally symmetric representation denoted by a set of 1x1 matrices: i.e. We need ALL symmetry operations! none are redundant.

C_{4v}	[E]	$[C_3$	$^3C_4]$	$[C_2 = {}^2C_4]$	$[\sigma_v$	$\sigma_v']$	$[\sigma_d$	$\sigma_d']$
Γ_1	[1]	[1	1]	[1]	[1	1]	[1	1]
Γ_2								
Γ_3								
Γ_4								
Γ_5								

- Previous Theorem says, that the sum of the squares of the dimensions of the group must be equal to the order of the group:

C_{4v}	E	C_4	$^3C_4$	C_2	σ_v	σ_v'	σ_d	σ_d'
Γ_1	1	1	1	1	1	1	1	1
Γ_2	1							
Γ_3	1							
Γ_4	1							
Γ_5	2							

Because $1^2 + 1^2 + 1^2 + 1^2 + 2^2 = 8$ = order of the group, i.e. one of the irreducible representations will be two-dimensional.

- Based on the theorems we also know that:
  - The characters of all operations in the same class are the same in each the irreducible representations.
  - The sum of the squares of each row must be = 8.
  - The point product of any two rows must be = 0 ⇐ Orthogonal.

C_{4v}	[E]	$[C_4$	$C^3{}_4]$	$[C_2]$	$[\sigma_v$	$\sigma'_v]$	$[\sigma_d$	$\sigma'_d]$
Γ_1	[1]	[1	1]	[1]	[1	1]	[1	1]
Γ_2	[1]	[1	1]	[1]	[–1	–1]	[–1	–1]
Γ_3	[1]	[–1	–1]	[1]	[–1	–1]	[1	1]
Γ_4	[1]	[–1	–1]	[1]	[1	1]	[–1	–1]
Γ_5	[2]	[	]	[]	[	]	[	]

Because of previous theorem we can actually simplify this a little:

C_{4v}	E	$2c_4$	C_2	$2\sigma_v$	$2\sigma_d$
Γ_1	1	1	1	1	1
Γ_2	1	1	1	–1	–1
Γ_3	1	–1	1	–1	1
Γ_4	1	–1	1	1	–1
Γ_5	2	a	b	c	d

- Using previous theorem we can now write down 4 equations that uniquely determine the remaining 4 unknown characters a, b, c, d:

$$\Gamma_1 * \Gamma_5 = 1*2+2*1*a+1*b*+2*1*c+2*1*d$$
$$= 2+2a+b+2c+2d = 0$$

$$\Gamma_2 * \Gamma_5 = 2+2a+b-2c-2d = 0$$

$$\Gamma_3 * \Gamma_5 = 2-2a+b+-2c+2d = 0$$

$$\Gamma_4 * \Gamma_5 = 2-2a+b+2c-2d = 0$$

From this we find: a = 0, b = -2, c = 0, d = 0

- Using the definitions of Mulliken symbols, we can complete the character table by naming the reducible representations:

C_{4v}	E	$2c_4$	C_2	$2\sigma_v$	$2\sigma_d$
A_1	1	1	1	1	1
A_2	1	1	1	−1	−1
B_1	1	−1	1	−1	1
B_2	1	−1	1	1	−1
E	2	0	−2	0	0

Using Character Tables

Again the C_{4v} character table – this time the "extended version":

C_{4v}	E	$2C_4$	C_2	$2\sigma_v$	$2\sigma_d$	Basis	Basis Function
A_1	1	1	1	1	1	z	x^2+y^2, z^2
A_2	1	1	1	−1	−1	R_z	
B_1	1	−1	1	1	−1		x^2-y^2
B_2	1	−1	1	−1	1		xy
E	2	0	−2	0	0	$(x,y)(R_x,R_y)$	(xz,xz)

Basis and Basis Functions

Consider the matrix representations of all the symmetry operations in C_{4v}:

$$C_4 = \begin{bmatrix} \cos\frac{2\pi}{4} & -\sin\frac{2\pi}{4} & 0 \\ \sin\frac{2\pi}{4} & \cos\frac{2\pi}{4} & 0 \\ 0 & 0 & 1 \end{bmatrix} = \begin{bmatrix} 0 & -1 & 0 \\ 1 & 0 & 0 \\ 0 & 0 & 1 \end{bmatrix}$$

$$C_4^3 = \begin{bmatrix} \cos\frac{3\pi}{2} & -\sin\frac{3\pi}{2} & 0 \\ \sin\frac{3\pi}{2} & \cos\frac{3\pi}{2} & 0 \\ 0 & 0 & 1 \end{bmatrix} = \begin{bmatrix} 0 & 1 & 0 \\ -1 & 0 & 0 \\ 0 & 0 & 1 \end{bmatrix}$$

$$C_2 = \begin{bmatrix} \cos\pi & -\sin\pi & 0 \\ \sin\pi & \cos\pi & 0 \\ 0 & 0 & 1 \end{bmatrix} = \begin{bmatrix} -1 & 0 & 0 \\ 0 & -1 & 0 \\ 0 & 0 & 1 \end{bmatrix}$$

$$\sigma_{yz} = \begin{bmatrix} -1 & 0 & 0 \\ 0 & 1 & 0 \\ 0 & 0 & 1 \end{bmatrix} \quad \sigma_{xz} = \begin{bmatrix} 1 & 0 & 0 \\ 0 & -1 & 0 \\ 0 & 0 & 1 \end{bmatrix}$$

$$\sigma_d = \begin{bmatrix} 0 & 1 & 0 \\ 1 & 0 & 0 \\ 0 & 0 & 1 \end{bmatrix} \quad \sigma'_d = \begin{bmatrix} 0 & -1 & 0 \\ -1 & 0 & 0 \\ 0 & 0 & 1 \end{bmatrix}$$

z never "mixes" with x or y. x and y are "mixed" by C_4 and σ_d.

- Can separate to set of block diagonalized matrices:

	E	$2C_4$	C_2	σ_{xz}	σ_{yz}	σ_d	σ'_d
$\Gamma_{x,y}$	$\begin{bmatrix} 1 & 0 \\ 0 & 1 \end{bmatrix}$	$\begin{bmatrix} 0 & -1 \\ 1 & 0 \end{bmatrix}$	$\begin{bmatrix} -1 & 0 \\ 0 & -1 \end{bmatrix}$	$\begin{bmatrix} 1 & 0 \\ 0 & -1 \end{bmatrix}$	$\begin{bmatrix} -1 & 0 \\ 0 & 1 \end{bmatrix}$	$\begin{bmatrix} 0 & 1 \\ 1 & 0 \end{bmatrix}$	$\begin{bmatrix} 0 & -1 \\ -1 & 0 \end{bmatrix}$
Γ_z	[1]	[1]	[1]	[1]	[1]	[1]	[1]

The characters of the matrices of $\Gamma_{x,y}$ and Γ_z are: Γ_z: all = 1.

	E	$2C_4$	C_2	σ_{xz}	σ_{yz}	σ_d	σ'_d
$\Gamma_{x,y}$	2	0	−2	0	0	0	0

Comparing this result with the complete character table above, we can say:

- x and y form a basis for the irreducible representation E in C_{4v}.
- z forms a basis for the irreducible representation A_1 in C_{4v}.

What can we do with Basis Functions?

Consider a d_{x2-y2} orbital:

- Any symmetry operation on C_{4v} transforms d_{x2-y2} onto itself or it's negative.
- The orbital transforms in a one-dimensional representation, i.e. with no mixing of coordinates.

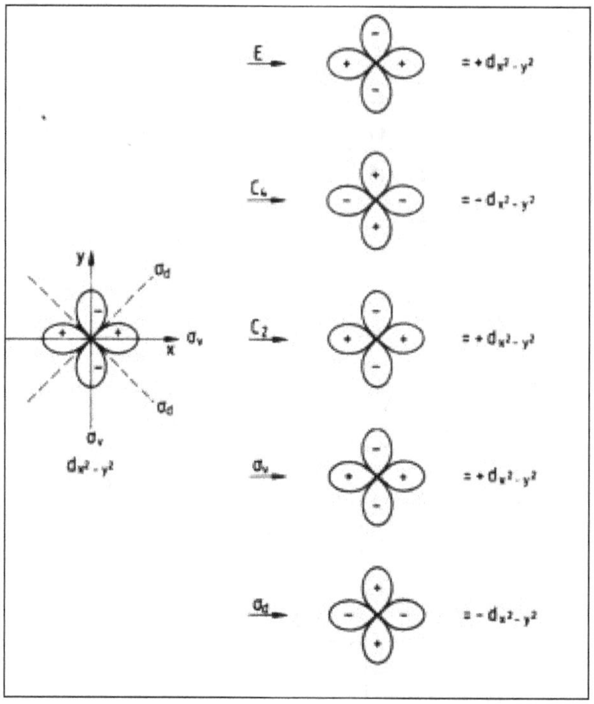

The correct irreducible representation for a $d_{x2\text{-}y2}$ under C_{4v} is thus:

C_{4v}	E	$2C_4$	C_2	$2\sigma_v$	$2\sigma_d$	Basis	Basis Function
B_1	1	−1	1	1	−1		$x^2 - y^2$

We say: "x^2 -y^2 transforms like B_1 under C_{4v}".

Next consider a rotation about the z-axis shown here as a curved arrow: Key: Look along the axis of rotation and determine the sense of rotation before and after the symmetry operation performed on the curved arrow.

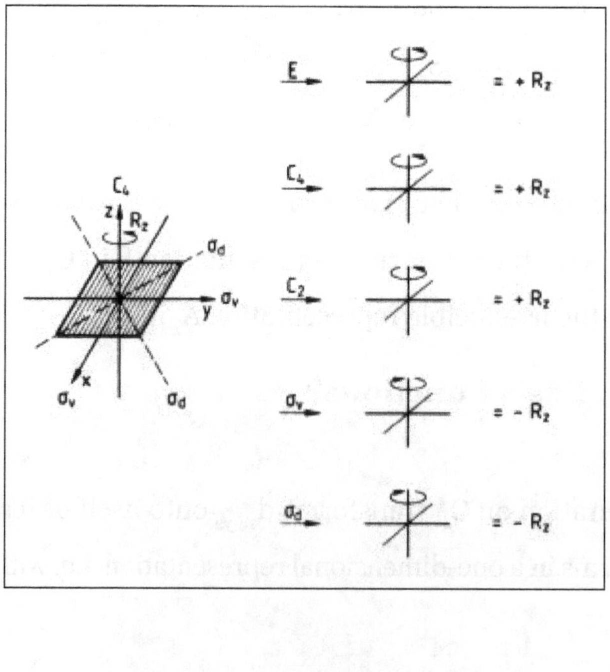

- Only the two σ operations have an effect leading to the following irreducible representation:

C_{4v}	E	$2C_4$	C_2	$2\sigma_v$	$2\sigma_d$	Basis	Basis Function
A_2	1	1	1	−1	−1	R_z	

"R_z transforms like A_2 under C_{4v}":

- Similarly one finds that rotations R_x and R_y mix and together form a basis for the E representation under C_{4v}.

In general the basis functions denoted by x, y, z, xz, xy, yz, $x^2 - y^2$, and z^2 directly relate to the symmetries of the orbitals and their transformations in the point group of the molecule under consideration.

- Many point groups (in particular the ones with primary axis of rotation with an odd #) have imaginary characters represented either by ± i or ε and ε*.
- For any point group with principal axis Cn: $\varepsilon = \exp(i2\pi/n)$.
- Using Euler's relationship:

$$e^{i\theta} = \cos\theta + i\sin\theta$$
$$e^{i2\pi/n} = \cos(2\pi/n) + i\sin(2\pi/n).$$

- Imaginary characters always appear in pairs of conjugated complex numbers – their occurrence is a mathematical necessity. e.g.: Point group C_3:

C_3	E C_3 C_3^2	$\epsilon = e^{2\pi i/3}$		
A E	1 1 1 $\left\{\begin{matrix} 1 & \epsilon & \epsilon^* \\ 1 & \epsilon^* & \epsilon \end{matrix}\right\}$	z, R_z $(x,y),(R_x,R_y)$	x^2+y^2, z^2 $(x^2-y^2, xy), (yz, xz)$	$z^3, x(x^2-3y^2), y(3x^2-y^2)$ $(xz^2, yz^2), [xyz, z(x^2-y^2)]$
$\Gamma_{x,y,z}$	3 0 0			

Since the principle axis is C_3, ε = exp(2πi/3):

- In order to use this character table on a real physical problem, we need real numbers; they are obtained by adding the pair-wise complex conjugated numbers to give real numbers, e.g. for C_3:

$$\text{using } e^{2\pi i/3} = \cos\frac{2\pi}{3} + i\sin\frac{2\pi}{3} = -\frac{1}{2} + i\frac{1}{2}\sqrt{3}$$

$$E = \begin{bmatrix} 1 & \varepsilon & \varepsilon^* \\ 1 & \varepsilon^* & \varepsilon \end{bmatrix} = \begin{bmatrix} 1 & \left(-\frac{1}{2}+i\frac{1}{2}\sqrt{3}\right) & \left(-\frac{1}{2}-i\frac{1}{2}\sqrt{3}\right) \\ 1 & \left(-\frac{1}{2}-i\frac{1}{2}\sqrt{3}\right) & \left(-\frac{1}{2}+i\frac{1}{2}\sqrt{3}\right) \end{bmatrix}$$

- $\{1+1\}\ \{\varepsilon+\varepsilon^*\}\ \{\varepsilon^*+\varepsilon\} = \{2\}\{-1\}\{-1\}$

A usable form of the C_3 table is thus:

C_3	E	C_3	C_3^2
A	1	1	1
E	2	−1	−1

Meaning: x and y form a basis for the E representation. Using the general rotation matrices developed earlier:

$$C_3 = \begin{bmatrix} \cos 2\pi/3 & -\sin 2\pi/3 \\ \sin 2\pi/3 & \cos 2\pi/3 \end{bmatrix} = \begin{bmatrix} -\frac{1}{2} & \sqrt{\frac{3}{2}} \\ -\sqrt{\frac{3}{2}} & -\frac{1}{2} \end{bmatrix}$$

$$C_3^2 = \begin{bmatrix} \cos 4\pi/3 & -\sin 4\pi/3 \\ \sin 4\pi/3 & \cos 4\pi/3 \end{bmatrix} = \begin{bmatrix} -\frac{1}{2} & \sqrt{\frac{3}{2}} \\ -\sqrt{\frac{3}{2}} & -\frac{1}{2} \end{bmatrix}$$

The traces of these matrices are both = -1. These are the characters shown in the usable table.

Deconstructing Reducible Representations

An infinite number of reducible representations are possible. Each reducible representation can be deconstructed (reduced) to a sum of a finite set of irreducible representations. E.g.: Consider an arbitrary reducible representation Γ_{red} in C_{2v} symmetry:

C_{2v}	E	C_2	σ_v	σ'_v
A_1	1	1	1	1
A_2	1	1	−1	−1
B_1	1	−1	1	−1
B_2	1	−1	−1	1
Γ_{red}	3	1	3	1

By inspection, we see that it is the sum of two A_1 and one B_1 irreducible representations:

$$\begin{array}{lcrrrr} 2\,A_1 & = & 2 & 2 & 2 & 2 \\ B_1 & = & 1 & -1 & 1 & -1 \\ \hline \Gamma_{red} = 2\,A_1 + B_1 & = & 3 & 1 & 3 & 1 \end{array}$$

Clearly, in more complicated case, the inspection method can be extremely difficult.

There exists a very simple formula that deconvolutes any reducible representation into its irreducible components:

$$a_1 = \frac{1}{h}\sum_R \chi^R \times \chi_i^R$$

- a_i = number of times that the irreducible representation Γ_i occurs in the reducible representation Γ_{red} under investigation.
- h = order of the point group (= number of symmetry operations).
- R = operation in the point group.
- χ^R = character of the operation R in Γ_{red}.
- χ_i^R = character of the operation R in Γ_i.

Let's apply this formula to the previous example in C_{2v}:

	E	C_2	σ_v	σ'_v
Γ_{red}	3	1	3	1

$$a_{A_1} = \frac{1}{4}\sum_R \chi^R \times \chi_{A_1}^R = \frac{1}{4}\left(\overset{E}{3\times1}+\overset{C_2}{1\times1}+\overset{\sigma_v}{3\times1}+\overset{\sigma'_v}{1\times1}\right) = 2 \quad \text{i.e. need } 2\times A_1$$

$$a_{A_2} = \frac{1}{4}\left(3\times1+1\times1+3\times(-1)+1\times(-1)\right) = 0$$

$$a_{B_1} = \frac{1}{4}\left(3\times1+1\times(-1)+3\times1+1\times(-1)\right) = 1$$

$$a_{B_2} = \frac{1}{4}\left(3\times1+1\times(-1)+3\times(-1)+1\times\right) = 0$$

$$\rightarrow \quad \Gamma_{red} = 2\,A_1 + B_1$$

The Direct Product

The direct product of two (ir)reducible representations is obtained by multiplying the respective characters of the representations. The result is again an (ir)reducible representation of the same group. e.g.:

D_3	E	$2\,C_3$	$3\,C_2$
A_1	1	1	1
A_2	1	1	−1
E	2	−1	0
$A_1 \times E$	2	−1	0
$A_2 \times E$	2	−1	0
$E \times E$	4	1	0
$A_2 \times A_2$	1	1	1

Construction of Character Table

Here we will take the example of $XeOF_4$. The molecule has one C_4 axis, $4\sigma_v$ ($2\sigma_v + 2\sigma_d$) and one C_2 axis along C_4 axis as shown in figure.

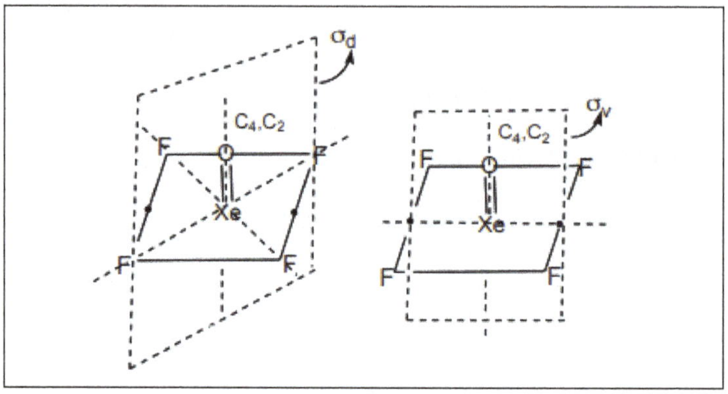

Symmetry elements in $XeOF_4$.

Molecule belongs to C_{4v} point group. The total number of symmetry operations generated by these symmetry elements are: E, $2C_4$, C_2, $2\sigma_v$, $2\sigma_d$. These have been dividing into five classes. So according rule from GOT number of irreducible representations in a group equals to number of classes. So there are five irreducible representations into C_{4v} point group character table. The total number of symmetry operation is eight and thus h=8 .The format of C_{4v} character table can be shown as:

C_{4v}	E	$2C_4$	C_2(along C_4 axis)	$2\sigma_v$	$2\sigma_d$
Γ_1					
Γ_2					
Γ_3					
Γ_4					
Γ_5					

The sum of squares of the dimensions of irreducible representations equals to order of the group. Let the dimensions of five irreducible representations be l_1, l_2, l_3, l_4, l_5.Therefore,

$$l_1^2 + l_2^2 + l_3^2 + l_4^2 + l_5^2 = 8$$

Also we know the relation $\sum_R |\chi_i(E)|^2 = h$.

This shows that the sum of the squares of characters for E in all the irreducible representations equals to order of the group. The above rule relating irreducible and reducible representations will only be satisfied when $l_1 = l_2 = l_3 = l_4 = 1$ and $l_5 = 2$ i.e. dimensions respectively are 1,1,1,1 and 2. For using above equation number of operations in a class must be taken care. Because dimension of irreducible representation equals to character of E ie $\chi(E)$. We write $\chi(E)$ under E as follows:

C_{4v}	E	$2C_4$	C_2	$2\sigma_v$	$2\sigma_d$
Γ_1	1				
Γ_2	1				

Γ_3	1				
Γ_4	1				
Γ_5	2				

We also know that in each point group character table there is always present a totally symmetric one dimensional irreducible representation for which each symmetry operation has the character +1. Let the irreducible representation be Γ. So C_{4v} character table with this information can be shown as:

C_{4v}	E	$2C_4$	C_2	$2\sigma_v$	$2\sigma_d$
Γ_1	1	1	1	1	1
Γ_2	1				
Γ_3	1				
Γ_4	1				
Γ_5	2				

Since Γ_1, Γ_2, Γ_3, Γ_4 are one dimensional, the characters for all operations in these irreducible representations will be ±1 only. Let the characters of these operations in Γ_5 be$\pm\chi C_4$, $\pm\chi C_2$, $\pm\chi\,\sigma_v$. And $\pm\chi\,\sigma_d$. So we can write the characters in the table below as:

C_{4v}	E	$2C_4$	C_2	$2\sigma_v$	$2\sigma_d$
Γ_1	1	1	1	1	1
Γ_2	1	±1	±1	±1	±1
Γ_3	1	±1	±1	±1	±1
Γ_4	1	±1	±1	±1	±1
Γ_5	2	$\pm\chi C_4$	$\pm\chi C_2$	$\pm\chi\,\sigma_v$	$\pm\chi\,\sigma_d$

In a class, the squares of the characters for the symmetry operation when multiplied by number of operations in that class and summed over all irreducible representations, equals to order of the group. Let us take C_2 class and apply this rule (down the column):

$$1^2+(\pm1)^2+(\pm1)^2+(\pm1)^2+(\pm\chi\,C_2)^2=8.$$

$$(1)+(1)+(1)+(1)+(\chi\,C_2)^2=8$$

Therefore, $(\chi C_2)^2=8-4=4$.

This is possible only when $\chi\,C_2$ in Γ_5 is ±2. Now take the class corresponding to C_4 and do the similar procedure i.e.

$$2(1)^2+2(\pm1)^2+2(\pm1)^2+2(\pm1)^2+2(\pm\chi C_4)^2=8$$

$$2+2+2+2+2(\chi C_4)^2=8$$

Therefore, $2(\chi C_4)^2=8-8=0$. This is possible when $\chi C4 = 0$ in $\Gamma 5$.

Now take the class of σ_v,

$$2(1)^2 + 2(\pm 1)^2 + 2(\pm 1)^2 + 2(\pm 1)^2 + 2(\pm \sigma_v)^2 = 8$$

$$2+2+2+2+2(\chi\sigma_v)^2 = 8$$

Therefore, $2(\chi\, \sigma_v)^2 = 8 - 8 = 0$. This is possible when $\chi\, \sigma_v = 0$ in Γ_5. Similarly $\chi\, \sigma_d = 0$ in Γ_5. Now the C_{4v} character table looks as:

C_{4v}	E	$2C_4$	C_2	$2\sigma_v$	$2\sigma_d$
Γ_1	1	1	1	1	1
Γ_2	1	±1	±1	±1	±1
Γ_3	1	±1	±1	±1	±1
Γ_4	1	±1	±1	±1	±1
Γ_5	2	0	±2	0	0

In order to find the character for C_2 symmetry operation in Γ_5. For this let us use orthogonality check by using the equation, $\Sigma\chi_i(R)\chi_j(R) = 0$ and let us take Γ_1 and Γ_5,

$$1.1.2 + 2.1.0 + 1.1(\pm 2) + 2.1.0 + 2.1.0 = 0$$

$$2 + 0 \pm 2 + 0 + 0 = 0$$

This is possible only if we take - sign before 2 so that the sum is zero 2+0-2+0+0=0. Let us now check the orthogonality of Γ_4, and Γ_5 irreducible representations,

$$1.1.2 + 2(\pm 1).0 + 1.(\pm 1)(-2) + 2(\pm 1).0 + 2(\pm 1).0 = 2 + 0 \pm 2 + 0 + 0 = 0$$

The product under C_2 must be of negative sign and χC_2 in Γ_5 is already -2. Therefore, χC_2 for C_2 in Γ_4 should be +1 so that the result is 2+0-2+0+0=0.

The C_{4v} character table now looks as:

C_{4v}	E	$2C_4$	C_2	$2\sigma_v$	$2\sigma_d$
Γ_1	1	1	1	1	1
Γ_2	1	±1	±1	±1	±1
Γ_3	1	±1	±1	±1	±1
Γ_4	1	±1	1	±1	±1
Γ_5	2	0	-2	0	0

Let us check orthogonality of Γ_2 and Γ_5 $1.2 + 2(\pm 1).0 + (\pm 1)(-2) + 2(\pm 1).0 + 2(\pm 1).0 = 2 + 0 + \pm 2 + 0 + 0.$

In order this equation is valid, the product under C_2 (highlighted) must be of negative sign and χC_2 in Γ_5 is already -2. Therefore, for χC_2 in Γ_2 should be +1 so that the result is 2+0-2+0+0=0. Similarly to χC_2 in Γ_3, is +1. Since there are eight operations in the group, four of these must have χ = +1 and other four must have $\chi = -1$ so that orthogonality rule is obeyed by these Γ_1, Γ_2, Γ_3, Γ_4. The remaining characters may be written as +1 or -1 in Γ_2, Γ_3, Γ_4 so that these have four +1 and four -1. This is possible only in this way,

C_{4v}	E	$2C_4$	C_2	$2\sigma_v$	$2\sigma_d$
$A_1 = \Gamma_1$	1	1	1	1	1
$A_2 = \Gamma_2$	1	1	1	-1	-1
$B_1 = \Gamma_3$	1	-1	1	-1	1
$B_2 = \Gamma_4$	1	-1	1	1	-1
$E = \Gamma_5$	2	0	-2	0	0

Where A_1, A_2, B_1, B_2, E are Mulliken's symbols for these irreducible representations.

Construction of Character Table for C_{2h} Point Group

Let us take 1,2 trans-dicholoroethylene. The various symmetry elements in this molecule are shown in Figure.

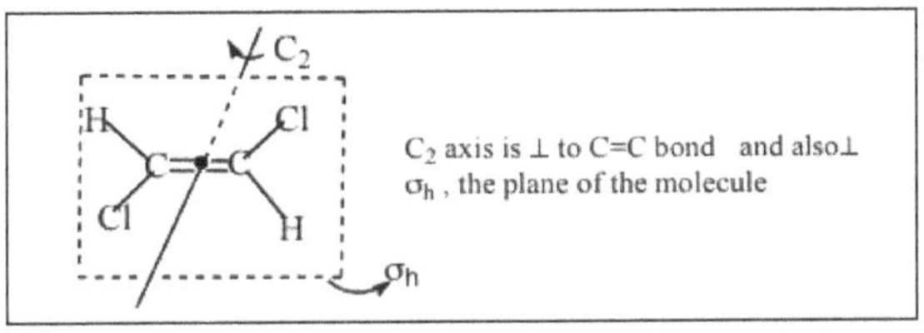

Symmetry elements in 1,2 trans-dichloroethylene.

Now let us work out characters for irreducible representations for C_{2h} point group. This point group has E, C_2, σ_h and i four symmetry operations. Therefore, the order of the group, h=4. Each operation belongs to a separate class. So according to rule of GOT there will be only four irreducible representations in C_{2h} point group. Let these be Γ_1, Γ_2, Γ_3, Γ_4.

Let the dimensions of these be l_1, l_2, l_3, l_4 respectively. According to GOT rules .these irreducible representations should satisfy the relation.

$$l_1^2 + l_2^2 + l_3^2 + l_4^2 = 4,$$

This is possible only when $l_1 = l_2 = l_3 = l_4 = 1$. All irreducible representations are of dimension one.

In a group there is always one totally symmetrical representation for which characters for all symmetry operations in a class are +1 only. The character of E gives the dimension of irreducible representation. So character for E in all these irreducible representations will be +1. The format of the character table of C_{2h} point group looks as:

C_{2h}	E	C_2	σ_h	$2\sigma_i$
Γ_1	1	1	1	1
Γ_2	1			
Γ_3	1			
Γ_4	1			

GOT demands that the remaining three irreducible representations will be orthogonal to this totally symmetrical irreducible representation. Thus, according to GOT all other irreducible representations Γ_2, Γ_3, Γ_4, must be orthogonal to Γ_1. These remaining Γ_2, Γ_3, Γ_4, representations will have characters as +1 for two symmetry operations and -1 character for other two symmetry operations of the C_{2h} point group to satisfy the orthogonality condition. Thus for C_{2h} point group the four irreducible representations will have the following characters:

C_{2h}	E	C_2	σ_h	I
$A_g = \Gamma_1$	1	1	1	1
$A_u = \Gamma_2$	1	1	-1	-1
$B_g = \Gamma_3$	1	-1	-1	1
$B_u = \Gamma_4$	1	-1	1	-1

That these characters are correct or not can be verified by checking the orthogonality of Γ_1, Γ_2, Γ_3, Γ_4.

- Orthogonality of Γ_1 and Γ_2,

  $$1.1+1.1+1.(-1)+1.(-1)=0$$

- Orthogonality of of Γ_1 and Γ_3,

  $$1.1+1.(-1)+1.(-1)+1.1=0$$

- Orthogonality of of Γ_1 and Γ_4,

  $$1.1+1.(-1)+1.1+1.(-1)=0$$

- Orthogonality of of Γ_2 and Γ_3,

  $$1.1+1.(-1)+(-1).(-1)+(-1).1=0$$

- Orthogonality of of Γ_4 and Γ_3,

  $$1.1+(-1).(-1)+(-1).1+1.(-1)=0$$

Thus these characters obtained are correct.

Reduction of Representations II

By making maximum use of molecular symmetry, we often greatly simplify problems involving molecular properties. For example, the formation of chemical bonds is strongly dependent on the atomic orbitals involved having the correct symmetries. To make full use of group theory in the applications we will be considering, we need to develop a little more 'machinery'. Specifically, given a basis set (of atomic orbitals, for example) we need to find out:

- How to determine the irreducible representations spanned by the basis functions.

- How to construct linear combinations of the original basis functions that transform as a given irreducible representation/symmetry species.

It turns out that both of these problems can be solved using something called the 'Great Orthogonality Theorem' (GOT for short). The GOT summarizes a number of orthogonality relationships implicit in matrix representations of symmetry groups, and may be derived in a somewhat qualitative fashion by considering these relationships in turn.

General Concepts of Orthogonality

You are probably already familiar with the geometric concept of orthogonality. Two vectors are orthogonal if their dot product (i.e. the projection of one vector onto the other) is zero. An example of a pair of orthogonal vectors is provided by the x and y Cartesian unit vectors.

$$x, y = 0$$

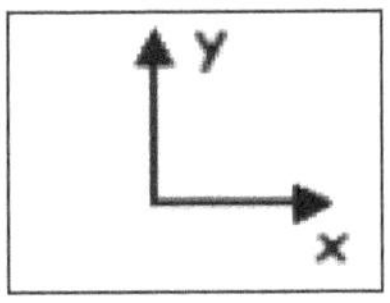

A consequence of the orthogonality of x and y is that any general vector in the xy plane may be written as a linear combination of these two basis vectors.

$$r = ax + by$$

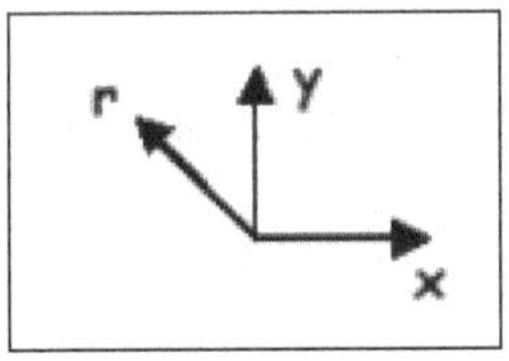

Mathematical functions may also be orthogonal. Two functions, $f_1(x)$ and $f_2(x)$, are defined to be orthogonal if the integral over their product is equal to zero i.e.

$$\int f_1(x) f_2(x) dx = \delta_{12}$$

This simply means that there must be 'no overlap' between orthogonal functions, which is the same as the orthogonality requirement for vectors, above. In the same way as for vectors, any general function may be written as a linear combination of a suitably chosen set of orthogonal basis functions. For example, the Legendre polynomials $P_n(x)$ form an orthogonal basis set for functions of one variable x.

$$f(x) = \sum_n c_n P_n(x)$$

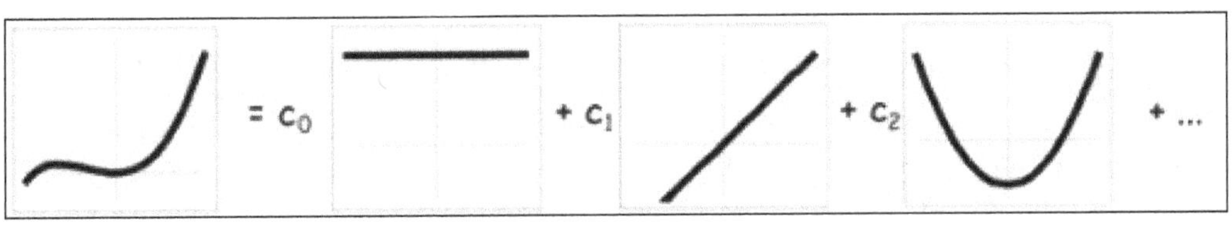

Orthogonality Relationships in Group Theory

The irreducible representations of a point group satisfy a number of orthogonality relationships:

1. If corresponding matrix elements in all of the matrix representatives of an irreducible representation are squared and added together, the result is equal to the order of the group divided by the dimensionality of the irreducible representation. i.e.

$$\sum_g \Gamma_k(g)_{ij}\Gamma_k(g)_{ij} = \frac{h}{d_k}$$

where k labels the irreducible representation, i and j label the row and column position within the irreducible representation, h is the order of the group, and dk is the order of the irreducible representation. E.g. The order of the group C_{3v} is 6. If we apply the above operation to the first element in the 2×2 (E) irreducible representation, the result should be equal to $\frac{h}{d_k} = \frac{6}{2} = 3$. Carrying out this operation gives:

$$(1)^2 + \left(-\frac{1}{2}\right)^2 + \left(-\frac{1}{2}\right)^2 + (1)^2 + \left(-\frac{1}{2}\right)^2 + \left(-\frac{1}{2}\right)^2 = 1 + \frac{1}{4} + \frac{1}{4} + 1 + \frac{1}{4} + \frac{1}{4} = 3$$

2. If instead of summing the squares of matrix elements in an irreducible representation, we sum the product of two different elements from within each matrix, the result is equal to zero. i.e.

$$\sum_g \Gamma_k(g)_{ij}\Gamma_k(g)_{i'j'} = 0$$

where $i \neq i'$ and/or $j \neq j'$. E.g. if we perform this operation using the two elements in the first row of the 2D irreducible representation used in 1, we get:

$$(1)(0) + \left(-\frac{1}{2}\right)\left(\frac{\sqrt{3}}{2}\right) + \left(-\frac{1}{2}\right)\left(-\frac{\sqrt{3}}{2}\right) + (1)(0) + \left(-\frac{1}{2}\right)\left(\frac{\sqrt{3}}{2}\right) + \left(-\frac{1}{2}\right)\left(-\frac{\sqrt{3}}{2}\right) = 0 + \frac{\sqrt{3}}{4} - \frac{\sqrt{3}}{4} + 0 - \frac{\sqrt{3}}{4} + \frac{\sqrt{3}}{4} = 0$$

3. If we sum the product of two elements from the matrices of two different irreducible representations k and m, the result is equal to zero. i.e.

$$\sum_g \Gamma_k(g)_{ij}\Gamma_m(g)_{i'j'} = 0$$

where there is now no restriction on the values of the indices i, i', j, j' (apart from the rather obvious restriction that they must be less than or equal to the dimensions of the irreducible representation). E.g. Performing this operation on the first elements of the A_1 and E irreducible representations we derived for C_{3v} gives:

$$(1)(1) + (1)\left(-\frac{1}{2}\right) + (1)\left(-\frac{1}{2}\right) + (1)(1) + (1)\left(-\frac{1}{2}\right) + (1)\left(-\frac{1}{2}\right) = 1 - \frac{1}{2} - \frac{1}{2} + 1 - \frac{1}{2} - \frac{1}{2} = 0$$

We can combine these three results into one general equation, the Great Orthogonality Theorem.

$$\sum_g \Gamma_k(g)_{ij}\Gamma_m(g)_{i'j'} = \frac{h}{\sqrt{d_k d_m}}\delta_{km}\delta_{ii'}\delta_{jj'}$$

For most applications we do not actually need the full Great Orthogonality Theorem. A little mathematical trickery transforms above equation into the 'Little Orthogonality Theorem' (or LOT), which is expressed in terms of the characters of the irreducible representations rather than the irreducible representations themselves.

$$\sum_{g} \chi_k(g)\chi_m(g) = h\delta_{km}$$

Since the characters for two symmetry operations in the same class are the same, we can also rewrite the sum over symmetry operations as a sum over classes.

$$\sum_{C} n_C \chi_k(C)\chi_m(C) = h\delta_{km}$$

where n_C is the number of symmetry operations in class C.

In all of the examples we've considered so far, the characters have been real. However, this is not necessarily true for all point groups, so to make the above equations completely general we need to include the possibility of imaginary characters. In this case we have:

$$\sum_{C} n_C \chi_k^*(C)\chi_m(C) = h\delta_{km}$$

where $\chi_k^*(C)$ is the complex conjugate of $\chi_k(C)$. Above equation is of course identical to equation $\sum_{C} n_C \chi_k(C)\chi_m(C) = h\delta_{km}$ when all the characters are real.

Using the LOT to Determine the Irreducible Representations Spanned by a Basis

We can often carry out a similarity transform on a general matrix representation so that all the representatives end up in the same block diagonal form. When this is possible, each set of submatrices also forms a valid matrix representation of the group. If none of the submatrices can be reduced further by carrying out another similarity transform, they are said to form an irreducible representation of the point group. An important property of matrix representatives is that their character is invariant under a similarity transform. This means that the character of the original representatives must be equal to the sum of the characters of the irreducible representations into which the representation is reduced. E.g. if we consider the representative for the C_3^- symmetry operation in our NH_3 example, we have:

$$\begin{pmatrix} 1 & 0 & 0 & 0 \\ 0 & 0 & 0 & 1 \\ 0 & 1 & 0 & 0 \\ 0 & 0 & 1 & 0 \end{pmatrix} \xrightarrow{\text{similarity transform}} \begin{pmatrix} 1 & 0 & 0 & 0 \\ 0 & 1 & 0 & 0 \\ 0 & 0 & -\frac{1}{2} & -\frac{\sqrt{3}}{2} \\ 0 & 0 & \frac{\sqrt{3}}{2} & -\frac{1}{2} \end{pmatrix} = (1)\otimes(1)\otimes\begin{pmatrix} -\frac{1}{2} & -\frac{\sqrt{3}}{2} \\ \frac{\sqrt{3}}{2} & -\frac{1}{2} \end{pmatrix}$$

$$\chi = 1 \qquad\qquad \chi = 1 \qquad\qquad \chi = 1+1-1 = 1$$

It follows that we can write the characters for a general representation $\Gamma(g)$ in terms of the characters of the irreducible representations $\Gamma_k(g)$ into which it can be reduced.

$$\chi(g) = \sum_k a_k \chi_k(g)$$

where the coefficients a_k in the sum are the number of times each irreducible representation appears in the representation. This means that in order to determine the irreducible representations spanned by a given basis. All we have to do is determine the coefficients a_k in the above equation. This is where the Little Orthogonality Theorem comes in handy. If we take the LOT in the form of above equation, and multiply each side through by a_k, we get,

$$\Sigma_g a_k \chi_k(g) \chi_m(g) = h a_k \delta_{km}$$

Summing both sides of the above equation over k gives,

$$\Sigma_g \Sigma_k a_k \chi_k(g) \chi_m(g) = h \Sigma_k a_k \delta_{km}$$

We can use equation $\chi(g) = \sum_k a_k \chi_k(g)$ to simplify the left hand side of this equation. Also, the sum on the right hand side reduces to am because δkm is only non-zero (and equal to 1) when k = m,

$$\Sigma_g \chi(g) \chi_m(g) = h a_m$$

Dividing both sides through by h (the order of the group), gives us an expression for the coefficients am in terms of the characters $\chi(g)$ of the original representation and the characters $\chi_m(g)$ of the mth irreducible representation.

$$a_m = \frac{1}{h} \Sigma_g \chi(g) \chi_m(g)$$

We can of course write this as a sum over classes rather than a sum over symmetry operations.

$$a_m = \frac{1}{h} \Sigma_C n_C \chi(g) \chi_m(g)$$

Cartesian Coordinates as Base

The matrices behave very similar to symmetry operations of the point group. These combine in similar fashion. How to quantify the effect of symmetry operations so that we can apply group theory more effectively and in systematic manner? This can be achieved if we can represent each and every symmetry operation of the point group by a matrix generated by taking, Cartesian coordinates, orbitals, base vectors, wave functions etc. as the bases. Effect of symmetry operations on these bases provide matrices corresponding to these symmetry operations. These matrices for each and every operations of the point group form a homomorphic group. These matrices behave very similar to symmetry operations of the point group. This means the effect of symmetry operation parallels the effect of matrix (corresponding to that symmetry operation generated by taking certain base) on a particular point of the object.

Symmetric and Antisymmetric Orbitals and Representation

Let us take p_x orbital and perform various symmetry operations and see the effect of these symmetry operations and let us quantify these effects in terms of mathematics. The various symmetry operations and their effects are shown as:

1. Identity operation E: According to E operation we are not doing anything on the p_x orbital or we are rotating by 3600 degree. The p_x orbital remains unchanged E $p_x \rightarrow P_x$ and p_x goes to p_x and we can say that E has (+1) effect on P_x orbital i.e.

$$E\ p_x \rightarrow P_x = (+1)p_x$$

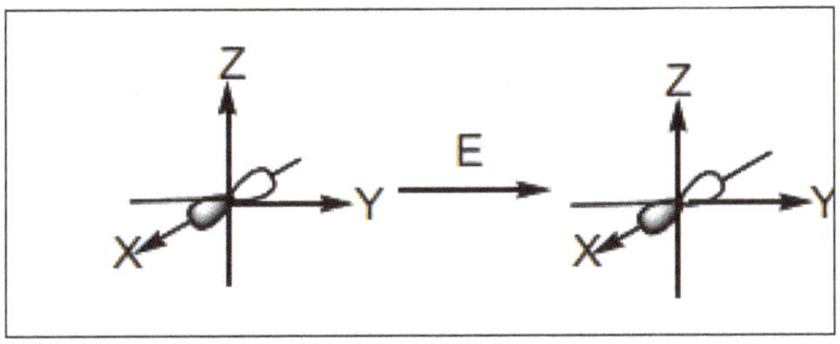

If there is no change the effect is written as +1 effect of this symmetry operation on the p_x orbital.

2. C_2 operation: Let us perform C_2 rotations about X,Y,Z axes on p_x orbitals. These are depicted as follows in figure. Effect of C_{2x} on p_x gives the following results i.e. $C_{2x}p_x$ = + p_x ie $(+1)p_x$ or the effect of C_2x on p_x is (+1). Thus P_x is symmetric with respect to C_{2x}.

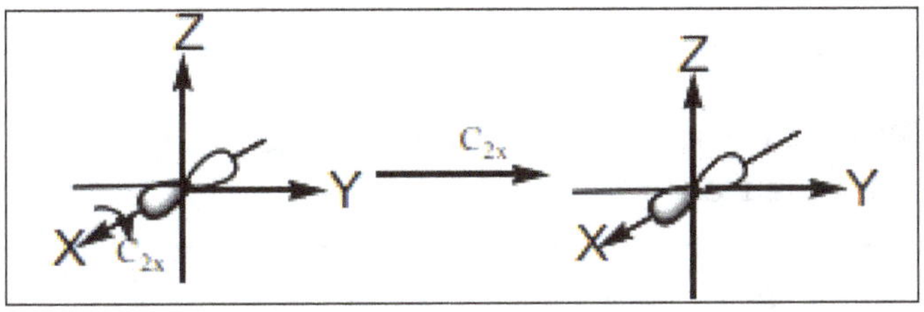

Effect of C_2 symmetry operation on p_x orbital.

And the effect of other C_2 axes C_{2y} and C_{2z} is shown in figure:

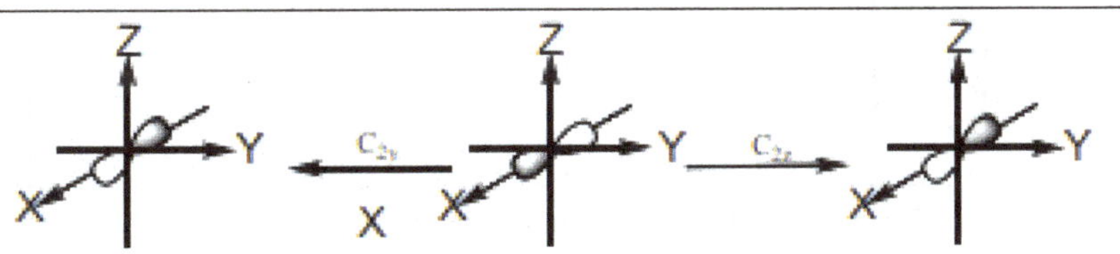

C_{2y} p_x = $-p_x$ ie $(-1)p_x$ or the effect of C_{2y} on p_x is (-1)

C_{2z} p_x = $-p_x$ ie $(-1)p_x$ or the effect of C_{2z} on p_x is (-1)

Effect of C_{2y} and C_{2z} symmetry operations on p_x orbital.

-1 effect means that the p_x orbital is antisymmetric with respect to C_{2y} and C_{2z} symmetry operations on p_x orbital.

3. σ_{yz} operation: When σ_{yz} operates on p_x one gets the result as, $\sigma_{yz}\,p_x = -p_x = (-1)p_x$ and this effect of symmetry operation is are shown in figure.

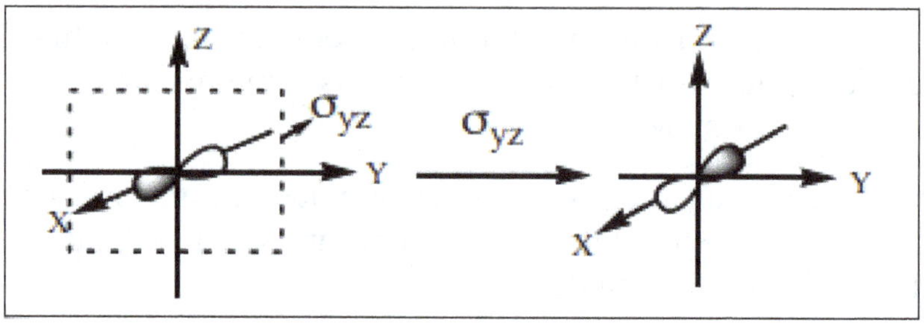

Effect of σ_{yz} on p_x orbital.

Thus P_x is antisymmetric with respect to σ_{yz} ie -1 effect.

4. σ_{xz} operation: σ_{xz} operates on p_x one gets the result as, $\sigma_{xz}\,p_x = p_x = (+1)px$ and this effect of symmetry operation is shown in figure.

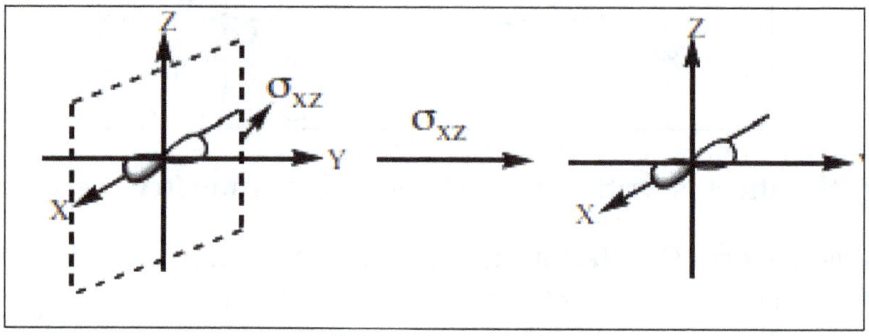

Effect of σ_{xz} on p_x orbital.

Thus P_x is symmetric with respect to σ_{xz} ie +1 effect.

5. Inversion operation: When i operates on p_x orbital one gets the result as: $i(p_x) = -p_x = (-1)p_x$ and the effect of i is shown in figure.

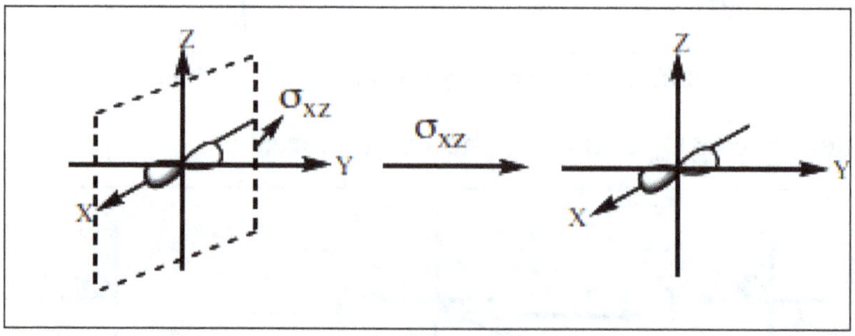

Effect of i on p_x orbital.

Thus P_x is antisymmetric with respect to i i.e. -1 effect.

Here +1 and -1 represent symmetric and antisymmetric with respect to particular symmetry operation.

Table: The effect of all the symmetry operations on p_x.

p_x	E	C_{2y}	C_{2z}	σ_{yz}	σ_{xz}	i	C_{2x}
	+1	-1	-1	-1	+1	-1	+1

Thus by taking ±1 as the effect of symmetry operations on p_x orbital we get a representation for E, C_2, σ_{xz}, σ_{yz}, and i using p_x orbital as the basis. Similarly we get the effect of these symmetry operations on p_y orbitals and the results are shown in figure.

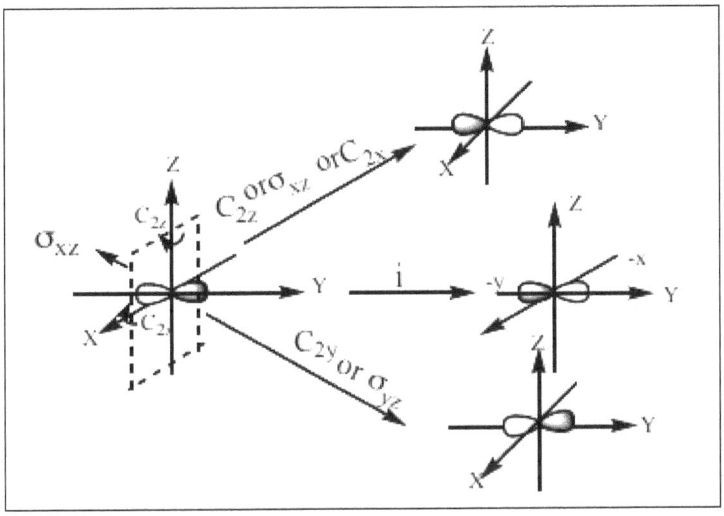

Effect of various symmetry operations on p_y orbital.

Table: Effects of symmetry operations on p_y orbital.

P_y	E	C_{2x}	C_{2z}	σ_{yz}	σ_{xz}	I
	+1	-1	-1	+1	-1	-1

The effects of various symmetry operations on p_z orbital are summarized in the table below. P_z orbital is perpendicular to x_y plane as shown,

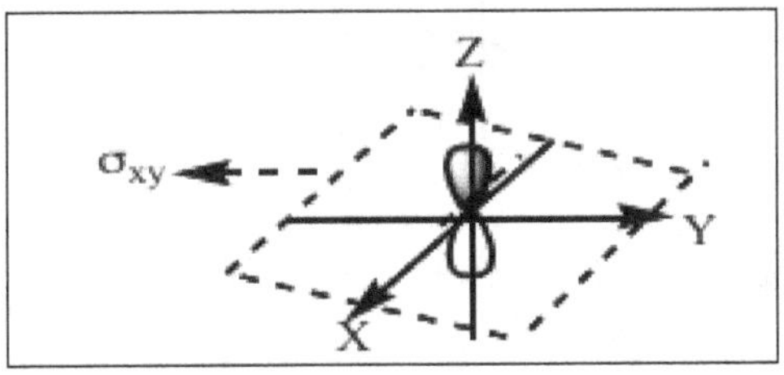

P_z	E	C_{2z}	C_{2x}	C_{2y}	σ_{yz}	σ_{xy}	σ_{xz}	I
	+1	1	-1	-1	+1	-1	+1	-1

These type of representations in terms of ± 1, using p_x, p_y, p_z orbitals as the basis, are known as non – degenerate representations.

Representation of C_{2v} point group using p_x, p_y and p_z orbitals as bases: In C_{2v} point group there are E,C_2,σ_{xz} and σ_{xz} symmetry operations. These p_x, p_y and p_z orbitals on O atom in H_2O molecule is shown in figure.

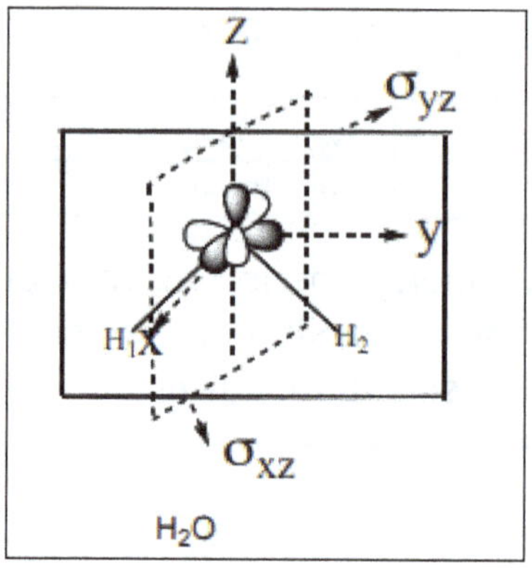

The effects of these symmetry operations on p_x, p_y and p_z orbitals is shown in table.

Table: Effects of symmetry operations of C_{2v} point group on p_x, p_y and p_z orbitals.

C_{2v}	E	C_2	σ_{xz}	σ_{yz}
p_x	1	-1	1	-1
p_y	1	-1	-1	1
p_z	1	1	1	1

Thus the effects of symmetry operations E, C_2, σ_{xz} and σ_{yz} of C_{2v} point group on p_x, p_y and p_z orbitals has been quantified in terms of ± 1.

Unit Vectors along Cartesian Coordinates as the Bases for Representation

A set of unit vectors x, y, z lying along X, Y, Z axes is shown in figure.

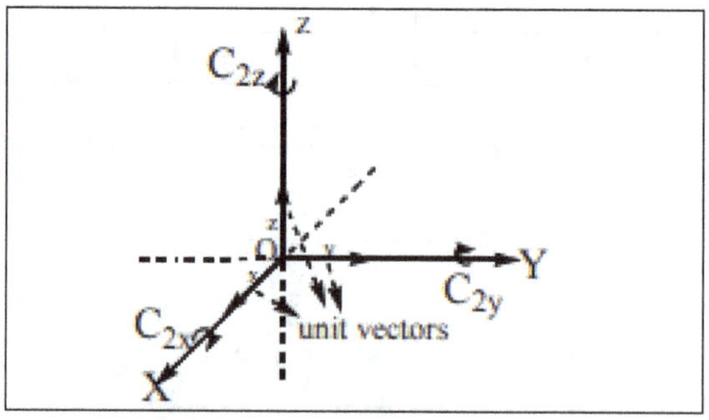

Set of unit vectors x, y, z in Cartesian coordinate system.

Now let us see what is the effect of E, C_{2x}, C_{2y}, C_{2z}, σ_{xz}, σ_{yz}, σ_{xy} and i on these unit vectors x, y, z lying along Cartesian axes.

1. Effect of E on unit vectors x, y, z. This is represented as:

$$E\begin{bmatrix} x \\ y \\ z \end{bmatrix} = \begin{bmatrix} x' \\ y' \\ z' \end{bmatrix}$$

ie x→x', y→y', z→z'. The E operation is shown in figure.

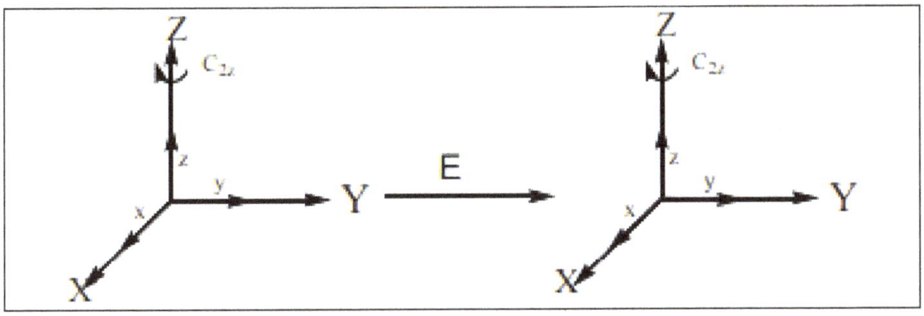

Effect of E on x, y, z unit vectors.

Since E does not do anything on these unit vectors these remain unchanged; (i) x = x' = (+1)x, (ii) y = y′ = (+1)y, (iii) z = z′ = (+1)z. Since x, y, z are orthogonal to each other, these do not contribute to other coordinates on performing symmetry operation. We can write three equations for operation E as:

$$E\,x = x' \rightarrow (+1)x + (0)y + (0)z$$
$$E\,y = y' \rightarrow (0)x + (+1)y + (0)z$$
$$E\,z = z' \rightarrow (0)x + (0)y + (+1)z$$

In matrix form the effect of E on x, y, z unit vectors can be written as:

$$E = \begin{bmatrix} 1 & 0 & 0 \\ 0 & 1 & 0 \\ 0 & 0 & 1 \end{bmatrix}$$

or we can represent the whole processes in terms of matrices,

$$E \begin{bmatrix} x \\ y \\ z \end{bmatrix} = \begin{bmatrix} x' \\ y' \\ z' \end{bmatrix}$$

$$\begin{bmatrix} 1 & 0 & 0 \\ 0 & 1 & 0 \\ 0 & 0 & 1 \end{bmatrix} \begin{bmatrix} x \\ y \\ z \end{bmatrix} = \begin{bmatrix} x' \\ y' \\ z' \end{bmatrix}$$

Here 3×3 matrix,

$$\begin{bmatrix} 1 & 0 & 0 \\ 0 & 1 & 0 \\ 0 & 0 & 1 \end{bmatrix}$$

corresponds to identity operation performed on unit vectors lying along X, Y, Z axes.

2. Effect of C_{2z} on unit vectors x, y, z. This is represented in figure as,

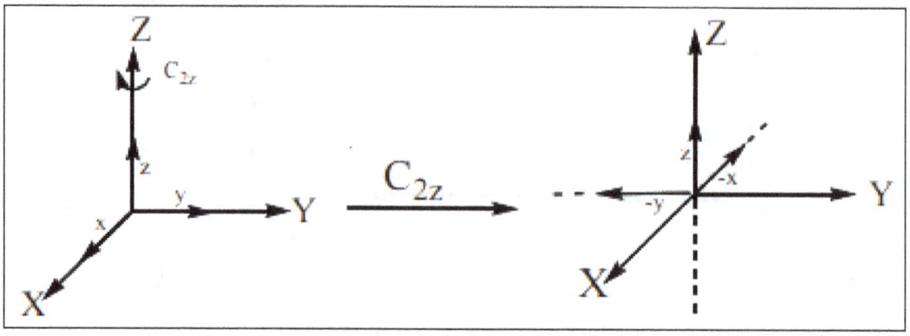

Effect of C_{2z} on unit vectors x,y,z.

$$x' = -x = (-1)x \rightarrow (-1)x + (0)y + (0)z$$
$$y' = -y = (-1)y \rightarrow (0)x + (-1)y + (0)z$$
$$z' = z = (+1)z \rightarrow (0)x + (0)y + (+1)z$$

$$\begin{bmatrix} x' \\ y' \\ z' \end{bmatrix} = \begin{bmatrix} -1 & 0 & 0 \\ 0 & -1 & 0 \\ 0 & 0 & +1 \end{bmatrix} \begin{bmatrix} x \\ y \\ z \end{bmatrix}$$

Thus matrix corresponding to operation C_{2z} on unit vectors x, y, z is,

$$\begin{bmatrix} -1 & 0 & 0 \\ 0 & -1 & 0 \\ 0 & 0 & 1 \end{bmatrix}$$

3. Effect of σ_{xz} on unit vectors x, y, z. This is represented as;

$$\sigma_{xz} \begin{bmatrix} x \\ y \\ z \end{bmatrix} = \begin{bmatrix} x' \\ y' \\ z' \end{bmatrix}$$

Figure below shows this σ_{xz} effect on unit vectors x,y,z,

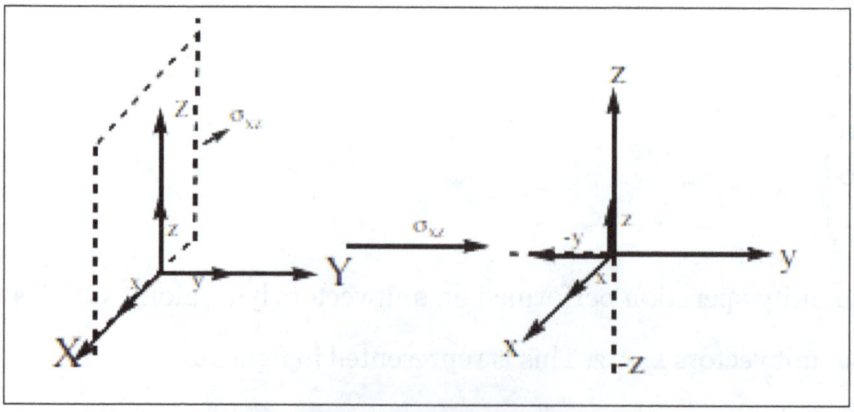

The σ_{xz} effect on unit vectors x,y,z.

Therefore, corresponding equations are;

$$x' = x = (+1)x \rightarrow (+1)x + (0)y + (0)z$$
$$y' = -y = (-1)y \rightarrow (0)x + (-1)y + (0)z$$
$$z' = z = (+1)z \rightarrow (0)x + (0)y + (+1)z$$

$$\begin{bmatrix} x' \\ y' \\ z' \end{bmatrix} = \begin{bmatrix} 1 & 0 & 0 \\ 0 & -1 & 0 \\ 0 & 0 & 1 \end{bmatrix} \begin{bmatrix} x \\ y \\ z \end{bmatrix}$$

Thus matrix corresponding to σ_{xz} is,

$$\begin{bmatrix} 1 & 0 & 0 \\ 0 & -1 & 0 \\ 0 & 0 & 1 \end{bmatrix}$$

4. Effect of σ_{yz} operation on x, y, z unit vectors: It can be written as:

$$\sigma_{yz} \begin{bmatrix} x \\ y \\ z \end{bmatrix} = \begin{bmatrix} x' \\ y' \\ z' \end{bmatrix}$$

Figure below shows this σ_{yz} effect on unit vectors x,y,z.

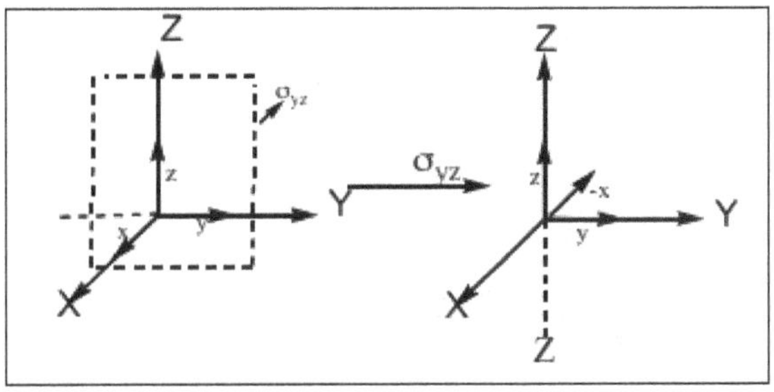

The σ_{yz} effect on unit vectors x,y,z.

Therefore, corresponding equations are;

$$x' = x = (-1)x \rightarrow (-1)x + (0)y + (0)z$$
$$y' = y = (+1)y \rightarrow (0)x + (+1)y + (0)z$$
$$z' = z = (+1)z \rightarrow (0)x + (0)y + (+1)z$$

Therefore,

$$\begin{bmatrix} x' \\ y' \\ z' \end{bmatrix} = \begin{bmatrix} -1 & 0 & 0 \\ 0 & 1 & 0 \\ 0 & 0 & 1 \end{bmatrix} \begin{bmatrix} x \\ y \\ z \end{bmatrix}$$

and matrix for σ_{yz} is,

$$\begin{bmatrix} -1 & 0 & 0 \\ 0 & 1 & 0 \\ 0 & 0 & 1 \end{bmatrix}$$

5. Effect of i operation on x,y,z unit vectors: it can be represented as;

$$i\begin{bmatrix} x \\ y \\ z \end{bmatrix} = \begin{bmatrix} x' \\ y' \\ z' \end{bmatrix}$$

Figure below shows this i effect on unit vectors x, y, z.

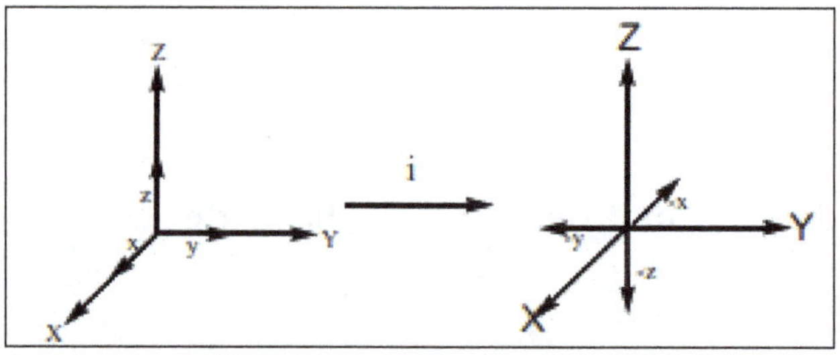

The i effect on unit vectors x, y, z.

Therefore, corresponding equations for operation i are;

$$x' = x = (-1)x \rightarrow (-1)x + (0)y + (0)z$$
$$y' = -y = (-1)y \rightarrow (0)x + (-1)y + (0)z$$
$$z' = z = (-1)z \rightarrow (0)x + (0)y + (-1)z$$

Therefore,

$$\begin{bmatrix} x' \\ y' \\ z' \end{bmatrix} = \begin{bmatrix} -1 & 0 & 0 \\ 0 & -1 & 0 \\ 0 & 0 & -1 \end{bmatrix}\begin{bmatrix} x \\ y \\ z \end{bmatrix}$$

Therefore, matrix for 'i' is,

$$\begin{bmatrix} -1 & 0 & 0 \\ 0 & -1 & 0 \\ 0 & 0 & -1 \end{bmatrix}$$

6. σ_{xy} operation on x, y, z unit vectors; It can be written as:

$$\sigma_{xy}\begin{bmatrix} x \\ y \\ z \end{bmatrix} = \begin{bmatrix} x' \\ y' \\ z' \end{bmatrix}$$

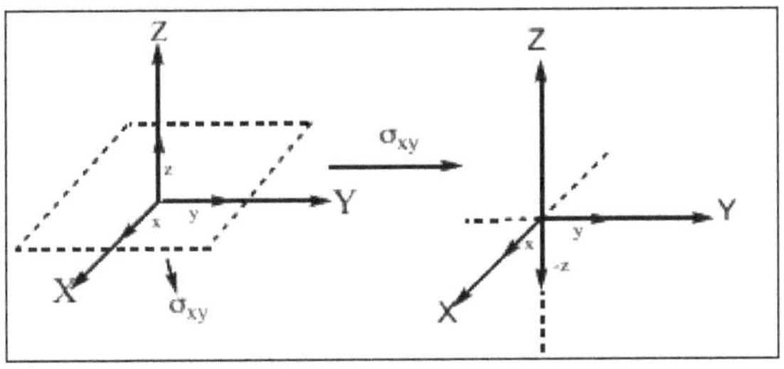

The σ_{xy} effect on unit vectors x, y, z.

The corresponding equations are:

$$x' = x = (+1)x \rightarrow (+1)x + (0)y + (0)z$$
$$y' = y = (+1)y \rightarrow (0)x + (+1)y + (0)z$$
$$z' = z = (-1)z \rightarrow (0)x + (0)y + (-1)z$$

Therefore,

$$\begin{bmatrix} x' \\ y' \\ z' \end{bmatrix} = \begin{bmatrix} 1 & 0 & 0 \\ 0 & 1 & 0 \\ 0 & 0 & -1 \end{bmatrix} \begin{bmatrix} x \\ y \\ z \end{bmatrix}$$

Matrix corresponding to σ_{xy} is,

$$\begin{bmatrix} 1 & 0 & 0 \\ 0 & 1 & 0 \\ 0 & 0 & -1 \end{bmatrix}$$

We have seen that x, y, z unit vectors transform into ± 1 only i.e. these remain same or their directions are changed. In all these cases we get 3 × 3 size matrices for each and every operation.

References

- Molecular-symmetry-group-theory-&-applications: chem.ox.ac.uk, Retrieved 06, March 2020
- A-Reduction-of-representations-II, Physical-and-Theoretical-Chemistry-Textbook-Maps: libretexts.org, Retrieved 21, June 2020

4

Chemical Reactions and Orbital Symmetry

The molecular transformations which are engaged in chemical reactions depend on the symmetries of the orbitals involved. The possibility of a particular chemical reaction mechanism can be determined through the symmetry properties of molecular orbitals and reaction coordinates. This chapter discusses in detail the symmetry rules for chemical reactions.

Symmetry Rules for Chemical Reactions

A very powerful mathematical tool has been developed that enables us to exploit symmetry in an exact manner. This tool is group theory. For most chemical problems dealing with molecules, it is point group theory that is used. The nuclei of the molecule are represented by a set of points, which are interchanged by the various symmetry operations. Each molecule belongs to one of the point groups, for which symbols such as T_d, C_v, O_h and so on, are used. Each point group is defined by the number and kind of symmetry elements that it possesses. These elements are planes, axes, centers and improper axes of symmetry.

Each point group is also characterized by a certain number of symmetry species associated with it. The species are represented by symbols such as A_1, E, T_{2g} and so on. Things or functions associated with a molecule must be resolvable into these various symmetry species.

There are two important conclusions that can be drawn by the application of group theory to a molecule. One has to do with the wave function ψ which is a solution to the Schrodinger equation,

$$H\psi_i = E_i\psi_i,$$

where H is the molecular Hamiltonian. A series of wave functions, or eigenfunctions ψ_i exists as solutions to above equation. Correspondingly, there are a series of energy values, or eigenvalues E_i. ψ_o and E_o are the ground state quantities. Now it can be shown that each wave function ψ_i must belong to one of the symmetry species of the group. This applies to the exact solutions of above equation, which are usually unknown. Fortunately, it also applies to molecular orbital theory, where a molecular orbital (MO) is written as a linear combination of atomic orbitals (LCAO). Each MO φ_i satisfies the equation,

$$h\phi_i = \epsilon_i \phi_i$$

where h is the averaged one electron Hamiltonian. Each ϕ_i must be a proper symmetry orbital. This puts strong restraints on the LCAO's that may be used.

The second important conclusion that may be drawn from group theory has to do with nuclear motions. These motions are usually classified as translation and rotations of the molecule as a whole, and vibrations within the molecule. Translation along the three Cartesian coordinates and rotations about the two or three axes of rotation each have a symmetry label. The vibrational motions are resolvable into a set of normal modes of vibration, each with a definite symmetry. A super molecule is formed which still belongs to one of the point groups. Now chemical reaction between two molecules requires changes in the initial nuclear positions, characteristic of the reactants, to new nuclear positions, characteristic of the products. These changes can always be resolved into normal modes of vibration of the super molecule, each belonging to a definite symmetry species.

At the same time the wave function must change. Since ψ^2 is the electron density function, it must change from that of the reactants to that of the products. In MO theory, the valence shell orbitals of all of the atoms involved will change from one set of LCAO's to another. The same set of AO's will be used as building blocks for both reactants and products.

Since chemical bonds are broken and new bonds are formed in a reaction, and since MO's are bonding, anti-bonding or non-bonding between various atoms, it is convenient to think of reaction as occurring by a partial transfer of electrons from certain filled MO's in the reactants to certain empty MO's, also in the reactants. This is a mechanism for generating the new LCAO's of the products. Woodward and Hoffman are chiefly responsible for the realization of the important role played by orbital symmetry in chemical changes. They introduced much of the language that is used today. However they did not use point group theory as such in their papers.

Application of Perturbation Theory

We will describe two apparently quite different, but still equivalent, methods for developing symmetry based rules for chemical reactions ($h\phi_i = \epsilon_i \phi_i$). First we will apply quantum mechanical perturbation theory to a collection of nuclei and electrons in the act of chemical change. Figure below is a plot of the potential energy of this system (electronic energy plus nuclear repulsion energy) as a function of the so called "reaction coordinate." This is a collective name given to the set of changing nuclear positions which correspond to going from reactants on the left to products on the fight. En route one or more potential energy maxima, such as B, may occur. At these points the system is said to be an activated complex, existing in a transition state. Local minima may also exist such as C, defining reactive intermediates if the minimum is shallow, or stable products, if the minimum is deep.

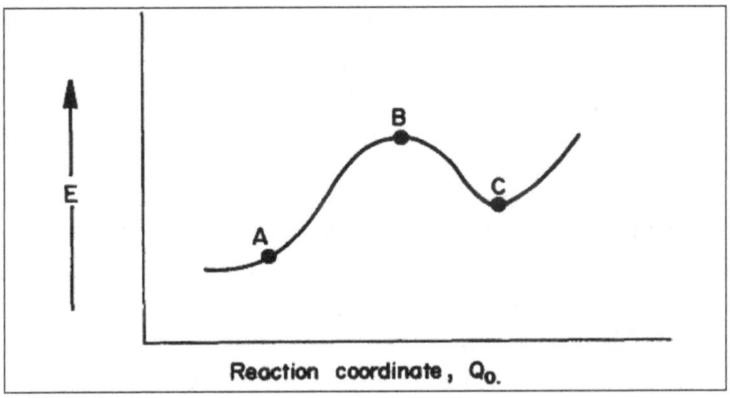

Plot of potential energy as a function of the reaction coordinate.

We now use quantum mechanics in the form of perturbation theory to relate the potential energy E to the reaction coordinates. First the Hamiltonian is expanded in a Taylor-Maclaurin series about the point Q_o, corresponding to the original configuration with Hamiltonian H_o:

$$H = H_o + \left(\frac{\delta U}{\delta Q}\right)Q + \frac{1}{2}\left(\frac{\delta^2 U}{\delta Q^2}\right)Q^2.$$

Here Q represents the reaction coordinate and also the magnitude of the small displacement from Q_o. For convenience, we consider only one normal mode at a time. Since the Hamiltonian must be invariant to all the symmetry operations of the pseudomolecule, it follows that Q and $(\delta U/\delta Q)$ have the same symmetry. Their direct product is totally symmetric. Since Q^2 is symmetric, it follows next that $(\delta^2 U/\delta Q^2)$ is also symmetric. U is the nuclear-electronic and nuclear-nuclear potential energy. The kinetic energy of the electrons and electron-electron potential energy are not functions of the nuclear coordinates, to the first order.

The last two terms in above equation represent the perturbation. Using standard second-order perturbation theory, we now solve for the new wave functions and energies. For the ground electronic state, the energy becomes,

$$E = E_o + Q\left\langle \psi_o \left| \frac{\delta U}{\delta Q} \right| \psi_o \right\rangle + \frac{Q^2}{2}\left\langle \psi_o \left| \frac{\delta^2 U}{\delta Q^2} \right| \psi_o \right\rangle + Q^2 \sum_k \frac{\left[\left\langle \psi_o \left| \frac{\delta U}{\delta Q} \right| \psi_k \right\rangle\right]^2}{E_o - E_k}$$

where E_o is the energy at Q_o, the next two terms are the first-order perturbation energy, and the last term is the second-order perturbation energy. While above equation is valid only for Q very small, we can select Q_o anywhere on Figure. Hence above equation is general for the purpose of displaying symmetry properties.

The symbol ⟨...⟩ represents integration over the electron coordinates, covering all space. We can now use a group theory rule to decide whether the integrals in above equation are exactly zero or not. The rule is that the direct product of three functions must contain the totally symmetric species, or the integral over all space is zero.

Let us consider the term in above equation which is linear in Q. At any maximum or minimum in the potential energy curve, $\delta E/\delta Q = 0$ and therefore the integral must be identically zero, independent of symmetry. At all other points this term must be the dominant one, since Q is small. If ψ_o belongs to a degenerate symmetry species (E or T), the term usually leads to the firstorder Jahn-Teller effect which removes the degeneracy. We will assume that ψ_o is nondegenerate.

Since the direct product of a nondegenerate species with itself is always totally symmetric, we derive our first symmetry rule: all reaction coordinates belong to the totally symmetric representation. That is $(\delta U/\delta Q)$ and also Q, must be totally symmetric, otherwise its product with ψ_o^2 will not be symmetric and the integral will be zero. However, it must be nonzero for all of the rising and falling parts of Figure. This means that once a reaction embarks on a particular reaction path it must stay within the same point group until it reaches an energy maximum or minimum. A totally symmetric set of nuclear motions can change bond angles and distances, but it cannot change the point group.

We now consider point A in Figure. The integral $\langle \psi_o | \delta Q | \psi_o \rangle$ has a positive value since the reaction has a positive activation energy. Instead of trying to evaluate the integral we accept that its value is the slope of Figure at the point A. The terms in Q^2 in above equation now become important. Their sum determines the curvature of the potential energy plot. For a reaction with a small activation energy, the curvature should be as small as possible (or negative).

The integral $\langle \psi_o | \delta^2 U / \delta Q^2 | \psi_o \rangle$ has a nonzero value by symmetry since $(\delta^2 U / \delta Q^2)$ is totally symmetric. Furthermore, it will be positive for all molecules. It represents the force constant which resists the move of any set of nuclei away from an original configuration for which ψ_o^2 is the electron density distribution. The last term in above equation represents the change in energy that results from changing the electron distribution to one more suited to the new nuclear positions determined by Q. Its value is always negative since E_o - E_k is a negative number.

This can be seen more easily if the equation for the wave function is written down from perturbation theory:

$$\psi = \psi_o + \sum_k \frac{\left\langle \psi_o \left| \frac{\delta U}{\delta Q} \right| \psi_k \right\rangle}{E_o - E_k} \psi_o.$$

The summations in above two equations are over all excited states. Each excited-state wave function is mixed into the ground-state wave function by an amount shown in above equation. The wave function is changed only because the resulting electron distribution ψ^2 is better suited to the new nuclear positions. Now we can use group theory to show that only the excited-state wave functions ψ_k that have the same symmetry as ψ_o can mix in and lower the potential energy barrier. This follows because $(\delta U/\delta Q)$ must be totally symmetric. Hence the direct product of ψ_o and ψ_k must be totally symmetric, but this requires that they have the same symmetry. We can conclude that, for a chemical reaction to occur with a reasonable activation energy, there must be low-lying excited states for the reacting system of the same symmetry as the ground state. Such a reaction is said to be symmetry allowed. A symmetry-forbidden reaction is simply one which has a very high activation energy because of the absence of suitable excited states.

Previous two equations are exact, as are the symmetry rules derived from them. For practical applications, some rather drastic assumptions must now be made. One is that LCAO-MO theory will be used in place of the exact wave functions ψ_o and ψ_k. Since we are interested only in the symmetry properties, this creates no serious error, since MO theory has the great virtue of accurately showing the symmetries of the various electronic states.

The second assumption is more serious, since we will replace the infinite sum of excited states in equations $H\psi_i = E_i\psi_i$, and $h\phi_i = \epsilon_i \phi_i$ by only a few lowest lying states. This procedure will work because we are not trying to evaluate the sum but only to decide if it has a substantial value. It can be shown that the various states contributing to equations $\Gamma\phi_i \times \Gamma\phi_f \subset \Gamma Q$ and

$$E = E_o + Q\left\langle \psi_o \left| \frac{\delta U}{\delta Q} \right| \psi_o \right\rangle + \frac{Q^2}{2}\left\langle \psi_o \left| \frac{\delta^2 U}{\delta Q^2} \right| \psi_o \right\rangle + Q^2 \sum_k \frac{\left[\left\langle \psi_o \left| \frac{\delta U}{\delta Q} \right| \psi_k \right\rangle\right]^2}{E_o - E_k}$$

fall off very rapidly as the

difference, $|E_o - E_k|$, becomes large. This is because the integral $\langle \phi_o | \delta U / \delta Q | \phi_k \rangle$ decreases very rapidly for two wave functions of quite different energy.

We use MO theory to represent the ground and excited states that are needed. The symmetry of $\psi_o\psi_k$ is replaced by $\phi_i\phi_f$, where ϕ_i is the occupied MO in the ground state and ϕ^f is the MO occupied in its place in the excited state. Positions of special importance are occupied by the highest occupied and lowest unoccupied molecular orbitals, since excitation of an electron from HOMO to LUMO defines the lowest excited state.

It is helpful to point out that the requirement that two orbitals, ϕ_i and ϕ_f, have the same symmetry is the same as saying that they must have a net positive overlap. Two molecular, or atomic, orbitals of different symmetry species have exactly zero overlap. We now go to a consideration of points B and C in Figure. B refers to an activated complex and C to a single molecular species, which is unstable with respect to isomerization, or breakdown to other products.

The term linear in Q in equation $H = H_o + \left(\frac{\delta U}{\delta Q}\right)Q + \frac{1}{2}\left(\frac{\delta^2 U}{\delta Q^2}\right)Q^2$ now vanishes, since we are at an extremum in the potential energy plot. The first quadratic term is positive, and the second one is negative. Clearly at a maximum, point B, the second term is larger than the first. At a minimum, point C, the first term dominates, but the magnitude of the second term determines whether we lie in a deep potential well or a shallow one.

Again the existence of low-lying states ψ_k of the correct symmetry to match with ψ_o is critical. Now there is no restriction on the reaction coordinate which forces it to be totally symmetric. However, ψ_o, $(\delta U/\delta Q)$ and ψ_k are still bound by the symmetry requirement that their direct product must contain the totally symmetric representation. If we consider rather symmetrical molecules to begin with, it will usually be found that the reaction coordinate, and $(\delta U/\delta Q)$, are nonsymmetric. The reason for this is that maximum and minimum potential energies are usually found for nuclear arrangements with a high degree of symmetry. Any disturbance of the nuclear positions will now reduce the symmetry. However, this corresponds to a change in the point group, which can only come about by a nonsymmetric vibrational mode.

Conversely, it may be pointed out that a number of point groups depend upon a unique value of Q_o in Figure. For example, a tetrahedral molecule has uniquely determined bond angles. All such cases must correspond to either maxima or minima, if the reaction coordinate is taken either as the bond angles or relative bond distances. In molecular orbital theory the product $\psi_o\psi_k$ is again replaced by $\phi_i\phi_f$, where both the occupied and empty MO's must be in the same molecule. Electron transfer from ϕ_i, to ϕ_f results in a shift in charge density in the molecule. Electron density increases in the regions where ϕ_i and ϕ_f have the same sign (positive overlap) and decreases where they have opposite signs (negative overlap). The positively charged nuclei then move in the direction of increased electron density. The motion of the nuclei defines a reaction coordinate. The symmetry of Q is the same as that of the product $\phi_i \times \phi_f$. We may write this as,

$$\Gamma\phi_i \times \Gamma\phi_f \subset \Gamma Q$$

where Γ means symmetry species.

The size of the energy gap between ϕ_i and ϕ_f is critical. A small gap means an unstable structure, unless no vibrational mode of the right symmetry exists for the molecule capable of changing its structure. A large energy gap between the HOMO and the LUMO means a stable molecular structure. Reactions can occur, but only with a high activation energy.

For an activated complex (point B) there must necessarily be at least one excited state of low energy. The symmetry of this state and the ground state then determines the mode of decomposition of the activated complex. When a molecule lies in a shallow potential well (point C), the activation energy for unimolecular change is small. In this case we can again expect a low-lying excited state. The symmetry of this state and the ground state will determine the preferred reaction of the unstable molecule. For a series of similar molecules, we expect a correlation between the position of the absorption bands in the visible-uv spectrum and the stability.

As an example, consider the similar molecules ozone and sulfur dioxide. The former is blue in color and is highly unstable. Sulfur dioxide is colorless and is also much more stable toward dissociation into SO and O. We correlate the instability of O_3 with the fact that it absorbs light of lower energy than SO_2 does because of lower lying excited states.

The Orbital Correlation Method

Before giving forth examples of the perturbation method, it is convenient to give the details of our second method, that of orbital correlation. The phrase "conservation of orbital symmetry," is often used. The method grew out of earlier work on state correlation. Wigner and Witmer developed rules for predicting the possible states of a diatomic molecule formed by combining two atoms.

Attempts to extend these rules to the states of reacting molecules and their product states did not prove very useful. State correlation did not provide enough information. A more useful method was initiated as early as 1927 by Hund, and later by Mulliken. This was a correlation of the MO's of a diatomic molecule with the AO's of the separated atoms on the one hand, and with the AO's of the so-called united atom on the other.

First the AO's of the separated atoms were combined to form MO's of the diatomic molecule. This is simply done by taking sums and differences of equivalent, or identical, orbitals; $ls_A \pm ls_B$, etc. These MO's each have a symmetry label in the D_{xh} (like nuclei) or C_{xh} (unlike nuclei) point groups. The MO's were then correlated with the AO's of the united atoms (coalesced nuclei) to determine if their energies increased or decreased as the nuclei approached each other.

The correlation depended on reduction in symmetry tables, so that a P_x atomic orbital of the united atom could be clarified in the D_{xh} point group of the molecule. There are three correlation rules:

- The symmetry of each MO remains unchanged as it becomes an AO of the united atom.
- MO's of the same symmetry cannot cross.
- No orbitals are gained or lost.

Surprisingly, there was no attempt to use the orbital correlation method for the reactions of polyatomic molecules until 1965. The procedure is straight-forward. A reaction coordinate is selected which preserves one or more elements of symmetry as the nuclei rearrange from reactants to products. This generates a point group in which these MO's of the reactants, products and all intermediate steps may be classified (if there are identical MO's in two molecules, their sum and difference must be taken).

The same three correlation rules are used to connect the MO's of the reactants with those of the products. Energies are estimated by knowing whether a given MO is bonding, nonbonding or anti-bonding. The new information that is gained is an estimate of the energy of the transition state. If this energy is higher than that of the reactants, there will be an appreciable activation energy. The reaction will be slow because of this.

The reaction is said to be "forbidden by orbital symmetry." This is too strong a statement since all that is predicted is an appreciable barrier. The reaction may still occur at a reasonable rate for many purposes. A reaction "allowed by orbital symmetry" is one where no barrier is predicted. It may be slow for other reasons. Nevertheless, analysis of possible reaction mechanisms by orbital correlation has proved to be very useful.

We will give an example by considering the exchange reaction between two hydrogen molecules occurring by a broad-side collision:

$$H_2 + D_2 = 2HD.$$

Deuterium would be used to show that a reaction actually took place. For simplicity we will consider two H_2 molecules.

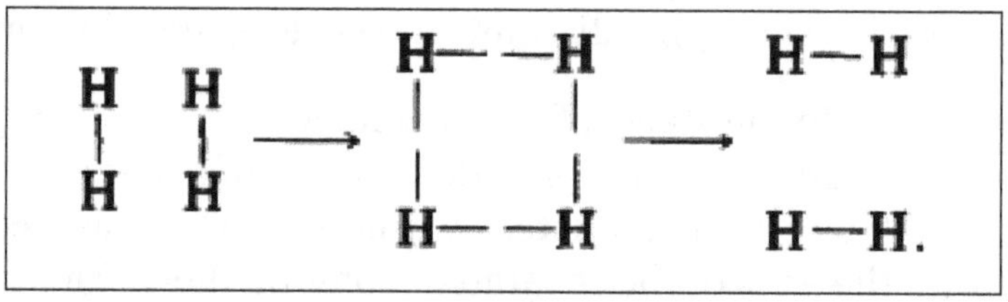

In such a mechanism three planes of symmetry are preserved, three twofold axes, and an inversion center. These elements define the D_{2h} point group. If the transition state is assumed to be a symmetric square, the point group is D_{4h}, but only at one point on the potential energy surface.

The starting MO's are the filled % orbitals of the hydrogen molecules, and the empty σu orbitals. In D_{2h}, we must take symmetry adapted linear combinations of these. These SALC's are simply the sums and differences of the MO's. Figure at the bottom, shows how the sums, $\sigma_g(1) \pm \sigma_g(2)$ generate species of a_g and b_{2u} symmetry in the D_{2h} point group. The combinations, $\sigma_u(1) \pm \sigma_u(2)$ generates species of b_{3u} and b_{1g} symmetry.

Figure shows how the energies of these four MO's change as we go from reactants to products. No calculations need be made. All we need to know is that at the start of the reaction, the electrons must be in the orbitals a_g and b_{3u} since we now bond together the two top H atoms and the two bottom H atoms of above molecule. For the products, then, a_g and b_{3u} are the low energy orbitals. This defines the orbital diagram. There is a crossing of the b_{2u} and b_{3u} orbitals. This crossing of an originally filled and an originally empty MO makes the reaction forbidden by orbital symmetry.

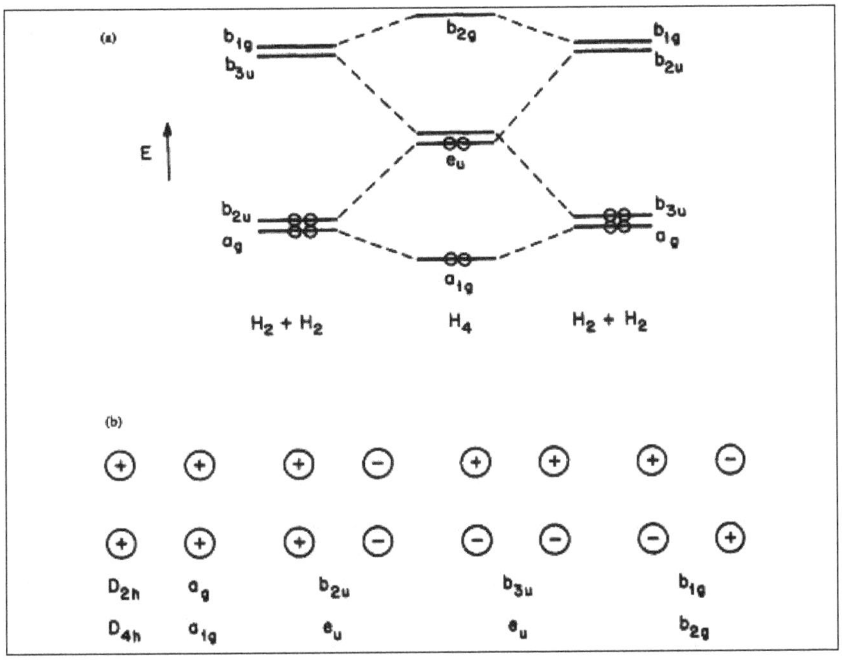

(a) Molecular orbital correlation diagram for reaction $H_2 + H_2 \rightarrow H_2 + H_2$ with square H_4 as the activated complex. (b) Linear combinations of four Is orbitals and their species in D_{2h} and D_{4h} point groups. The plane of the molecule is the x-y plane.

The meaning of this phrase, of course, is that the transition state is of high energy. The two electrons in the b_{2u} orbital must rise in energy as the nuclei move along the reaction coordinate. Notice that in the TS, the b_{2u} and b_{3u} orbitals become degenerate. This is the result of the higher D_{4h} symmetry assumed at this point. Symmetry produces degeneracies. After the TS, the electrons appear in the more stable b_{3u} orbital. This occurs because of the process known as configuration interaction. This mixes together the $(b_{2u})^2$ and $(b_{3u})^2$ configurations even before the TS is reached.

We can use the same example to illustrate the perturbation theory method. The filled MO's are of a_g and b_{2u} species in the D_{2h} point. The only low energy empty MO's are of b_{1g} and b_{3u} species. Therefore there are no low-lying excited states of the same symmetry as the ground state, and the energy barrier must be large.

Pericyclic Reactions

Pericyclic reaction is defined as a thermal or photochemical induced reaction involving a stereospecific, (rather highly stereoselective) concerted cyclic shift of the electrons. The reactant is converted to a product typically by a cyclic conjugated transition state (T.S) either thermally or photochemistry. The first key term concerted implies to no intermediate since bond breaking and bond making takes place simultaneously. The second key point cyclic shift of electrons signifies a kind of reaction that occurs around the periphery of a molecule following a typical involvement of π-orbital via a circular pathway. It rules out either homolytic/heterolytic cleavage or radical/ionic pathway.

There are three major classes of pericyclic reactions. Besides, Group transfer reaction is accounted as another class of pericyclic reactions:

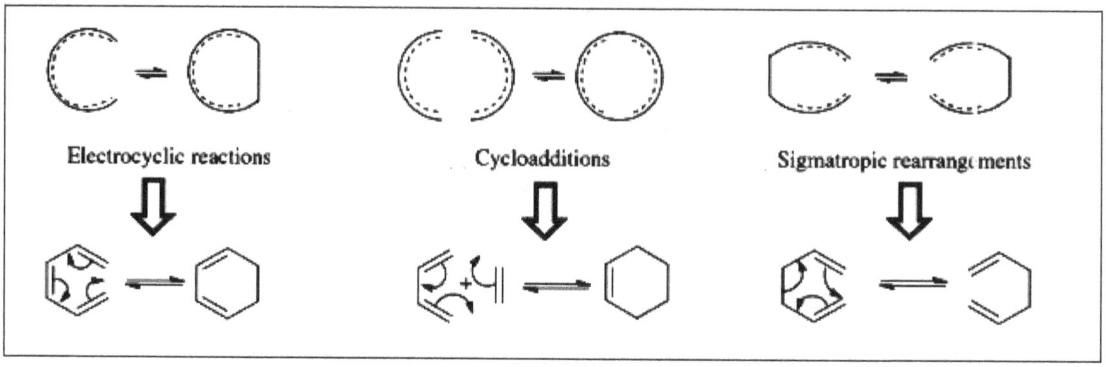

Characteristics Features of Pericyclic Reactions

- A concerted pathway via a single cyclic conjugated T.S where lack of evidence in isolation of intermediate like free radical, carbocation etc. rather the bond breaking and bond making takes place synchronously.
- Cyclic shift of electrons as a result of in general participation of p/π-orbital with some extent to σ-bond: A reorganization of electron pairwise within the close loop of interacting orbitals (around cyclic).
- Usually unaffected by acids or bases, the change of solvent polarity, catalyst (exception in Diels-Alder cycloaddition reaction accelerated by Lewis acid catalyst) and radical initiator or change in structural motif of the molecule.
- No nucleophilic or electrophilic reagents are required.
- No such ionic or radical pathway.
- Highly stereoselective in nature.
- Most of the reactions are equilibrium processes in which direction of equilibrium depends on the enthalpy and entropy of the reacting species. Large decrease in entropy or loss of degree of freedom illustrates the rigid transition state (e.g D.A reaction $\Delta S \neq = -150$ to -200 $JK^{-1}mole^{-1}$ and in sigmatropic reaction $\Delta S \neq = -50$ $JK^{-1}mole^{-1}$) and with a low enthalpy of activation reflecting the exothermic nature of the reaction.
- They follow Woodward-Hoffmann rule of "Conservation of Orbital Symmetry" which states that the orbital symmetry should be maintained throughout the reaction i.e. before and after the reaction the orbital symmetry of both the reactant and product should be conserved. Thus, it is an orbital symmetry controlled reaction. The pericyclic reaction excludes any usual nucleophilicelectrophilic selective transformation of stereoheterotopic faces and ligand discrimination by intramolecular symmetry.

Source/Influencing Factor/Mode of Activation

- Heat or Thermal (Δ).

- Light or Photochemical (hγ).

Non-Influencing Factor

- Polar/Non polar solvents.
- Radical initiator or inhibitor.
- Catalysis (except Diels-Alder cycloaddition reaction).

Criteria for Pericyclic Reaction

- Mode of activation. (Δ/hγ).
- Number of p/π-orbital participatiion [4nπ or (4n+2) π].
- Stereochemical outcome (stereospecific or stereoselective).

Advantages of Pericyclic Reaction

- Generation of variety of new carbon-carbon σ-bond (due to reorganization of p/π-bond).
- Less probabilities of formation of unwanted side products. (Due to concerted cyclic process and unreacted or unaffected by the influence of solvent or any reagent). So, it is straight forward, non-complicated and anticipated reaction.
- The reaction is highly stereospecific in nature. (Due to concerted orbital symmetry controlled process).
- Enormous application in biosynthesis.
- Non-conventional energy source e.g. the cycloreversal of cyclobutane to two ethelene molecule is accompanied by liberation of heat.

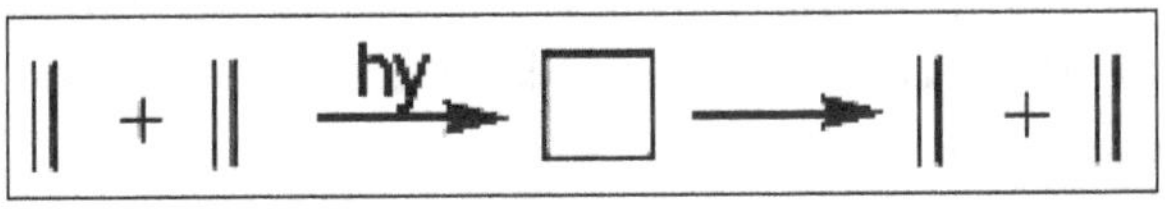

Evidence for the Connectedness of Pericyclic Reaction

1. Although small, an inverse secondary kinetic isotope effects was observed for the following diene and the dienophile when the deuterium substitution on the four carbon atoms changes from trigonal to tetrahedral as the reaction proceeds. It was experimentally found a geometrically related $(k_D/k_H)^2 = (k_{D2}/k_H)$ [not arithmetically related, $(k_{D2}/k_H) = 2(k_D/k_H) - 1$] to the isotope effects at both deuterated ends indicating both ends are formed simultaneously.

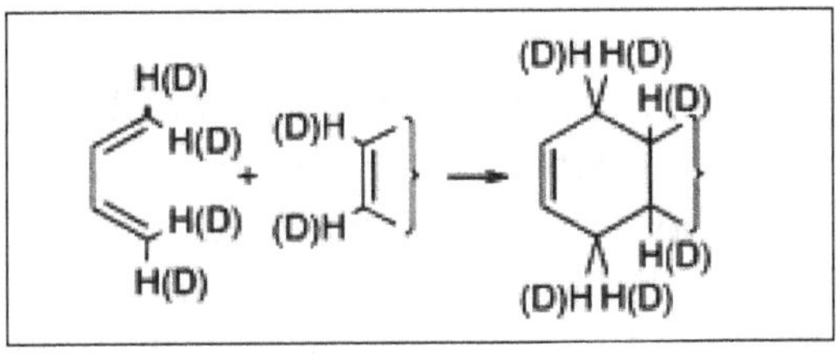

2. Bimolecular reactions account for inherently high negative entropies of activation ($\Delta S^{\neq}$). Exceptionally high negative $\Delta S^{\neq}$ (typically in the range of -150 to -200 J K^{-1} mol^{-1}) for Diels-Alder (D.A) reactions or cycloaddition reactions reveals a favourable alignment with compact transition structure of two components only if both bonds can form at the same time.

3. The rates of D.A reactions are slightly increased by a factor of only 10 upon the change in the polarity of the solvents or with large change in dipole moment from 2.3 to 39 in contrast to several orders of magnitude for stepwise ionic cycloaddition reactions.

4. The dienophile tetracyanoethylene reacts significantly much faster rate with butadiene than 1,1-dicyanoethylene which can't be rationalized by ionic stepwise pathway.

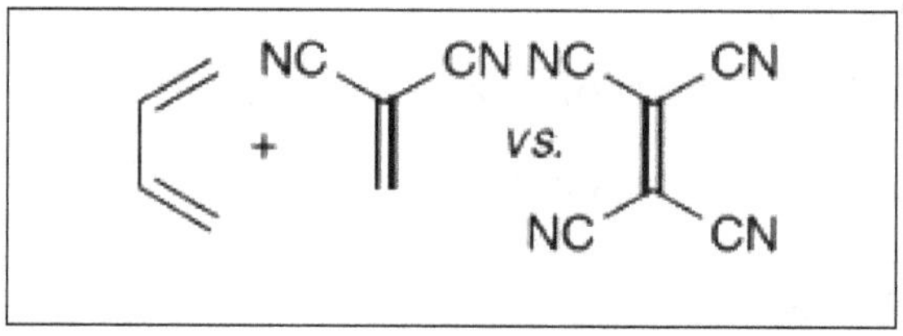

Difference between SN² Reaction and Pericyclic Reaction

SN² reaction constitutes a transition state of non-isolable intermediate that may be linear or cyclic transition state where the HOMO of nucleophilic approach must be exactly opposite to that of LUMO of leaving group. A stereospecific product having inversion of configuration is expected to result in SN² reaction. On the other side, Pericyclic reaction is an orbital symmetry controlled reaction with non-isolable intermediate via cyclic conjugated transition state which demands stereospecific product by virtue of suprafacial or anatarafacial or HOMO-LUMO interaction.

Woodward-Hoffmann Selection Rule

Let us take an example in which maleic anhydride react butadiene spontaneously but not with ethelene under thermal condition. In order to explain the phenomenon, we need to consider the Woodward-Hoffmann selection rule.

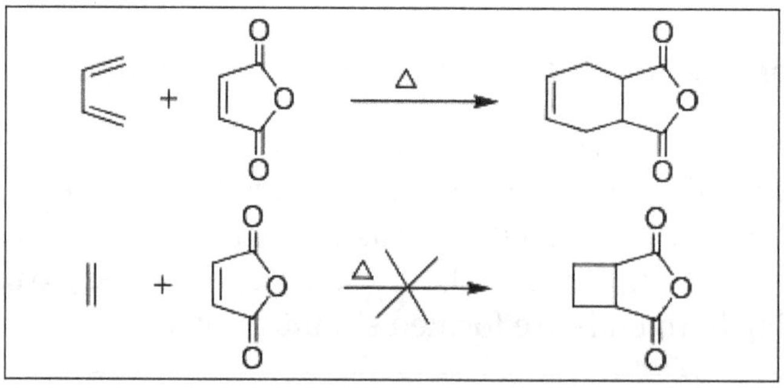

In a Pericyclic reaction, the set of in phase atomic orbitals (A.Os) to form molecular orbitals (M.Os) of reactants are transformed into corresponding in phase set of M.Os of the products by conserving the symmetry of the orbitals throughout the reaction in a concerted manner via symmetry allowed low energy transition state. This is called as woodward-Hoffmann rule of "Conservation of Orbital Symmetry". This clearly indicates that the orbital symmetry should be maintained throughout the reaction i.e. before and after the reaction the orbital symmetry of both the reactant and product should be conserved. Hence, Pericyclic reaction can be attributed as orbital symmetry controlled reaction.

Maintaining the principle of conservation of orbital symmetry, a Pericyclic reaction can be visualized as reorganization of electron (pairwise) with in a close loop of interacting orbitals through symmetry allowed concerted process. Selective transformation of organic reaction is based on:

- Stereoheterotopic faces.
- Stereoheterotopic ligand.

These are distinguished by intramolecular symmetry or overall symmetry of the molecule. But, the Pericyclic Reaction relies on the interactive molecular orbitals (M.Os), not as a whole molecular symmetry. The basic difference between pericyclic chemistry with heterocyclic or alicyclic reaction is that the pericyclic reaction is guided by "Rule of Conservation of Orbital Symmetry" proposed by Woodward and Hoffman. In addition to this, the former involves a cyclic concerted process with the help of p/π-orbital and the mode of activation must be either thermal or photochemical. No other reagents are associated with the pericyclic reaction.

To explain the stereochemistry of the product, three fundamental approaches have to be considered:

- Correlation of orbitals in relation to symmetry should be preserved (W. H principle).
- Aromaticity or anti aromaticity of T.S by Hückel Molecular orbital theory.
- Frontier Molecular Orbital (FMO) approach: HOMO-LUMO interactions.

Most of the pericyclic reactions, of course not all, are little influenced by Coulombic forces e.g. the rates of D.A reactions are little affected by polarity of the solvent. We can therefore expect that the major factor influencing the reactivity will be the size of the Frontier orbital interaction represented by energy expression from Klopman and Salem's perturbation of orbital symmetry. Thus, FMO approach among three fundamental approaches is most widely applicable in Pericyclic reaction.

Class of Pericyclic Reactions

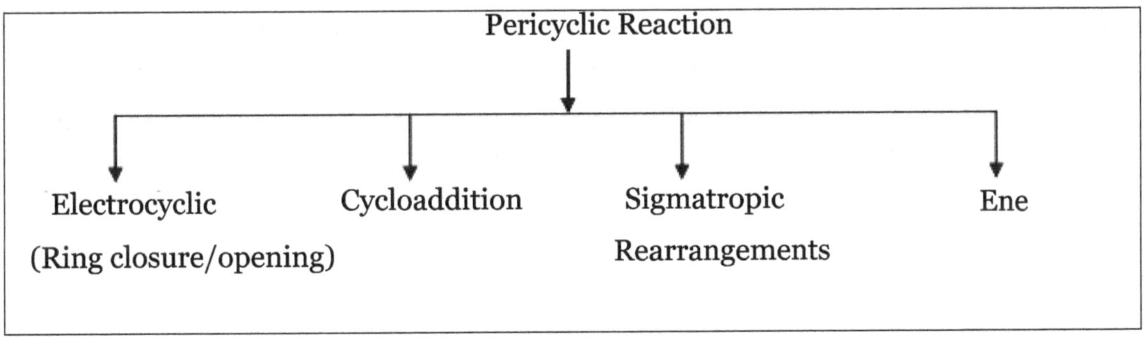

Electrocyclic Reaction

An intramolecular pericyclic reaction where reversible concerted cyclic shift of electrons occurs by virtue of ring closing with the formation of new σ-bonds across the ends of conjugated acyclic

π-electrons of polyene or breaking of ring to form acyclic analogue from its corresponding cyclic framework, is typically termed as Electrocyclic reaction. The reaction can be best way represented by the following diagram:

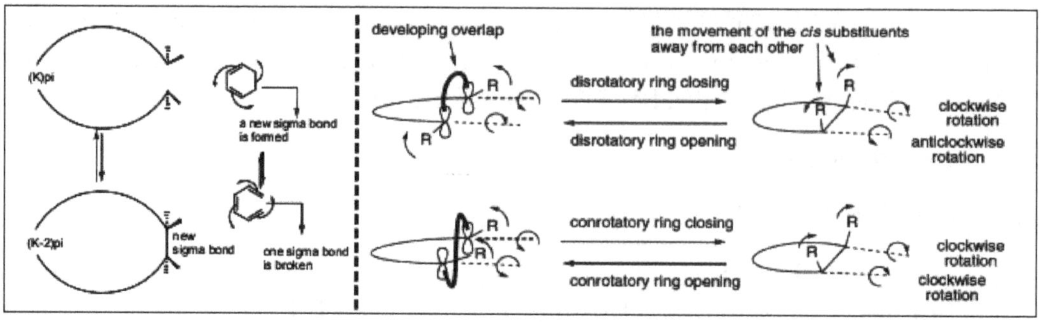

The reaction is activated either by heat or by UV-light. The reaction follows the principle of microscopic reversibility. The reaction sequence can be explained by three fundamental approaches:

- Correlation of orbital symmetry of terminal lobes of interacting orbitals.
- Perturbation of M.O theory or HMO theory of aromaticity or anti aromaticity in T.S.
- FMO approach of HOMO-LUMO interactions of orbital energy states.

Conditions for Electrocyclic Reaction

- Symmetry of terminal orbitals in conjugated π-system has to be considered.
- Terminal p-orbitals form σ-bond by maintaining phase distribution of lobes after rotation.
- Two possible ways of the rotation of lobes or p-orbitals: (a) Conrotation motion (b) Disrotation motion.

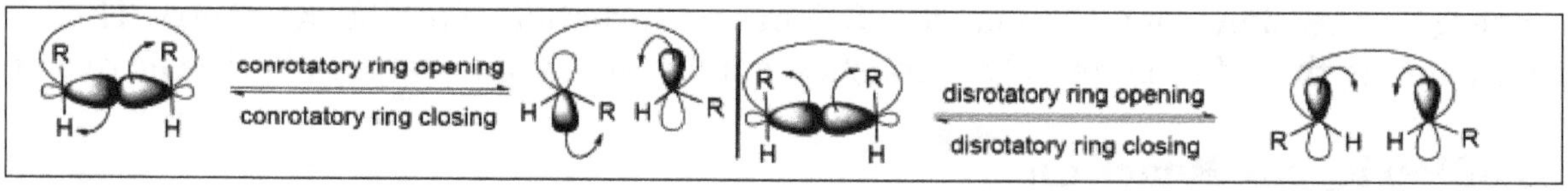

- Movement of the groups or substituents attached to terminal lobes according to the lobe's movements which bring about the stereochemistry of the resultant product.
- Conrotation or disrotation should follow Woodward–Hoffmann principle of orbital symmetry and will be favoured if the process is devoid of any steric crowding and generation of strain in the molecule.

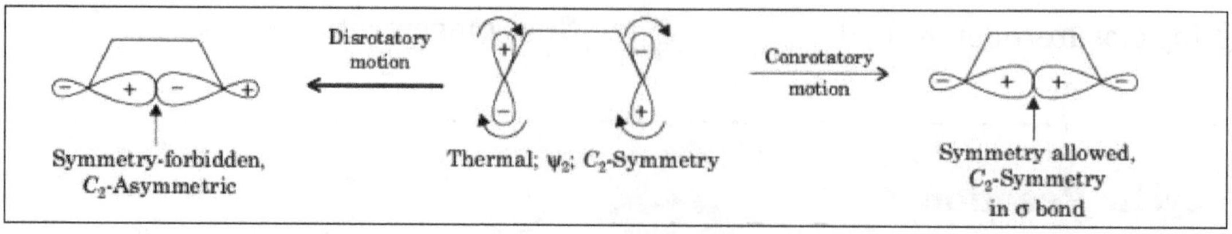

Example of electrocyclic reactions and their stereochemical significance of two modes of ring opening/closing are illustrated through thermal and photochemical transformations of isomeric

3,4-dimethylcyclobutenes and isomeric 5,6-dimethylcyclohexa-1,3-dienes. The stereochemistry of the substituents depends on the mode of activation and the mode of rotation of terminal orbitals.

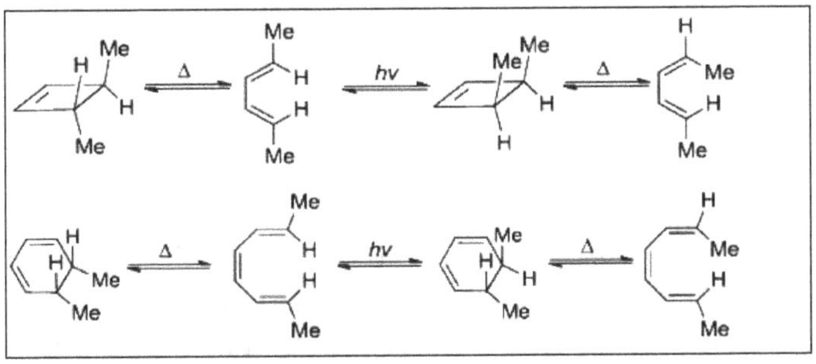

Let us consider a butadiene system (4nπ) as s-cis conformation where the terminal groups are considered as 'in' group (concave face of the arc) and 'out' group (convex face of the arc).

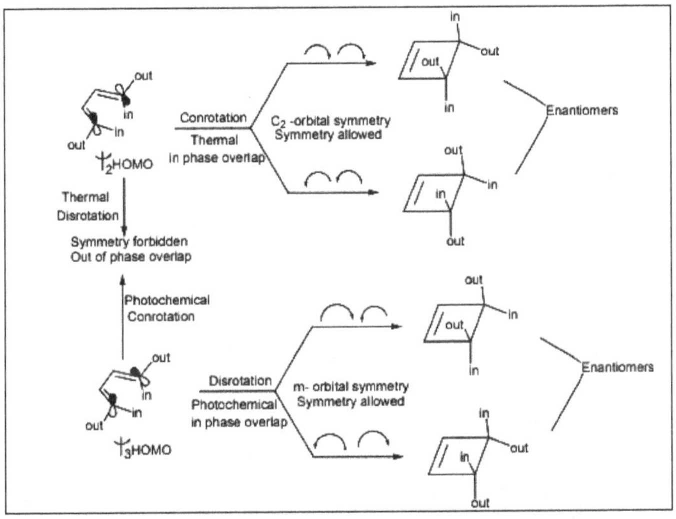

If the termini rotates in the same direction i.e. either both clockwise or both anticlockwise, is called Conrotatary motion. Here both 'in-in' and 'out-out' groups are in trans relationship. If the termini rotates in the opposite direction i.e. one in clockwise and another in anticlockwise or vice-versa, is called Disrotatary motion. Here both 'in-in' and 'out-out' groups are in cis relationship. In both cases the enantiomeric mixtures are formed. Enantiomers have equal energies and hence enantiomeric pathways have identical energies. The enantiomeric excess depends on the movements of the termini offering the in phase orbital overlap.

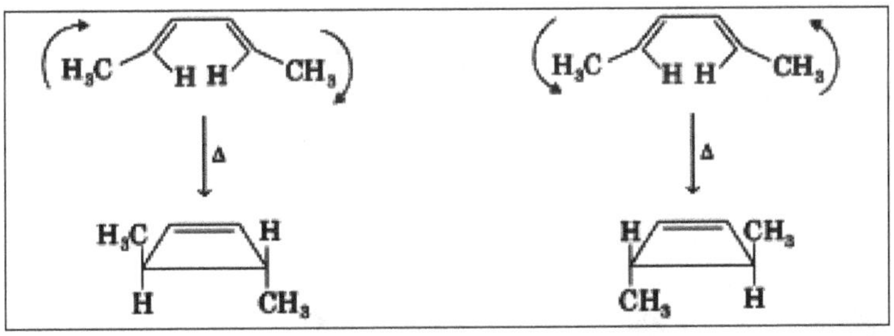

During conrotation C_2 symmetry has been maintained whereas for disrotation mirror plane (m) symmetry has been followed for 4nπ system. But the situation is exactly reversed in case of (4n+2)

π system. In both cases HOMO orbitals do participate in the conrotation or disrotation pathway which in turn regulates the stereochemical outcome.

Pictorial Representation of FMO Approach

According to Frontier Molecular Orbital approach, the valence π-electrons in the HOMO orbital of polyene system dictates the course of the Pericyclic Reaction. Similarly the valence π-electrons in the HOMO orbital of cyclo-polyene system also regulate the course of the pericyclic reaction. Therefore, combination of above HOMO orbital participation ultimately leads to a bonding interaction. The selection of HOMO participation obviously depends on the mode of activation. Let us consider a photochemical electrocyclic ring closure reaction of 1,3 butadiene to cyclobutene derivative. The pictorial representation is shown below:

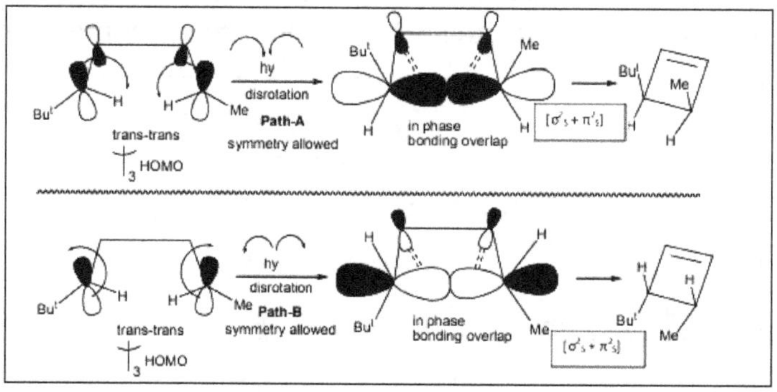

Which Pathway is more Favourable, Path-A or Path-B?

Both the Path A and Path B are symmetry allowed process and is represented by $\left[\sigma^2{}_S + \pi^2{}_S\right]$ interaction where 'S' stands for suprafacial interaction. The FMO suggests the both pathways are facilitated by suprafacial in phase bonding interaction leading to symmetry allowed process. Although enantiomeric mixture is formed but the Path-A is slightly energetically more favourable than Path-B since during T.S the bulky substituents ${}^tBu$ and Me groups tend to move towards inward direction in Path-B although both of the pathways results syn conformation. But such impact is even more pronounced while considering reverse reaction i.e. Electrocyclic ring opening. Then it will be observed that inward movements of arrows can be better pathways than outward movements of arrows because the later constitutes a bulk effect as a result of direct confrontation in T.S due to close proximity of tBu and Me groups. This makes Path-Y although symmetry allowed process but not observed due to thermodynamically almost impossible situation to take place.

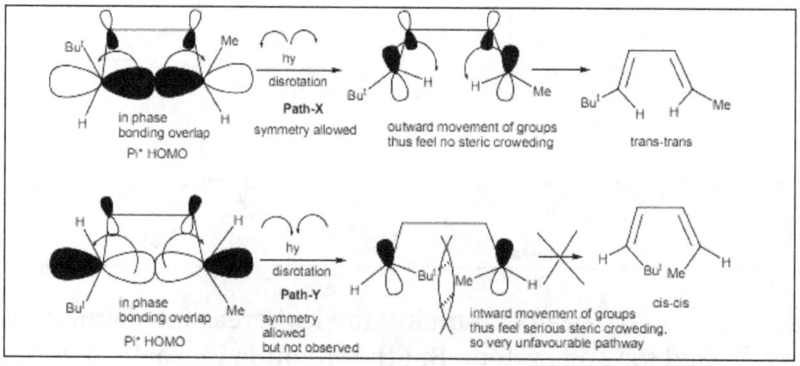

Thermal induced cyclization uses ground state ψ_3 as HOMO which has m-symmetry. So, the mode of thermal cyclisation is disrotatory motion. Photochemical induced cyclization uses excited state ψ_4 as HOMO which has C_2-symmetry. So, conrotatory motion is the mode of photo-induced cyclisation. For ring opening reaction, the similar energy of HOMO orbital, the symmetry element and mode of reaction would be followed.

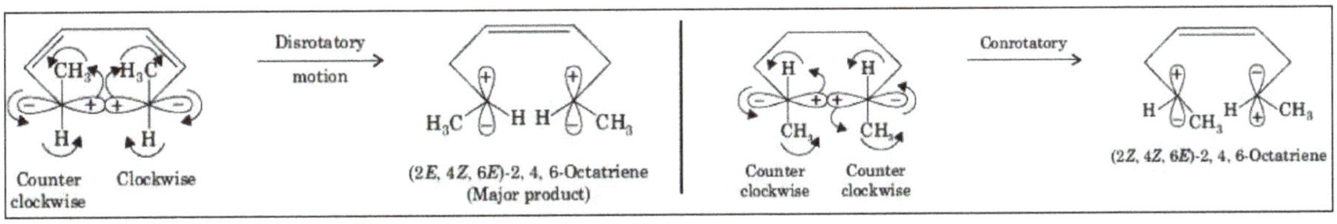

In case of 6π i.e [(4n+2)π] system, the same situation happens as that found in 4π system. Here both Path-A and Path-B have equal probabilities during thermal induced ring closing reaction. Because of similar type of steric interaction may occur at T.S, their formation of either pathway is statistically favourable. However, the photochemical pathway is both kinetically and thermodynamically unfavourable compared to that of thermal process because Path-D clearly shows a bulk atmosphere during the course of the reaction.

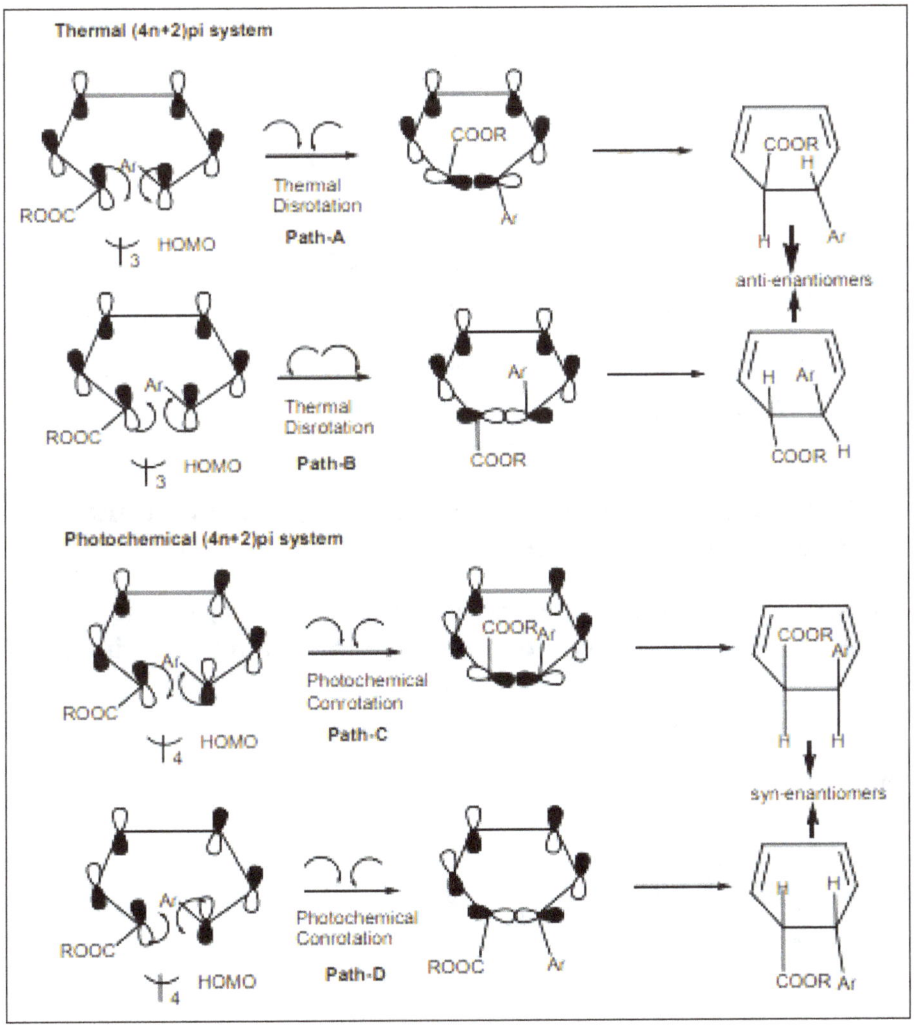

Allylic carbocationic system:

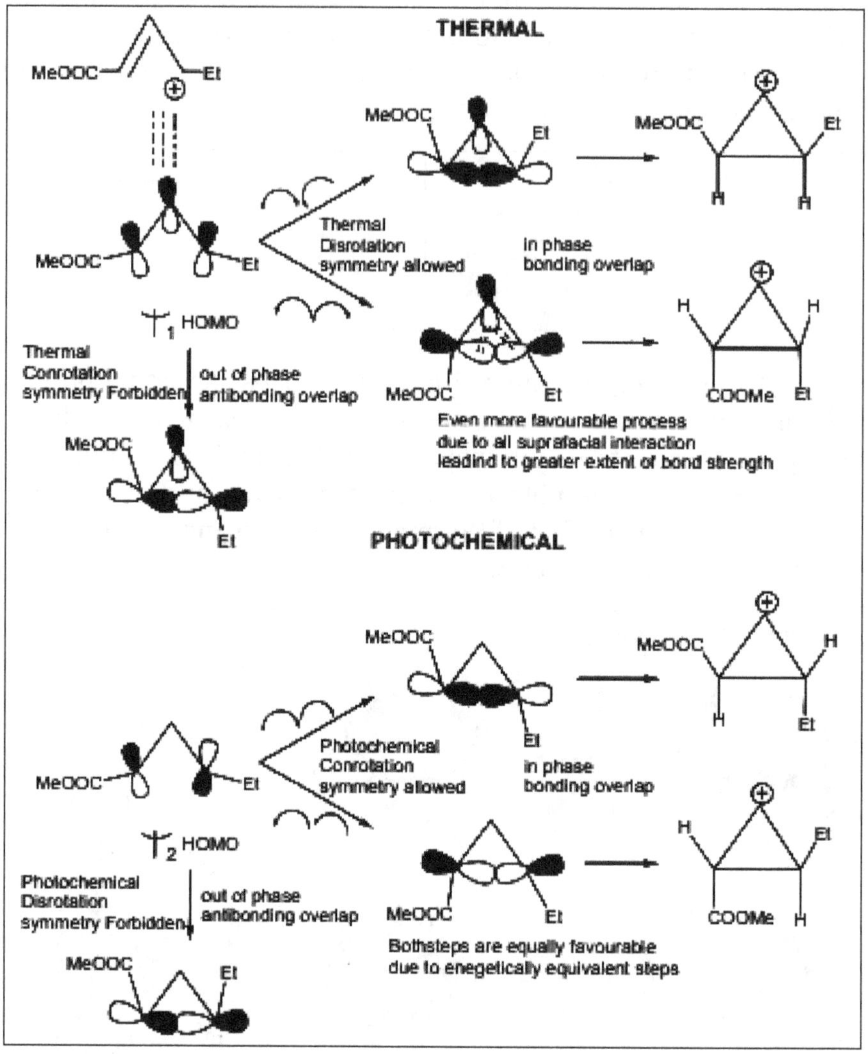

Driving Force for Allylic Cation or Pentadienyl Cation to Undergo Electrocyclic Ring Closure Reaction

Generally less conjugated carbocations undergo in Electrocyclic ring closing reaction. The formation of ring from allylic or pentadienayl cation increase the strain in the ring as well as decrease the stabilization of carbocation. In spite of that such reactions undergo due to gain in extra carbon-carbon σ-bond which compensate the destabilization energy.

FMO of butadiene:

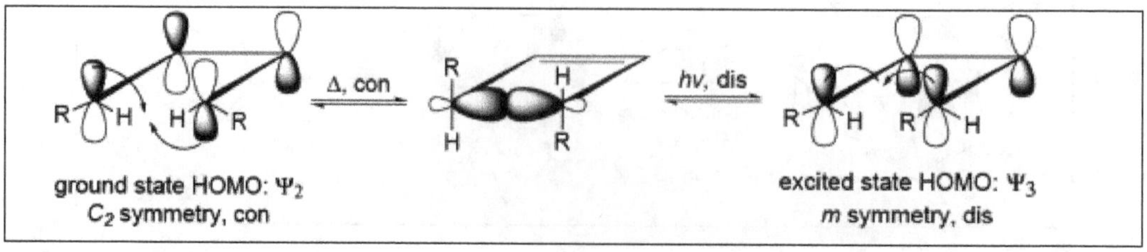

FMO of hexatriene:

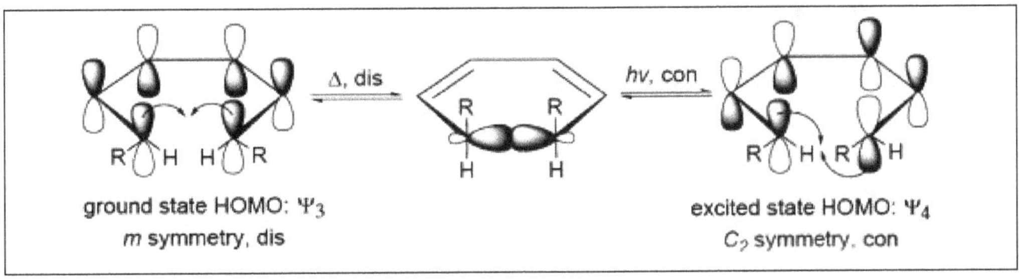

Table: A correlation among the entire factors associated with Elecrocyclic Reaction.

No. of π electrons	Symmetry	Mode of activation	Pathway	Reactants stereochem-istry	Product Stereo-chemistry
4nπ	C_2(s)	Thermal(Δ)	Conrotation	trans- trans cis- cis	trans
				trans- cis	cis
4nπ	m(s)	Photochemical (hγ)	Disrotation	trans- trans cis- cis	cis
				trans- cis	trans
(4n+2)π	m(s)	Thermal(Δ)	Disrotation	trans- trans cis- cis	cis
				trans- cis	trans
(4n+2)π	C_2(s)	Photochemical (hγ)	Conrotation	trans- trans cis- cis	trans
				trans- cis	cis

Thermal electrocyclic reactions involving a total number of electrons that can be expressed in the form (4nþ2) are disrotatory, and thermal electrocyclic reactions in which the total number of electrons can be expressed in the form (4n) are conrotatory. Some examples of electrocyclic reaction are shown below:

1. A representative of various types of electrocyclic reactions is illustrated:

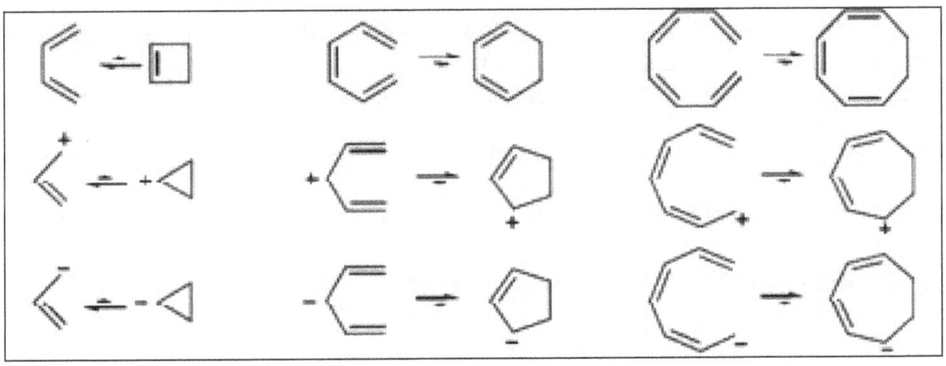

2. The reaction below is photo-induced conrotatory mode of reaction in which reactant comprises (4n + 2)π electrons with ψ_4* excited HOMO having C_2-symmetry.

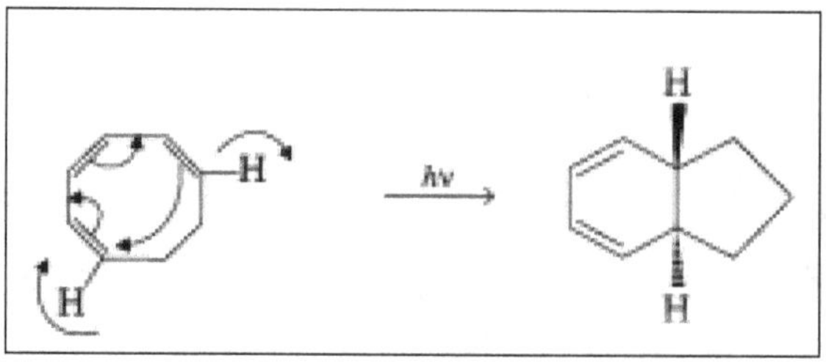

3. The following disrotatory mode of conversions of hexatriene [(4n + 2)π electrons system] are observed in thermal condition.

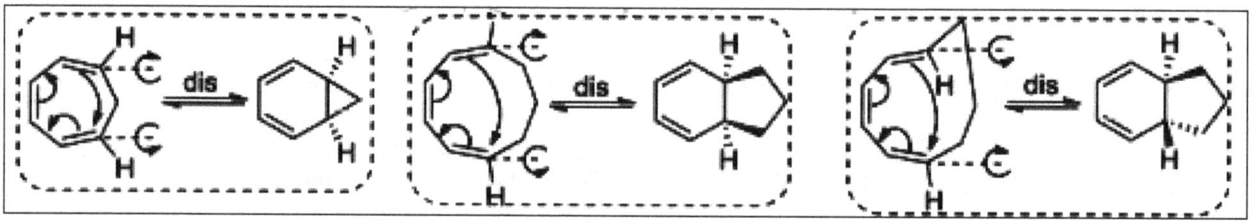

4. The following reactions represent double electrocyclic ring closure reaction. The first involves (4n) π electrons, resulting a conrotatory motion while the second cyclization has (4n + 2) π electrons showing disrotatory motion.

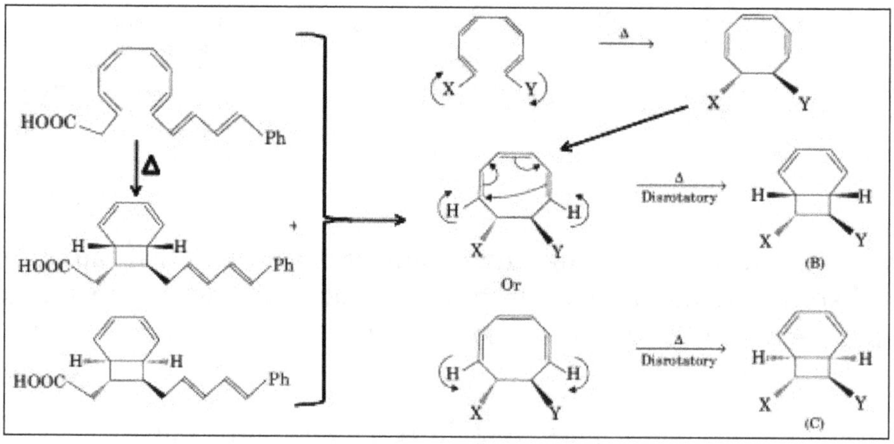

5. Octatetraene produces 7,8-dimethyl cycloocta-1,3,5-triene upon photochemical disrotatory ring closure followed by thermal disrotatory electrocyclization to yield fused ring.

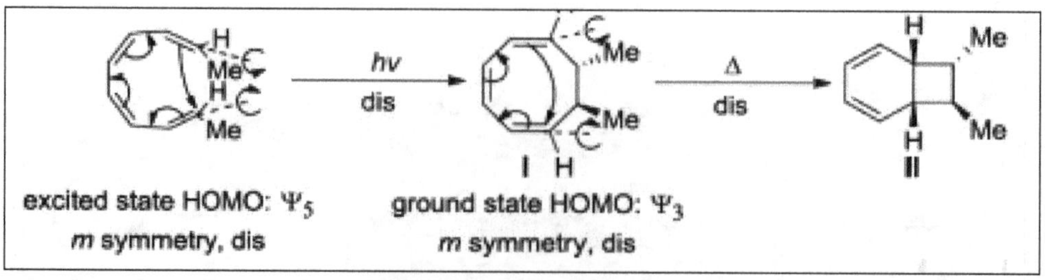

6. Under thermal conditions, cyclooctatetraene preferentially undergoes a 6π-electron disrotatory electrocyclization whereas under photochemical conditions, cyclooctatetraene preferentially undergoes a 4π-electron disrotatory electrocyclization. However, both form only the cis-isomer.

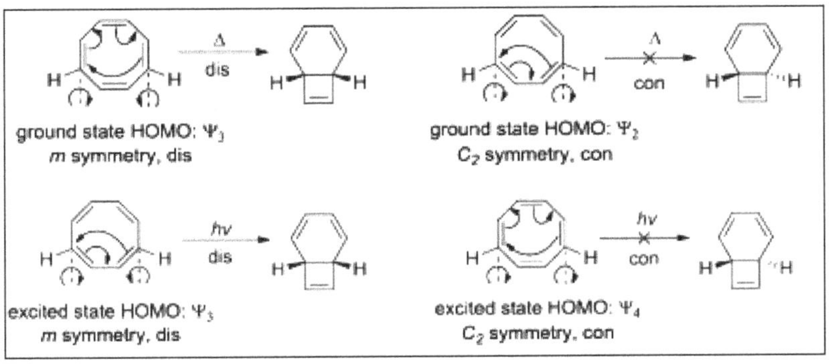

7. Compound II and III are formed by the ring opening via thermal conrotation and disrotatory electrocyclization respectively. Compound III undergoes [1,5] hydrogen shift to yield IV.

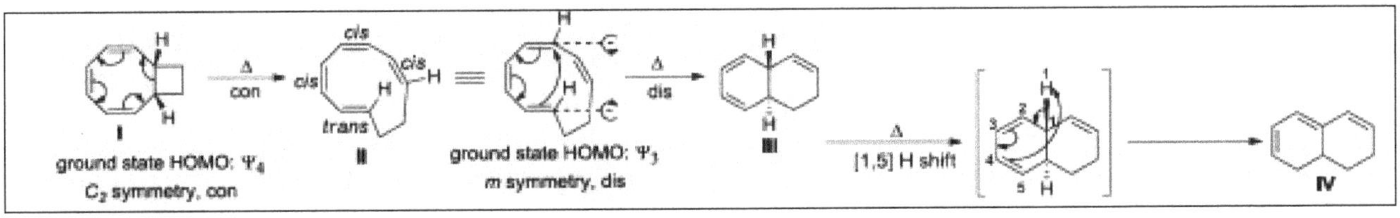

8. [16]-annulene isomerizes both thermally through two cis-fused 6-membered double disrotatory closure and photochemically through two trans-fused 6-membered double conrotatory closure.

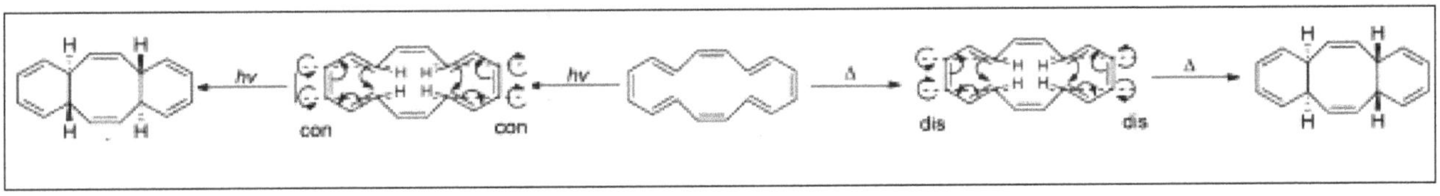

9. It is usual for a photochemical reaction to favour the less stable isomer of an equilibrium. For instance, the conjugated diene absorbs UV light, but the bicyclic compound does not, hence, the photochemical reaction favours the latter.

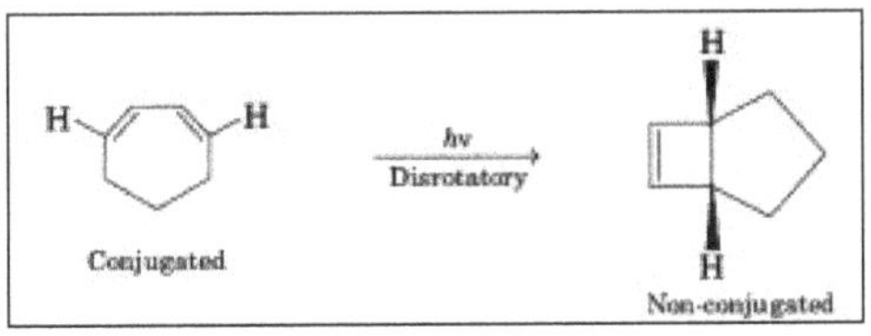

10. The allowed thermal conrotatory process would give a highly strain molecule containing a trans double bond within a small ring.

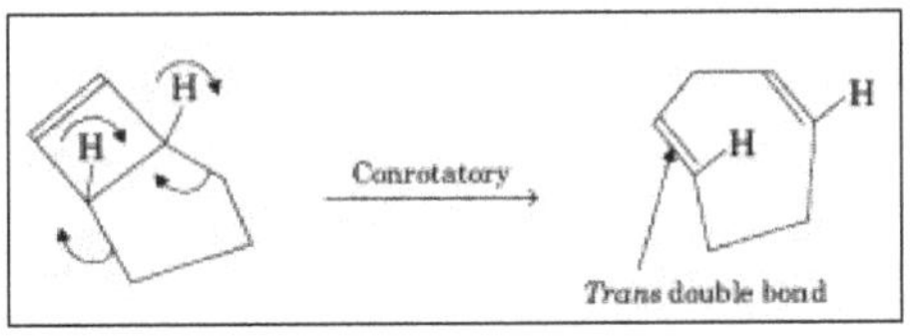

11. (2Z, 4Z, 6Z)-2, 4, 6-cyclooctatrione first isomerizes photochemically to (2E, 4Z, 6Z)-2, 4, 6-cyclooctatrienone, which then cyclises thermally to cis-bicyclo [4, 2, 0]octa-3, 6-dien-2-one by a conrotatory process.

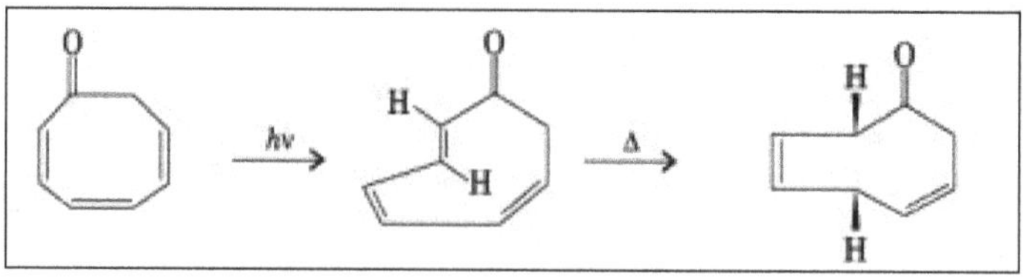

12. Thermal isomerization of cis-bicyclo [6, 2, 0] deca-2, 9-diene gives (1E, 3Z, 5Z)-1, 3, 5-cyclodecatriene.

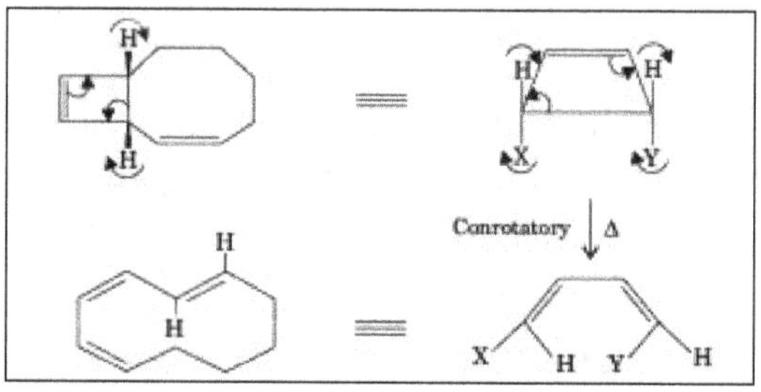

13. In bicycloheptane, a conrotatory reaction is virtually impossible since it would put trans double bond in the seven membered ring. Rather a forbidden disrotatory reaction occurs at high temperature (≈ 400 °C).

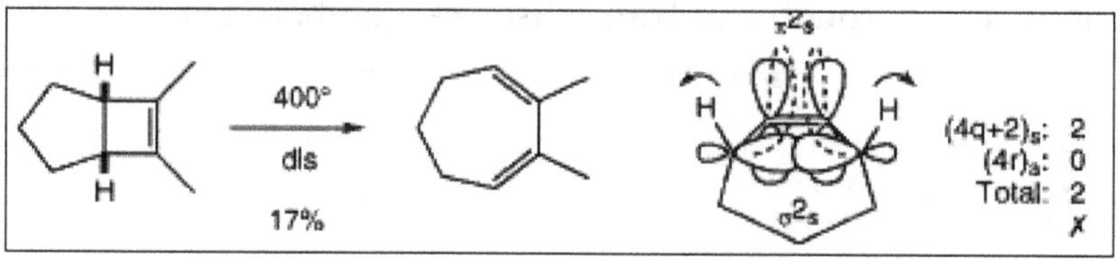

14. Since the ring fusion as well as double bond within the ring is the low energy cis arrangement, a conrotatory ring opening is expected to give strained trans double bond within 8-membered ring. It is thermodynamically unfavourable. The reaction thus needs a high temperature to resulting in the formation of another isomer.

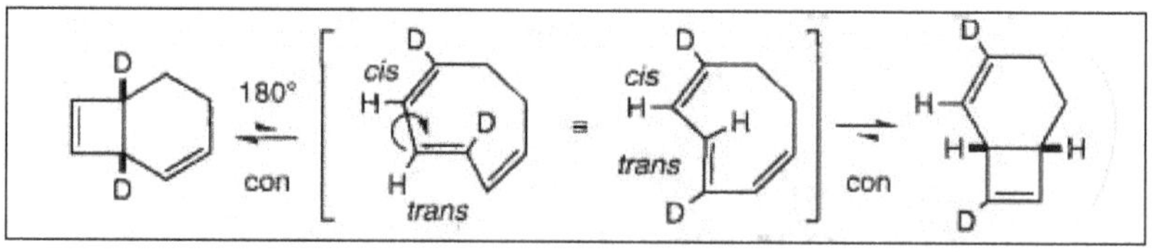

15. Stereoisomeric pentadienyl cations give stereoiosmeric cyclopentenyl cations in conrotatory reactions. The reactions are fully stereospecific. In contrary, for cyclopentenyl anion that disrotatory opens to allow two H's moving outward.

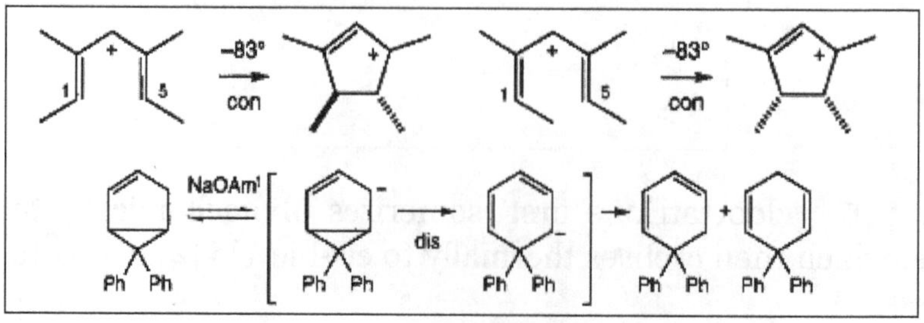

16. In case of constrained systems, the photochemical reaction can easily lead to a product thermodynamically much higher in energy than the starting material; but nevertheless they are thermally stable. Because the bicyclic ketone (first reaction) lost is not only strained but it lost the partial aromaticity of tropone and similar loss of aromaticity for dihydroanthracene from cis-stilbene.

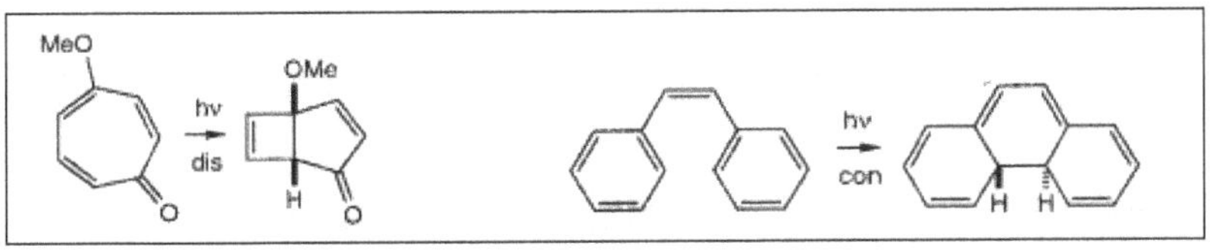

17. Disrotatory ring opening of various cyclopropyl halide results three different shapes of allyll cations: W-shaped, sickle shape and U-shaped which largely retain their conformation at low temperature as obvious from ¹H NMR spectroscopy. These reactions demonstrate torquoselectivity. It defines that the direction of disrotatory ring opening is determined by which side of the ring the halide leaves from as shown from contrast between the products from the cyclopropyl halides. The former opens the two $-CH_3$ groups moving outwards as thermodynamically preferred direction. However, later shun thermodynamics to open as U-shaped geometry.

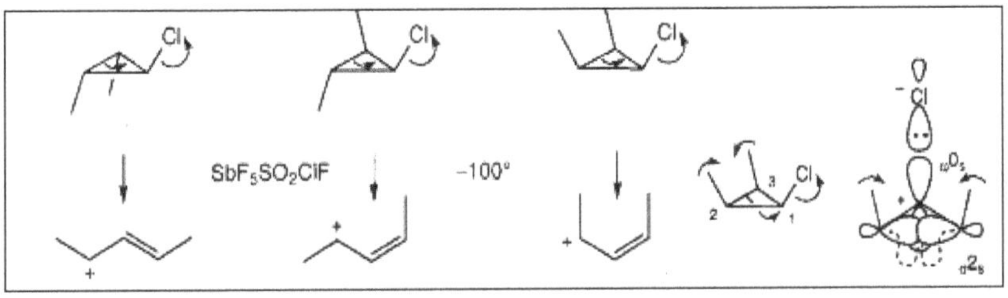

18. There is another example of Torquoselectivity as the predisposition of a given R substituent for a given conrotatory motion. Steric factors cause a preference for the larger group to move outward. It is observed that π-donor substituents prefer con-out mode whereas π-acceptor substituents prefer con-in mode.

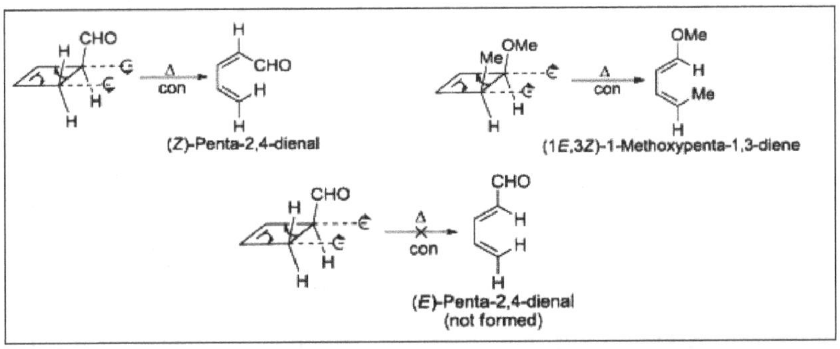

19. The trans isomer of 2,3-ditertiary butyl cyclopropanone readily isomerizes at 80 °C whereas the cis isomer does not because the disrotatory ring opening required for isomerization is sterically prevented for cis isomer.

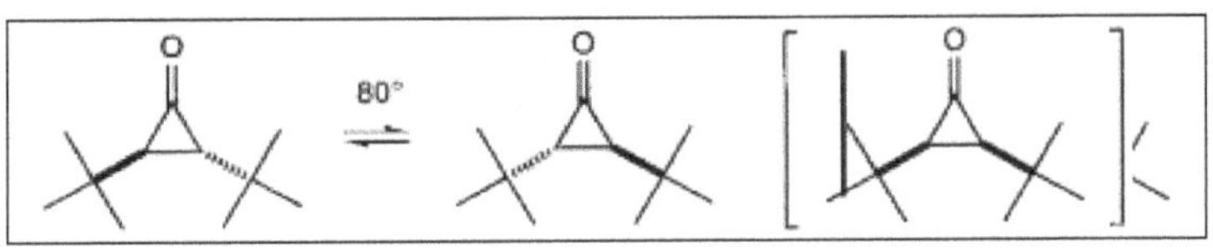

20. Thermal isomerization of Ph-substituted dienes occurs via cyclobutene intermediate following a thermally allowed conrotatory electrocyclization.

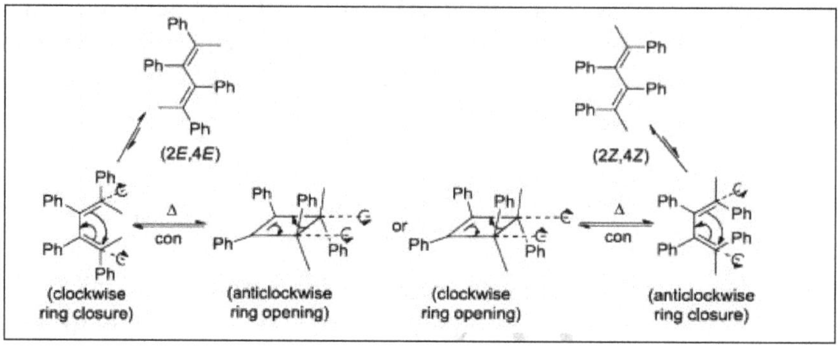

21. The trans-stilbine isomerizes to cis-stilbine (I) under photochemically. On irradiation of cis-stilbine provides dihydrophenanthrene (II) by conrotatory cyclization.

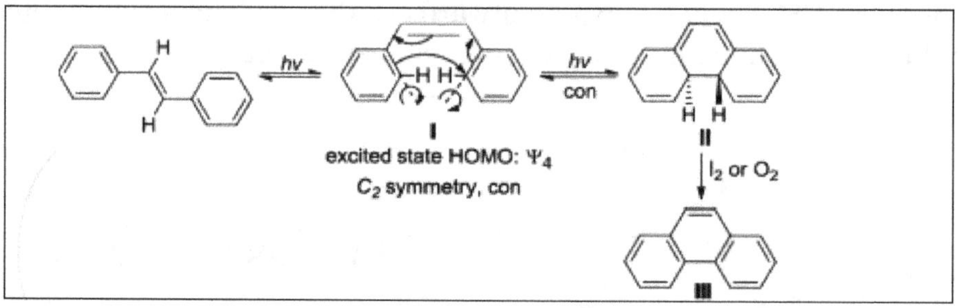

22. A beautiful illustration of the synthesis of ergosterol involves electrocyclic reaction under irradiation by light.

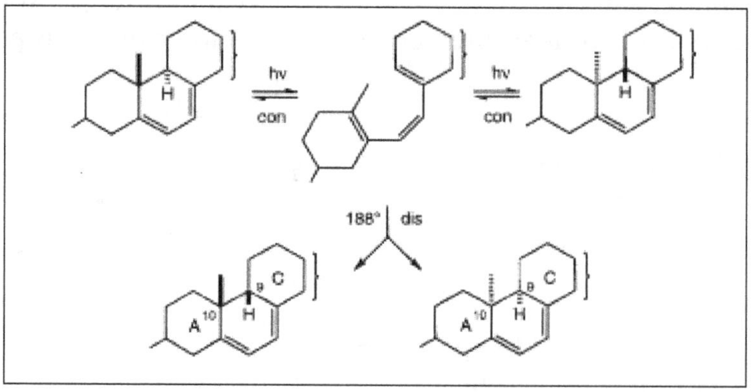

23. Vitamin D_3 (cholecalciferol) is produced through the action of UV irradiation on its precursor 7-dehydrocholesterol.

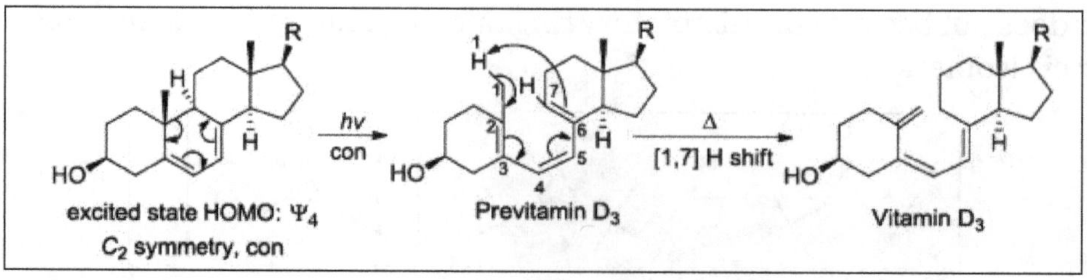

Typical Examples of Electrocyclic Reactions

1.
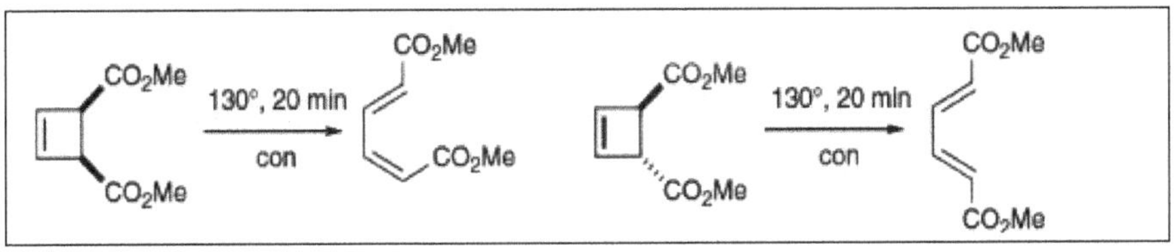

2.
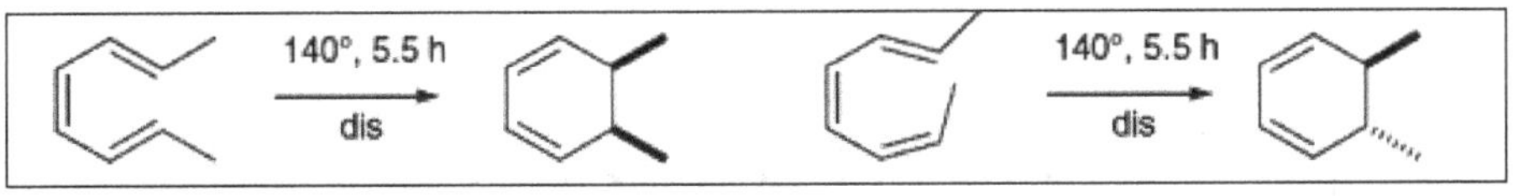

3.
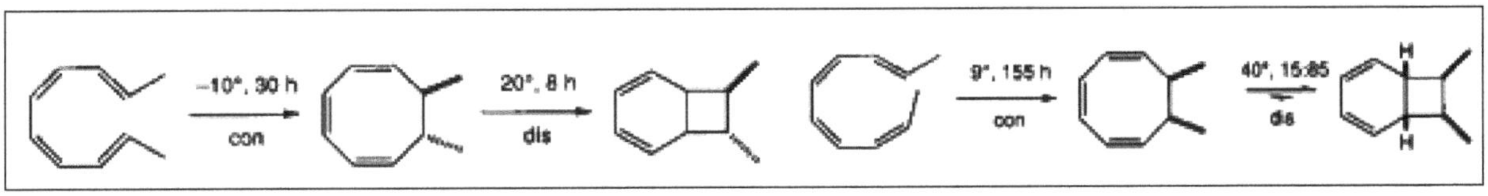

4.
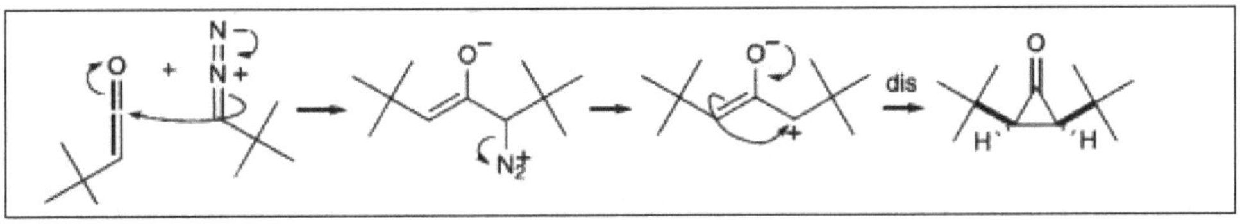

5.
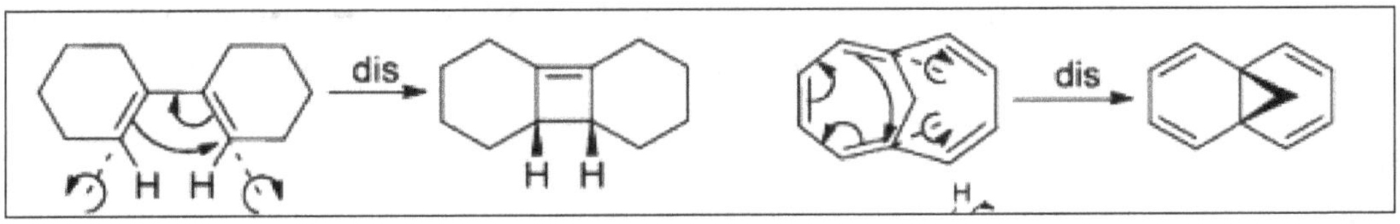

6.
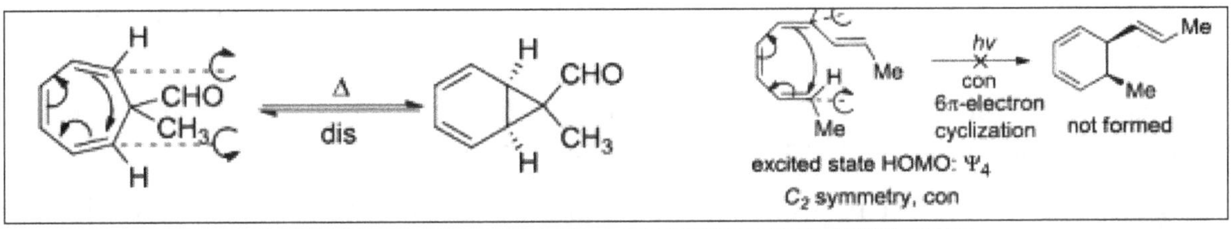

7.
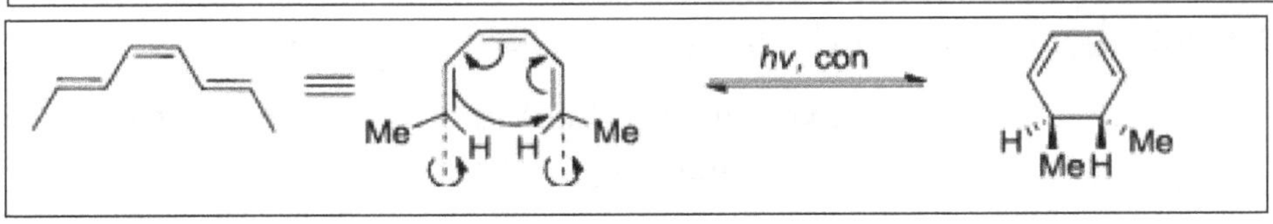

8.
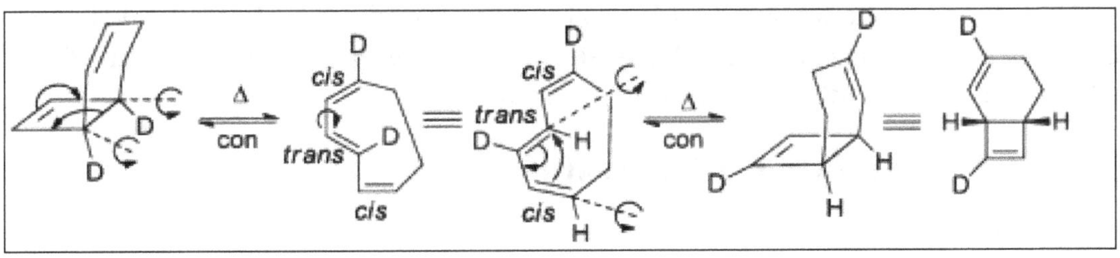

9.
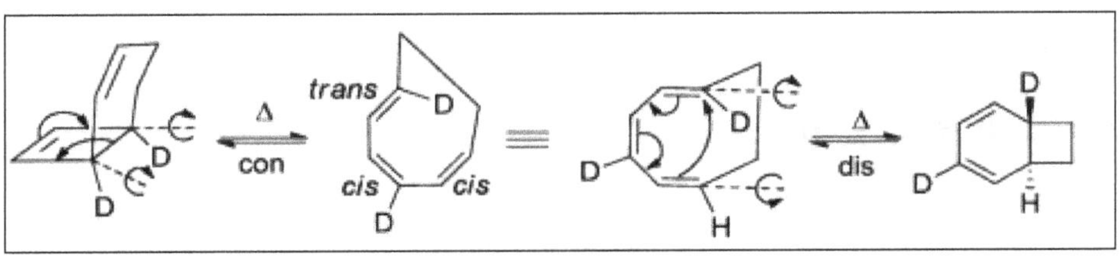

10. 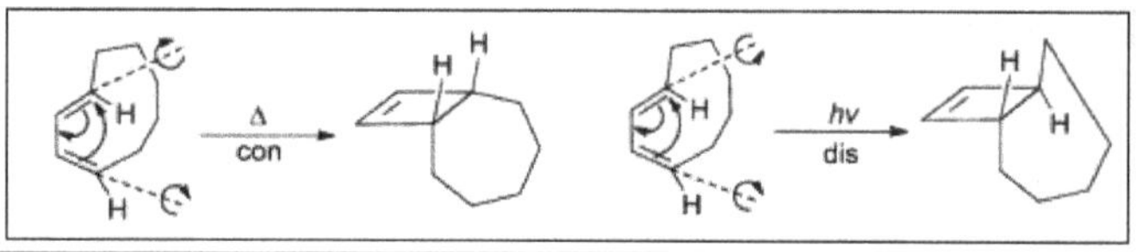

11. 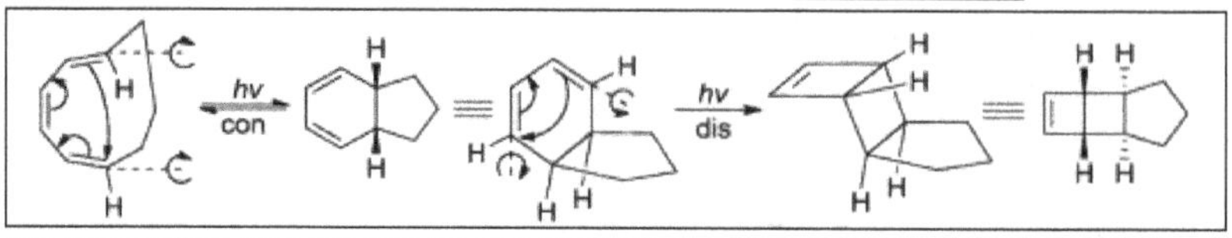

12. 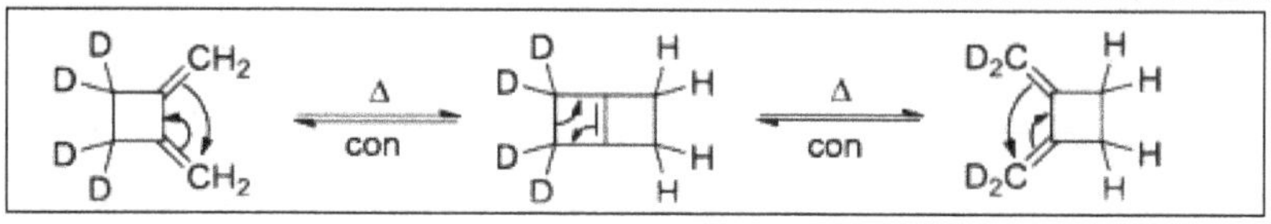

Cycloaddition Reactions

A concerted cyclic shift of two separate π-electron systems in which a ring is produced with the formation of two new σ-bonds is termed as Cycloaddition reaction. The reaction is activated either by heat or by UV-light. The reaction follows the principle of microscopic reversibility. The reaction sequence can also be explained by three fundamental approaches; (a) correlation of orbital symmetry of terminal lobe of interacting orbitals (b) perturbation of M.O theory or HMO theory of aromaticity or anti aromaticity in T.S (c) FMO approach of HOMO-LUMO interaction of corresponding orbital energy state. The reverse reaction of cycloaddition is called cycloreversion.

Conditions for Cycloaddition Reaction

- The symmetry of those orbitals w.r.t terminal conjugated π-system of two separate polyene system has to be considered.
- Always the terminal p-orbitals of both polyene which are going to form σ-bond must overlap by maintaining the in phase distribution of lobes after rotation.
- The groups or substituent attached to terminal lobes also moves according to the lobe's movements. This brings about the stereochemistry of the resultant product.
- HOMO of the one polyene interact effectively with LUMO of other polyene forming both endo and exo product where endo product is kinetically more stable.
- That cycloaddition will be favoured which follows the Woodward–Hoffmann principle of orbital symmetry. The involvement of any steric crowding or generation of strain in the product molecule will avoid the subsequent mechanism.

Source/Influencing factor/Mode of Activation

- Heat or Thermal (Δ).
- Light or Photochemical (hγ).

Criteria for Cycloaddition Reaction

- Mode of activation. (Δ/hγ).

- Number of p/π-orbital participates. [4nπ or (4n+2)π].
- Stereochemical outcome (whether stereospecific or stereoselective).
- The nature of orbitals undergoing change (σ or π).

Here Woodward-Hoffmann suggested that if the reaction would take place through same face of a π-system, the reaction is said to be Suprafacial (s) to that of π-system and if the reaction occurs through opposite face of a π-system, the reaction is said to be Antarafacial (a) to that of π-system. For example (supra)-ethylene and (supra) cis-2-butene yields cis-1,2-dimethyl cyclobutane whereas (supra)-ethylene and (antara) cis-2-butene yields trans-1,2-dimethyl cyclobutane.

Pictorial Representation

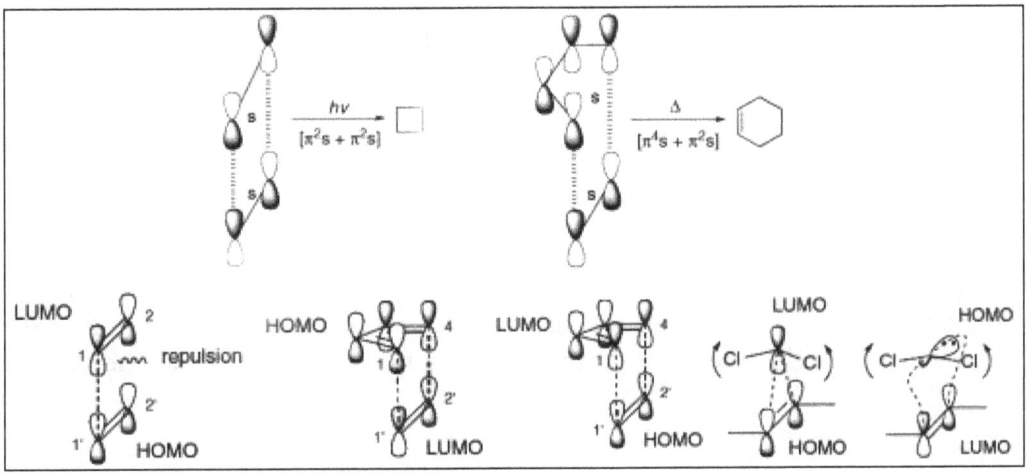

Diene in the s-cis conformation undergoes Diels-Alder reaction because C-1 and C-4 of the diene is close enough to react through a cyclic transition state. In s-trans conformation, the T.S would be strain.

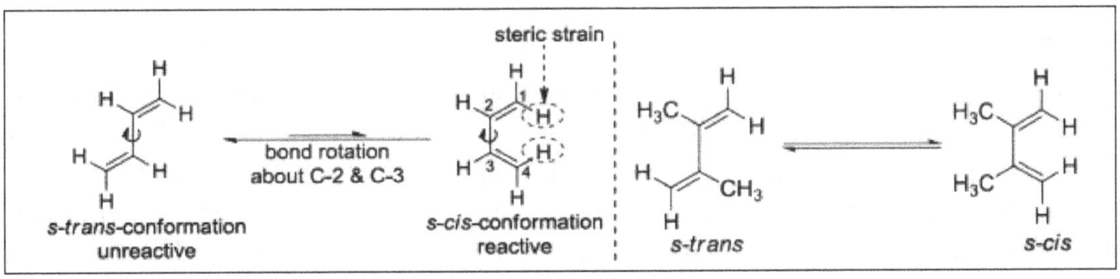

Dienes with an enforced coplanar s-cis conformation are exceptionally reactive in Diels-Alder reaction. The reactivity of diene depends on the rigidity (locked conformation in cyclic form), coplanarity and the concentration of the s-cis conformation in the equilibrium mixture.

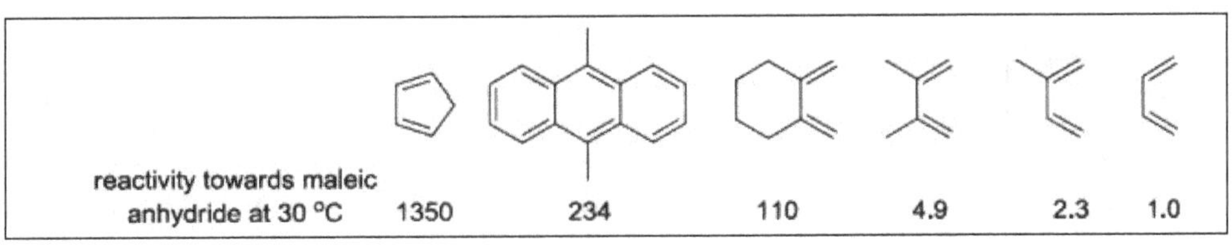

The following acyclic dienes are very unstable in s-cis form owing to severe steric hindrance. Cyclic dienes of permanent s-trans conformation are unable to adopt s-cis conformation and thus will not undergo Diels-Alder reaction.

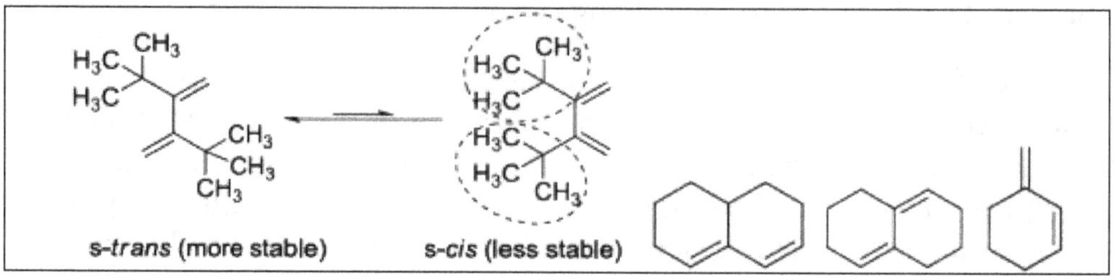

Approaching π—bonds and pair them up in a lower energy symmetric and a higher energy anti-symmetric combination can be classified based on symmetry of orbitals w.r.t an imaginary plane.

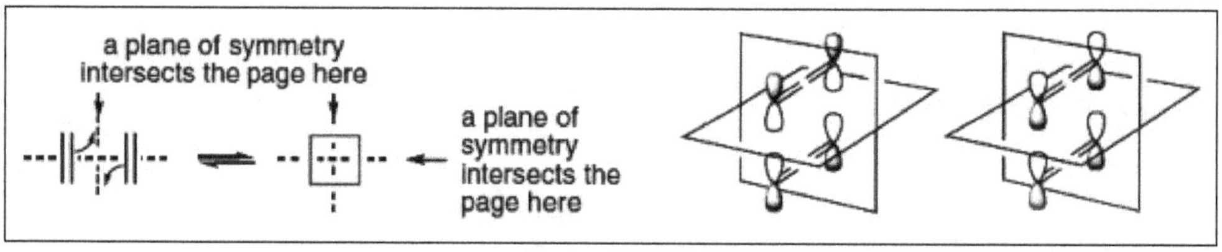

The Diels-Alder reaction which is suprafacial on both components, a plane of symmetry bisects the bond between C-2 and C-3 of the diene and the π bond of the dienophile. If any substituents cause the diene or dienophile unsymmetrical, they do not necessarily disturb the symmetry of the orbitals that are involved directly.

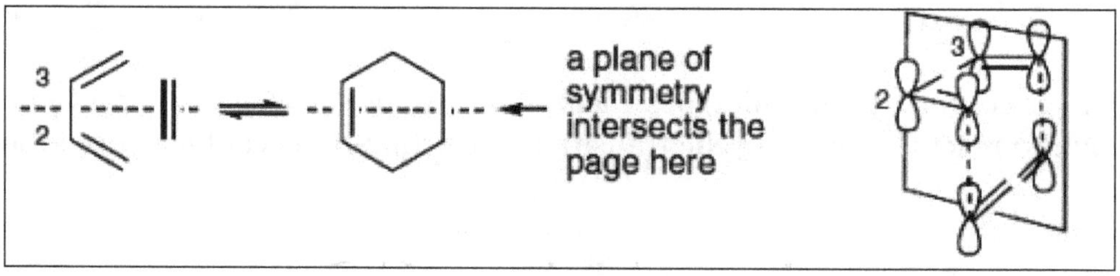

In the transition structure, the π bond between C-2 and C-3 has sharply formed (66%) and likewise, the σ bonds are little formed (22%). The π bonds in the dienophile and the diene seems not so weakened (32% and 22%, respectively). A late transition structure is accounted by the larger changes in the π-bonding in the diene and an early transition structure implies the small changes in the σ bonding and in the π-bonding of the dienophile. This asynchronicity reflects the sum of the HOMO(diene)/HOMO(dienophile) interactions and substantial contributions from both FOIs.

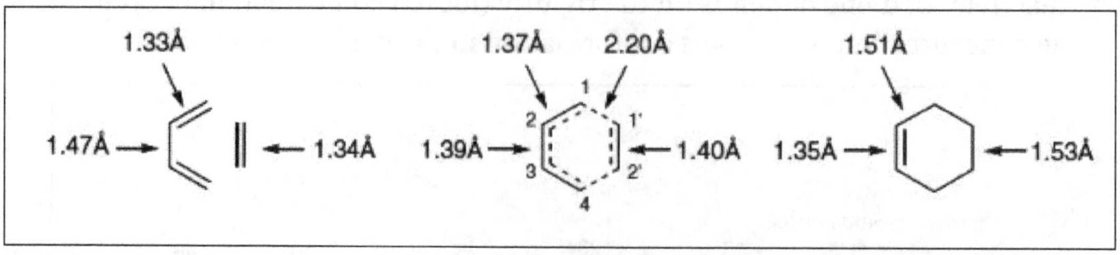

Molecular orbitals from combination of σ orbitals (left) and σ* (right):

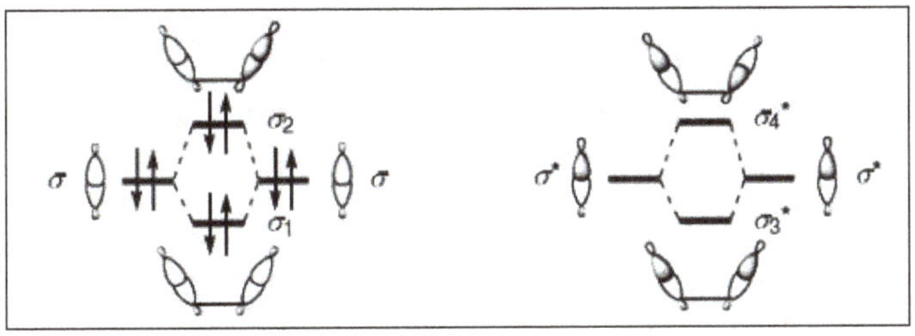

Primary and Secondary Orbital Interaction

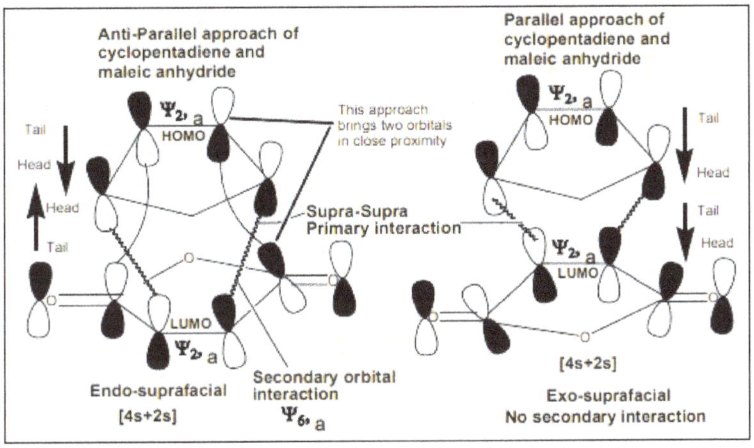

Both the cyclopentadiene and maleic anhydride can interact to each other w.r.t C_2 symmetry orbital under thermal condition. Their approach is either in parallel direction or in anti-parallel direction as shown in the figure. Both the approaches are basically supra-supra addition where in phase bonding overlap takes place. The antiparallel approach leads to endo product whereas parallel approach exerts exo product. The endo product is formed as kinetically controlled product (KCP) which occurs at low temperature while the exo addition is produced as thermodynamically controlled product (TCP) at comparatively higher temperature. The reason is that there is an additional orbital interaction between non-terminal lobes of both diene and carbonyl lobes of dienophile which strengthen the net overlap. This interaction enables lowering in activation barrier and faster the reaction rate.

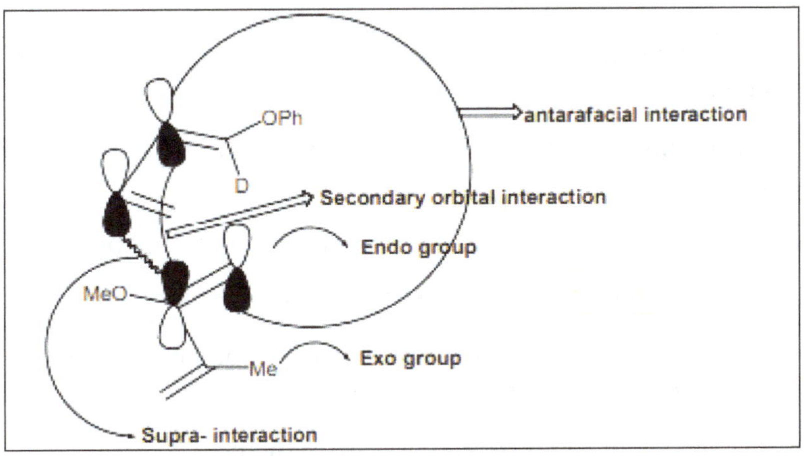

Supra-antara orbital interaction.

Therefore, such interaction which stabilizes the endo addition is known as secondary orbital interaction. Since exo product is formed due to parallel approach of two π-systems, so the lobes involving secondary orbital interaction are far apart from each other. Hence, the secondary orbital interaction is virtually absent in exo product. The endo product can be converted to exo product at high temperature (190 °C) but the reverse cannot be possible.

The above supra-antara interaction leads to a bonding with highly strained system. Hence the reaction is symmetry forbidden reaction. The exo product is more stable than the endo product for steric reasons. Since the reaction is concerted, the transition state for exo product formation must be less crowded than the transition state for endo product formation. In practice, it has been found that the endo product is more stable than the expected exo product. This clearly confirms that the transition state of endo product should be more stable than that of the exo product due to some other factors which overwhelms steric consideration. This factor is known as secondary orbital interaction. In endo transition state the double bonded carbon of dienophile is directed to the inside of diene ring where there can be interaction between back diene orbitals and orbitals on the substituents (C≡N) in dienophile. Such interaction is absent in case of exo orientation as dienophile is directed away from the cyclic diene conjugated system.

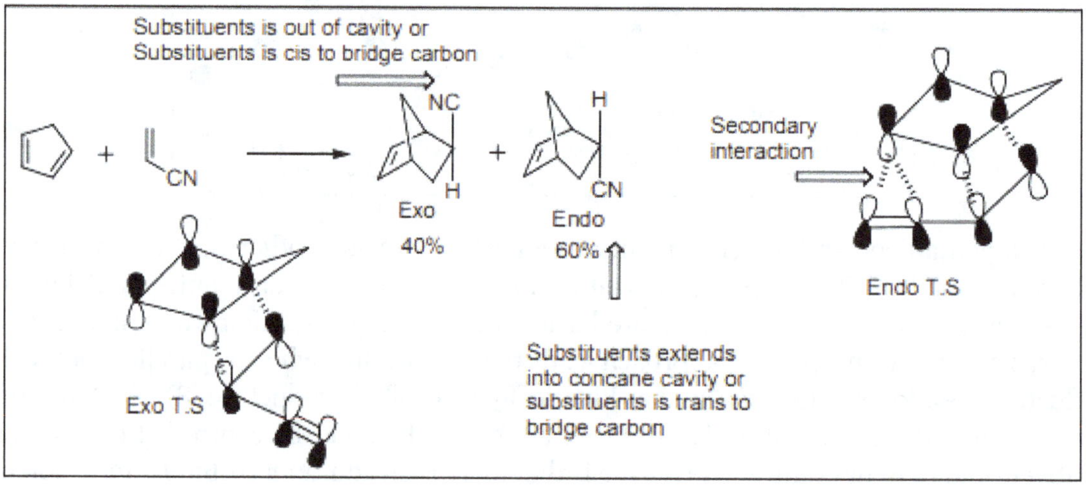

- For two component system the maximum number of mode of addition=$2^n=2^2=4$. (ss, sa, as, aa).
- ss mode means parallel approach of orbital whereas sa, as, aa indicates orthogonal approach of orbitals.
- Terminal group retains its configuration in suprafacial approach while it is inverted in antarafacial approach.
- Both endo and exo addition occurs through ss mode depending on the substitution pattern.
- Favourable as well as facile pathways is shown when $[\pi^4_s + \pi^2_s]$ is greater than $[\pi^4_a + \pi^2_a]$.
- Always s-cis conformation reacts in cycloaddition reaction because in s-trans conformation the terminal lobes are anti to each other.
- Any bulky group or ring at the termini will retard the rate.

- The rate of cycloaddition reaction depends on the energy difference between HOMO and LUMO orbital. The electron releasing group enhances the HOMO energy of diene and electron withdrawing group lowers the LUMO energy of dienophile except for cycloaddition reaction with reverse electron demand.

Energies of Frontier Orbitals

- C-(extra conjugation): Raises the energy of the HOMO and lowers the energy of the LUMO.
- Z-(an electron-withdrawing group): Slightly lowers the energy of the HOMO and substantially lowers the energy of the LUMO.
- X-(an electron-donating group): Substantially raises the energy of the HOMO and slightly raises the energy of the LUMO.

Frontier orbital energies (eV) of alkenes and dienes (1 eV = 23 kcal = 96.5 kJ):

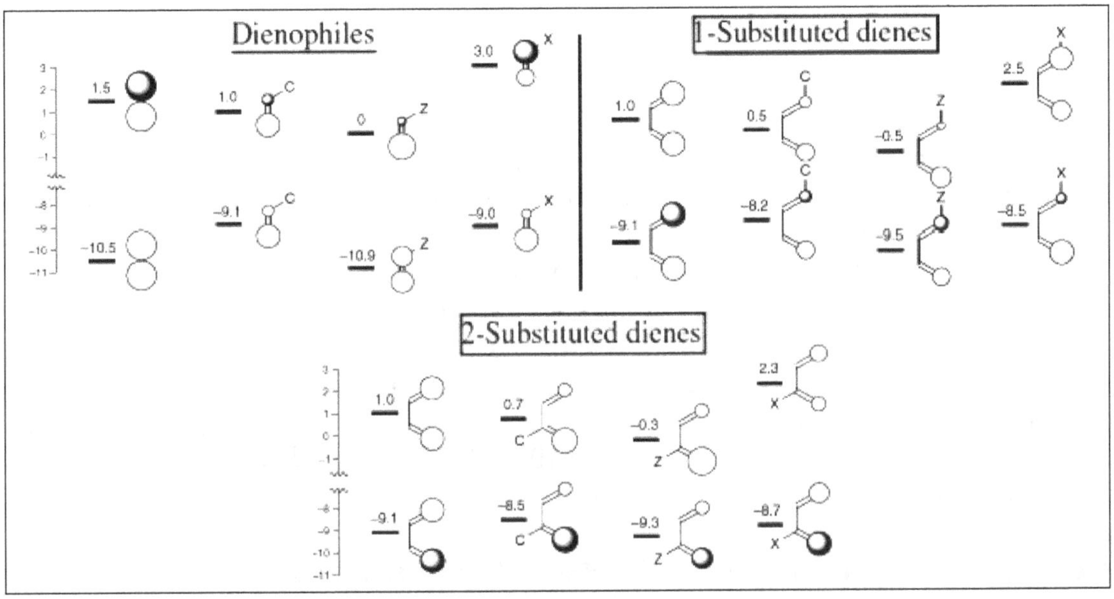

Frontier orbital energies are experimentally measured by photoelectron spectra for the occupied orbitals, and a combination of electron affinity measurements, charge transfer spectra and polarographic reduction potentials for the unoccupied orbitals. Frontier orbital energies provide a clear idea about the rate of cycloaddition reactions by considering lower energy difference between HOMO and LUMO. The size of the orbital coefficients seldom influences in the frontier orbitals energies. Rather the relative sizes of the coefficients are principally used to account for regioselectivity.

Butadiene reacts with ethylene at high temperature (165 °C), high pressure to give a low yield of cyclohexene. The rate becomes little faster with acrolein (at 150 °C) in terms taking less time at the same temperature. Introducing a second Z-substituent (EWG) such as methylenemalonate or maleic anhydride increases the rate even more that eventually reacts at room temperature.

An X-substituent (ERG) diene on C-1 or C-2 such as trans-piperylene, isoprene or methoxybutadiene increases the rate further with acrolein at 130 °C. In contrast to times and temperatures, the yield of the isolated product (here say 80 %) is a reliable way of measuring relative rates.

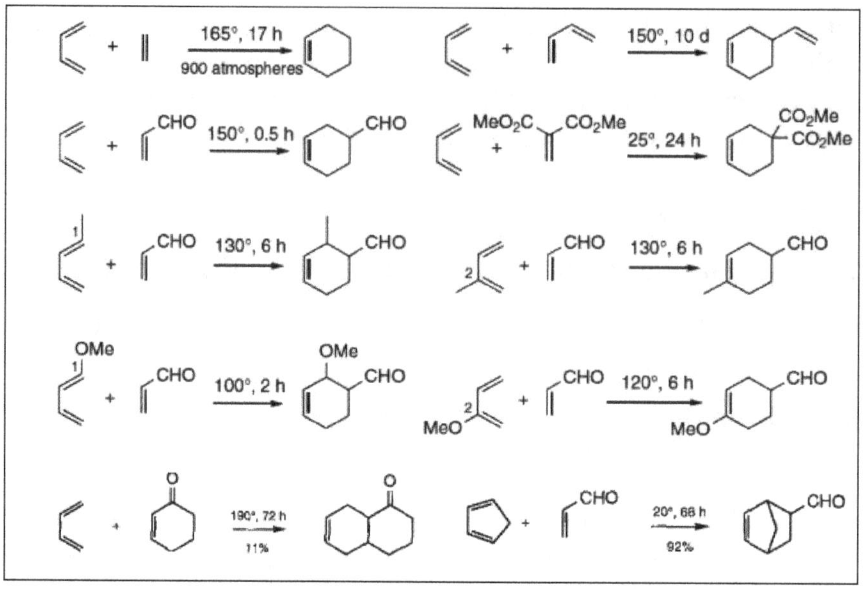

Cause: In order to explain the rate of the reaction, the proper selection of HOMO and LUMO for either of diene or dienophile is essential. Considering two pathways, the direction of the reaction and rate depend on the smaller energy difference between HOMO-LUMO either of diene or dienophile. Normally, a Diels-Alder reaction is characterized by the interaction of HOMO-diene (9.1 eV) and LUMO-dienophile (0 eV) with energy difference = 9.1 eV as compared to 11.9 eV arising from LUMO-diene (1.0 eV) and HOMO-dienophile (10.9 eV). Thus, a faster reaction is associated with more electron-withdrawing group in dienophile comprising low-lying LUMO such as tetracyanoethylene. This enables smaller the separation of the HOMO-diene and the LUMO-dienophile. The combination of X-substituent HOMO-diene (raising energy) and Z-substituted LUMO-dienophile produces relatively lower energy gap of 8.5 eV. That is why, the energy separation between the HOMO of butadiene and the LUMO of a Z-substituted dienophile is less than that between the HOMO of butadiene and the LUMO of ethylene.

Frontier Orbital Interactions for Diels-Alder Reactions

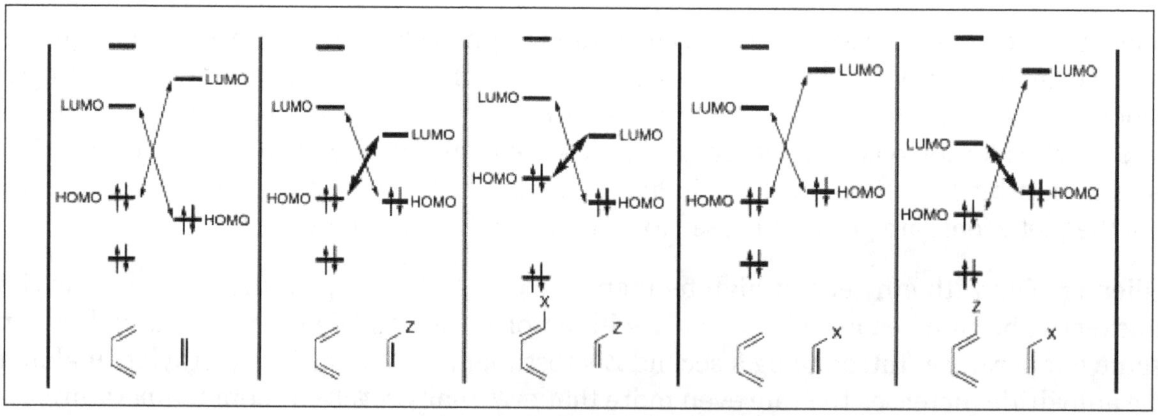

Diene with electron-withdrawing Z-substituents reacting with dienophiles with electron-donating X-substituents are said to be Diels-Alder reactions with inverse electron demand. Following examples represent this idea, although are not common.

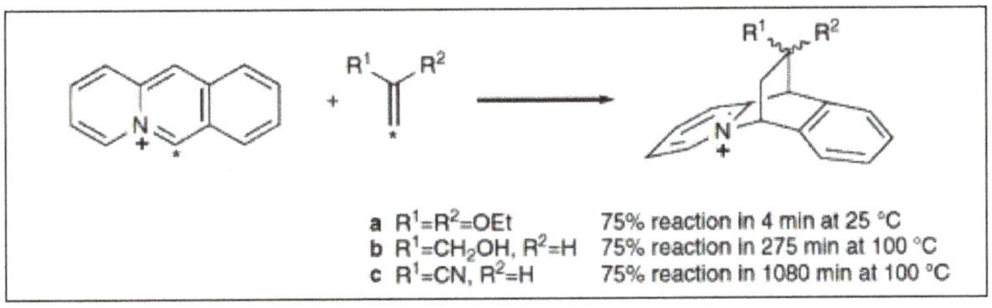

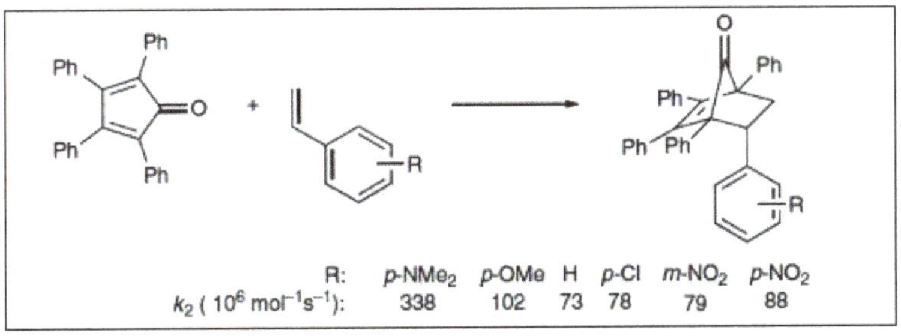

Cause: Suppose the electron density is supplied from from the diene. Since one node is there, the ψ_2 HOMO of diene ensures the flow of electrons that are concentrated at the sides where the σ–bonds are formed in lieu of concentrated in the centre of the diene (left). In case of inverse electron demand, the major supply of electrons comes from the ψ_1 HOMO of dienophile which consists of no node. Consequently, it interacts with ψ_3 diene LUMO even more strongly with lowe activation barrier.

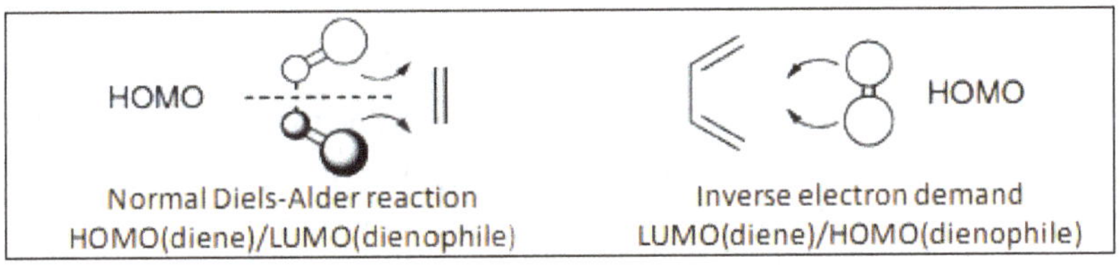

In both examples (above), R = ERG raises the HOMO of dienophile that becomes closer to LUMO of diene containing EWG. It helps strong HOMO-LUMO interactions. If two dienes one having a EWG and another with ERG are allowed to react, diene with EWG always acts as dienophile while diene with ERG behaves as diene.

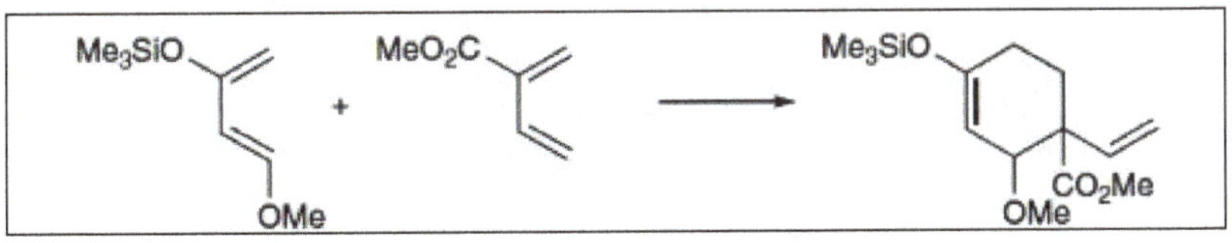

Regioselectivity of Diels-Alder Reactions

Regioselectivity refers to the orientation of a cycloaddition. Methoxybutadiene reacts with acrolein to give more of the 'ortho' adduct than of the 'meta' adduct.

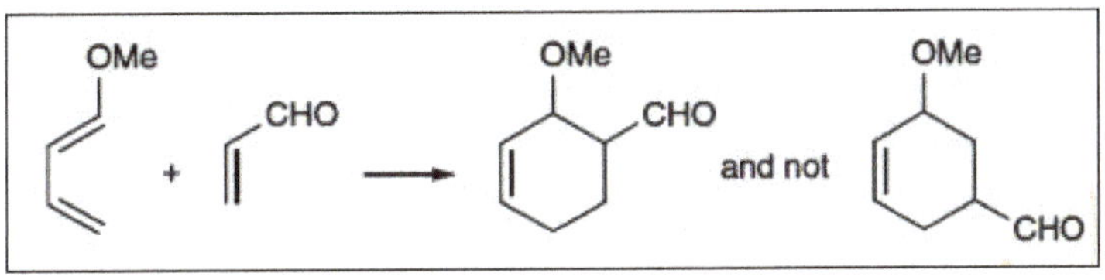

Cause: The coefficients of the atomic orbitals undergoing new σ—bond formation should be the concern of regioselectivity. The combination HOMO-diene and the LUMO-dienophile gives E_{act}, Er –Es = 8.5 eV (left) predominantly, as compared to the reverse where Er – Es = 13.4 eV (right).

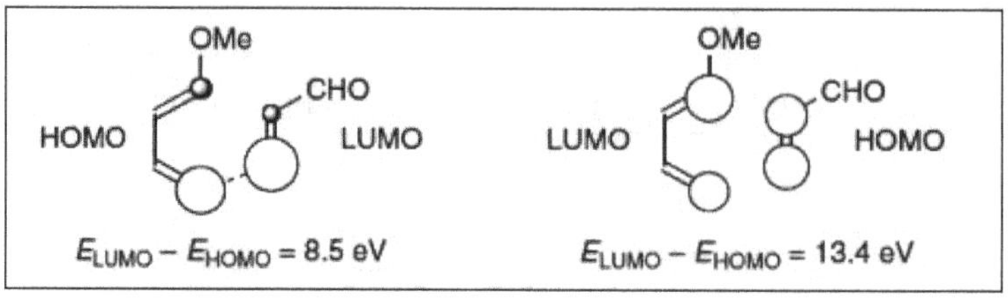

Let the square of the terminal coefficients on X be x and (x+n), and let the square of the coefficients on Y be y and (y+m). Large-large/small-small interaction [xy+(x+n)(y+m)] > Large-small/small-large interactions [x(y+m)+(x+n)y]. substracting gives the value nm. Therefore, former interaction is greater as long as n and m are of the same sign.

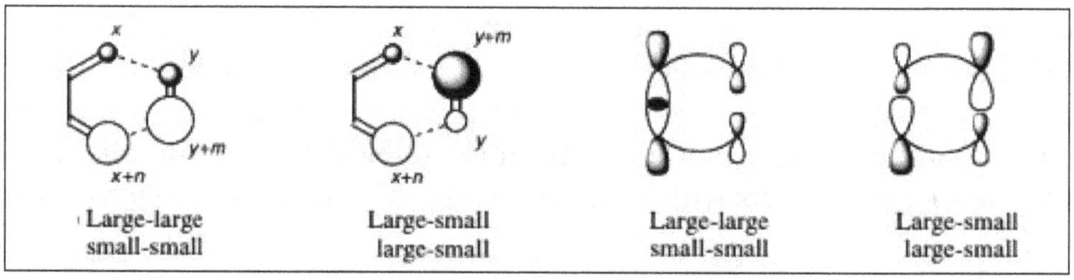

The large-large interaction proceeds over the small-small interaction (third figure). It indicates that the transition structure of an unsymmetrical Diels-Alder reaction is itself unsymmetrical where two σ bonds are formed to different extents. Despite the fact, the reaction is still concerted, with both bonds forming at the same time, but it is not synchronous.

Drawing curly arrows and creating canonical structures to express the logic that C-4 of the diene is a nucleophilic carbon and C-3′ of the acrolein an electrophilic carbon works well for the reaction between ERG-diene and EWG of dienophile. However, this is not applicable to all.

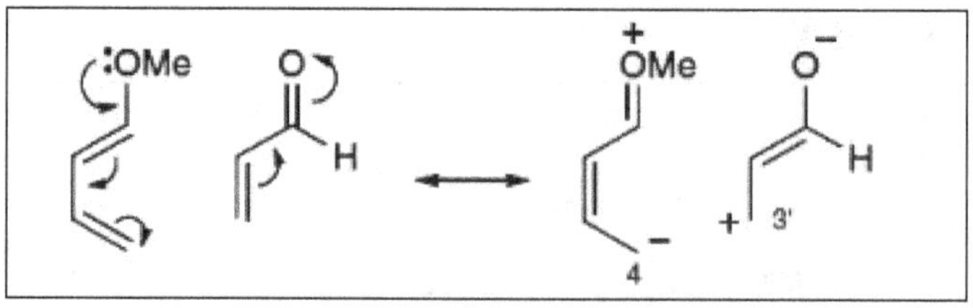

Clearly the valence bond argument is not good enough, rather the frontier orbital interactions. For instance, the curly arrows reflect that both C-4 of the diene carboxylic acid and C-3′ of acrylic acid are expected to be electrophilic which repels due to partial positive charges. The expected adduct would be the 'meta' adduct. In contrary, the mainly the 'ortho' adduct were isolated.

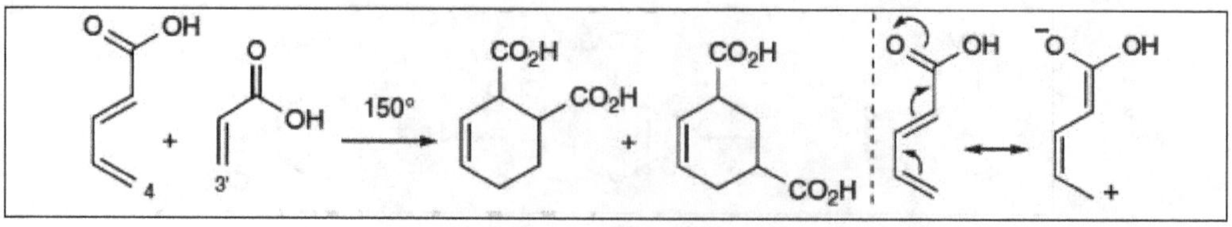

The 'ortho' adduct is also consistent with smaller HOMO-LUMO energy gap. It is assumed that the orbitals are polarized with marginally higher coefficient on C-4 in the diene-HOMO which stems from the hexatriene-like character of the conjugated system. As a result, the counter-intuitive combination with bonding between C-4 and C-3/ is achieved.

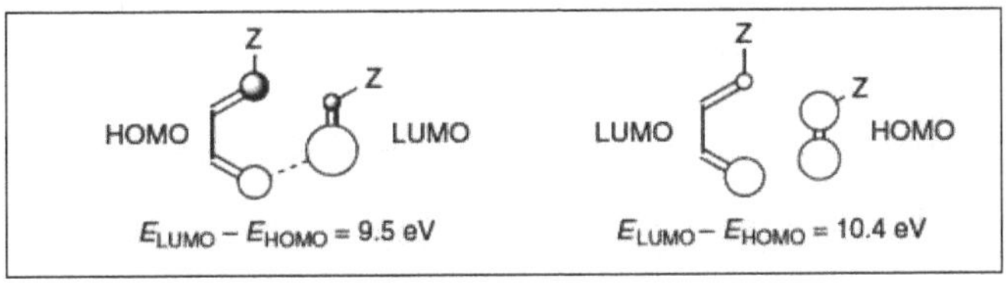

Examples that Obey 'Ortho-Para' Rule

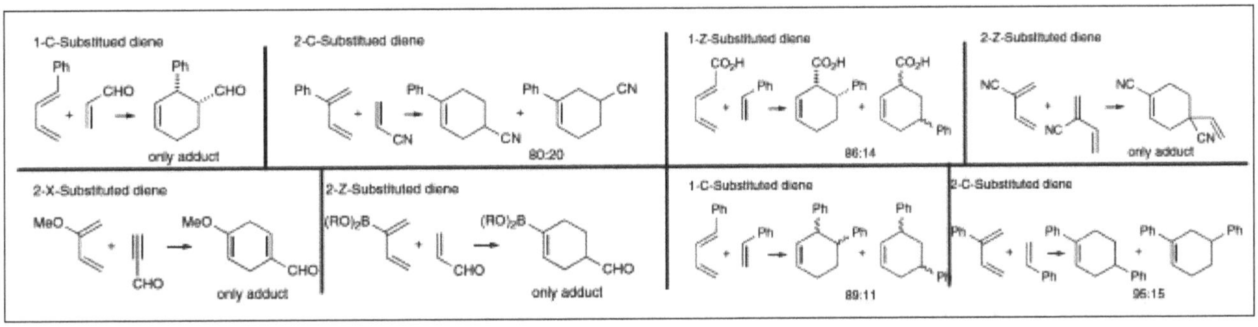

Exception of 'Ortho-Para' Rule

The rule breaks when both diene (at C-1 and C-2 position) and dienophile contain ERG (X substituent). It yields a substantial yield of 'meta' adduct with reasonable loss of regioselectivity partly because of high temperature reactions.

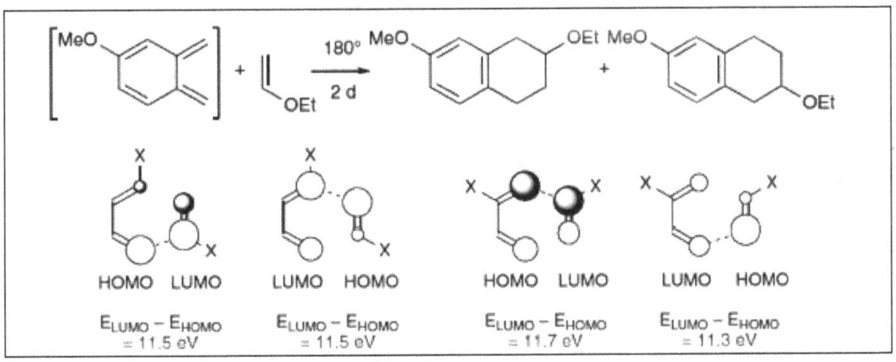

Hyperconjugation of the C-2 methyl group through the double bond with the carbonyl group attached to C-3 can reduce the electron-withdrawing effect of that carbonyl group. The C=C orbital is more polarised by the C-2 carbonyl which is the guiding factor than by the C-3 carbonyl.

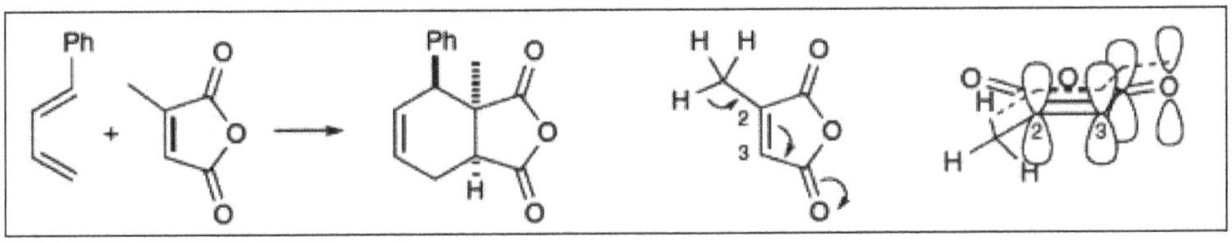

2-aryl substituent is usually less effective in polarising the coefficient at of frontier orbital at C-2 position than that at C-1 which eventually explains the observed regioselectivity.

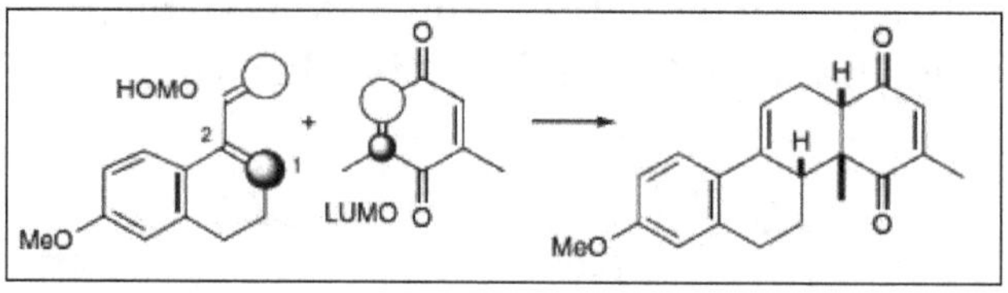

Based on the frontier orbital coefficients of tropone, the regioselectivity arises while reacts to various dienophiles.

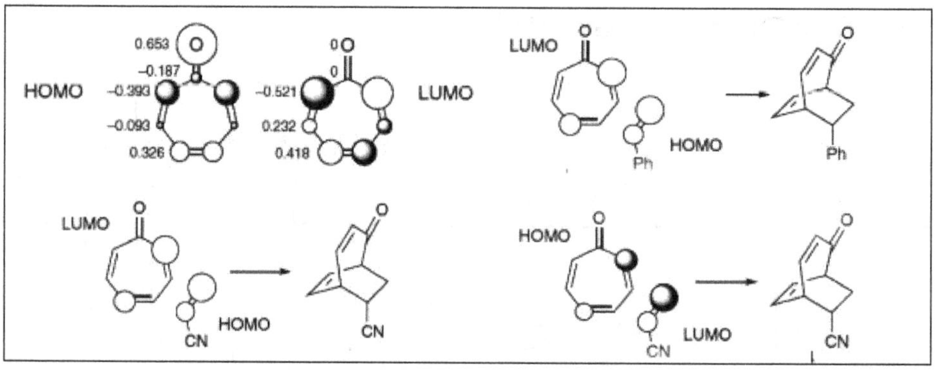

Typical [4+2] cycloadditions occur after electrocyclic ring closure.

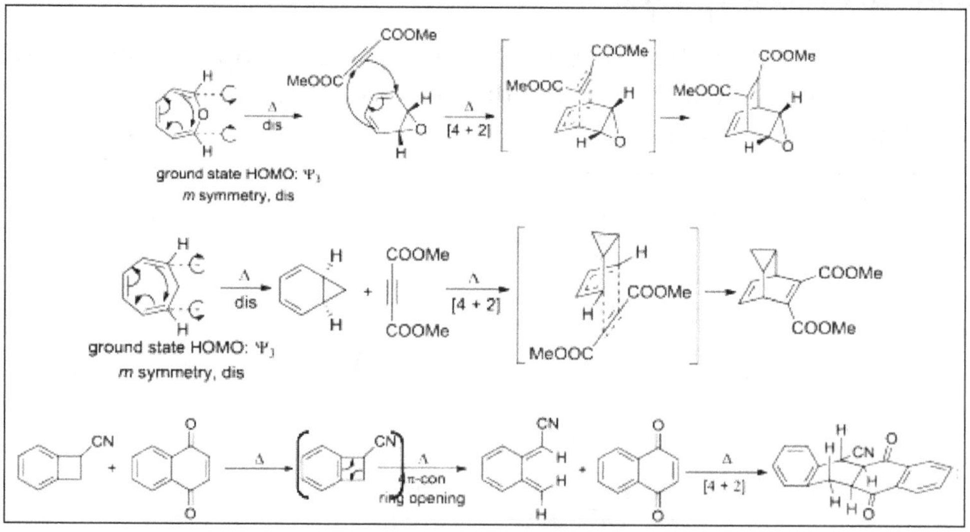

A transformation is shown using cycloaddition reaction as intermediate.

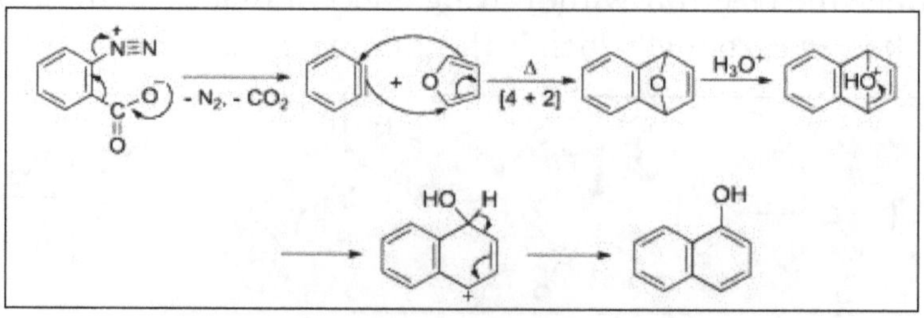

The Stereoselectivity of Diels-Alder Reactions

Three stereochemical issues related to product formation for cycloaddition:

- Cis/trans product formation w.r.t stereochemistry of starting diene and dienophile.
- Exo/endo products based on secondary HOMO-LUMO interactions.
- Regioselectivity measured by the relative size of terminal orbital lobes.

OUT-ENDO-CIS Rule

- OUT = Refers to the orientation of a group on the termini of diene.
- ENDO = Refers to the substituents (usually EWG) on the dienophile.
- CIS = Refers to stereochemical relationship of two groups in the product.

In order to determine product stereochemistry in D.A reaction Out-Endo-Cis rule is applied provided the interaction is supra-supra.

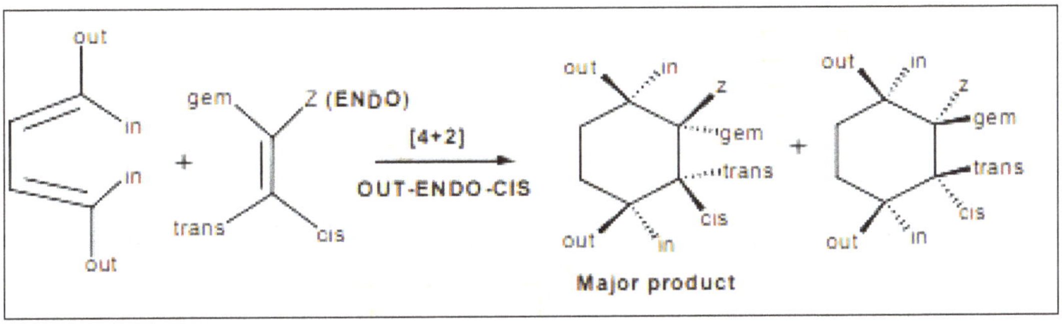

Therefore, out group on diene and endo group in dienophile are in cis in the product. Generally most electron withdrawing group in dienophile has been considered as endo group. Similarly OUT-EXO-TRANS, IN-ENDO-TRANS, INEXO-CIS are also can be applied.

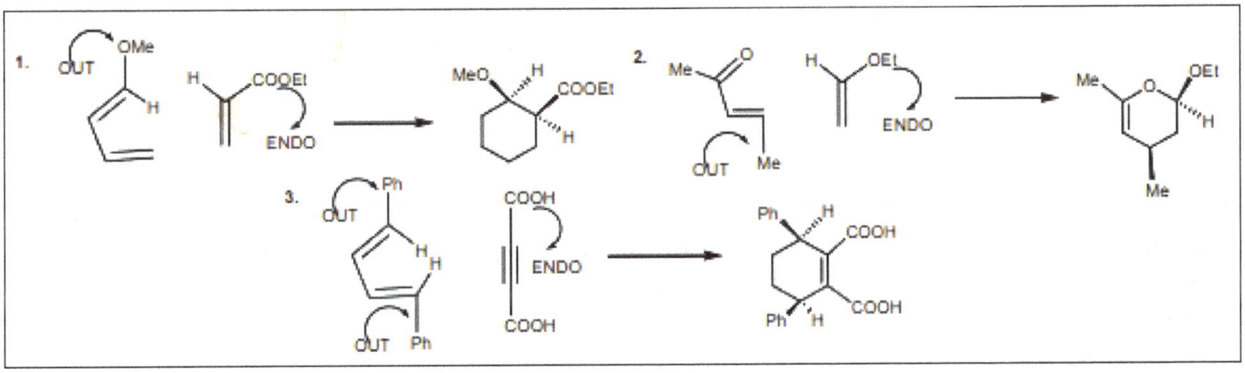

The stereochemical finding for Diels-Alder reactions relies on Alder's endo rule in which the favoured transition structure includes electron-withdrawing substituents in the more hindered environment under the diene unit. This gives kinetically favourable endo product at lower temperature with additional secondary orbital interactions. Long heating equilibrates this isomer with the thermodynamically favoured exo adduct by a retro-cycloaddition pathway.

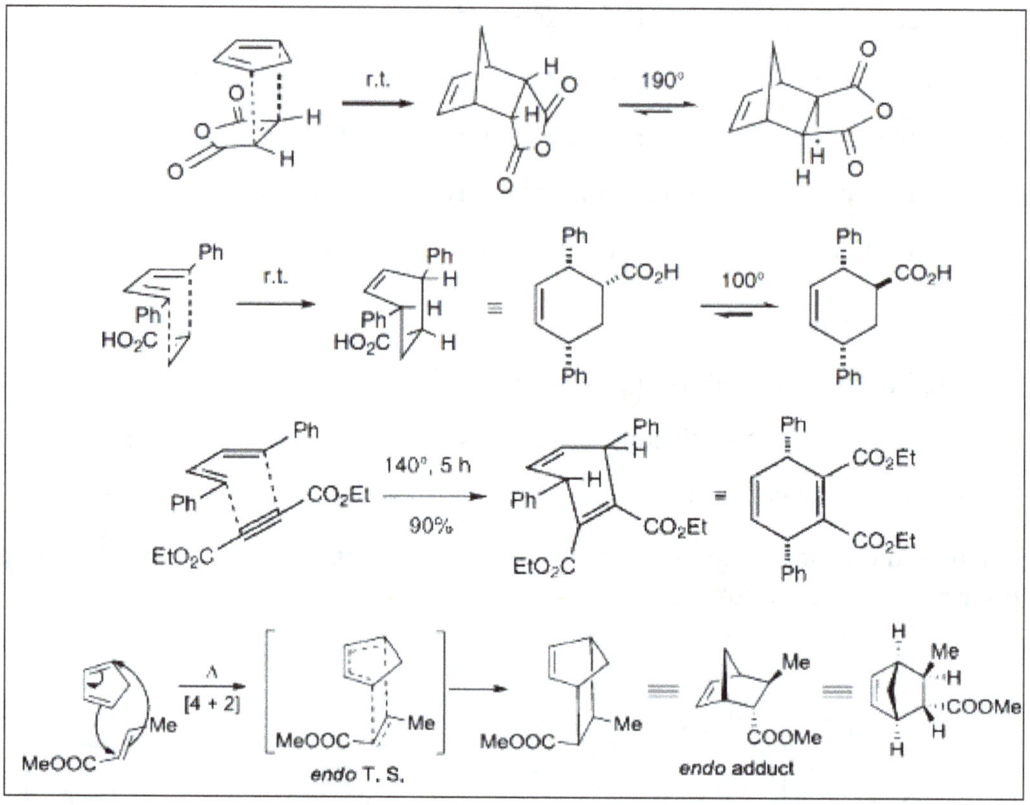

Secondary Interactions Determines the Regioselectivity of a Diels-Alder Reaction

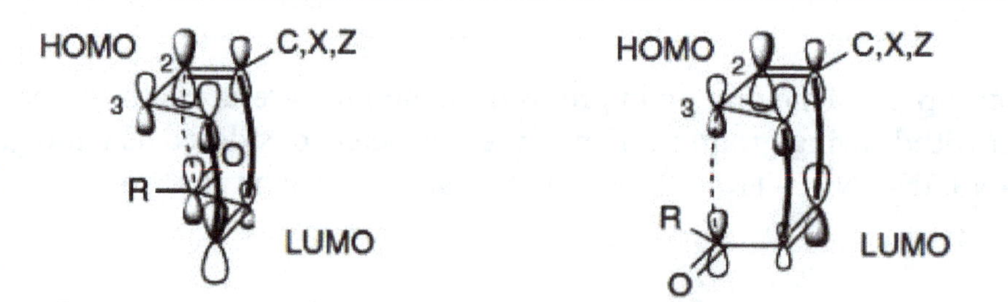

Transition structure for formation of an 'ortho' adduct

Transition structure for formation of a 'meta' adduct

The possibility for an electrostatic or Coulombic repulsive effect occurs in the exo transition structure. Such effect, in turn, has the greater propensity for forming the endo isomer. A Coulombic attraction favours exo attack both kinetically and thermodynamically in case of furan react with cyclopropenone.

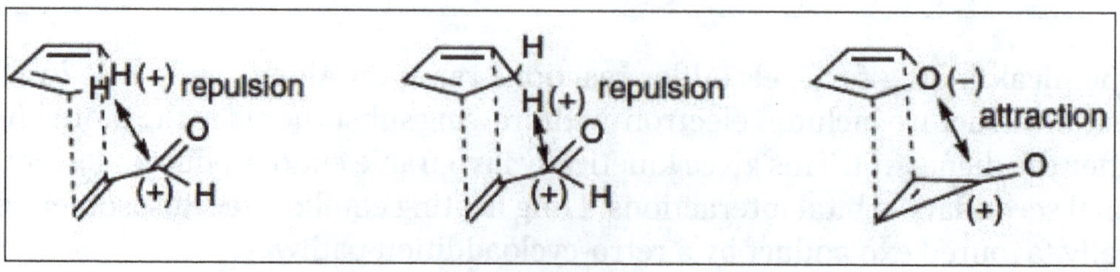

Energy profile diagram:

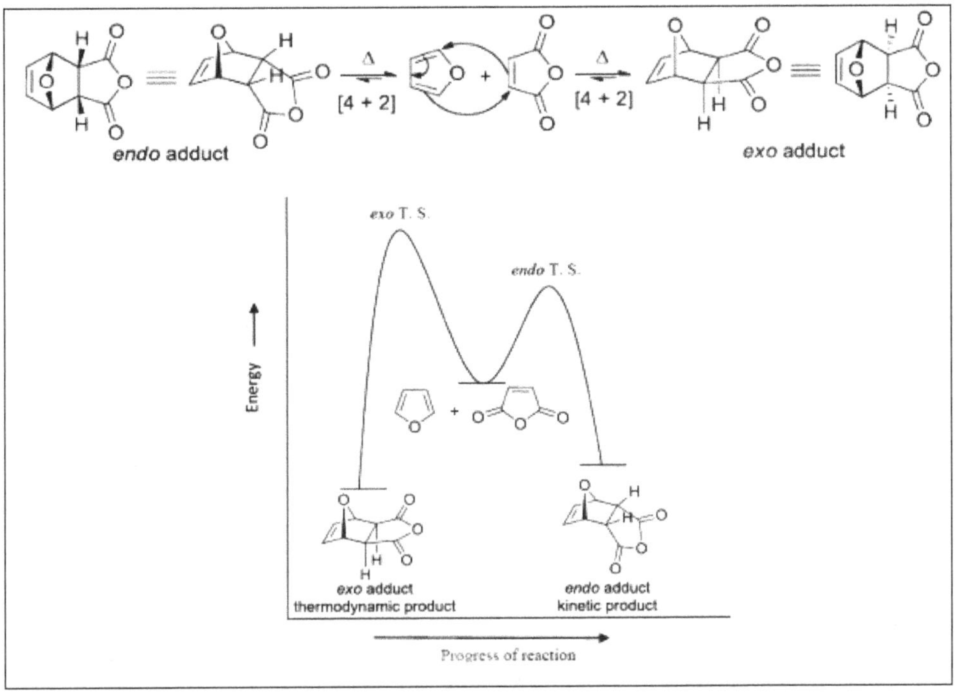

Effect of Lewis Acids on Diels-Alder Reactions

Lewis acids exert a strong catalysing effect in Diels-Alder reactions both as faster reaction rate and stereo/regioselectivity of the reactions.

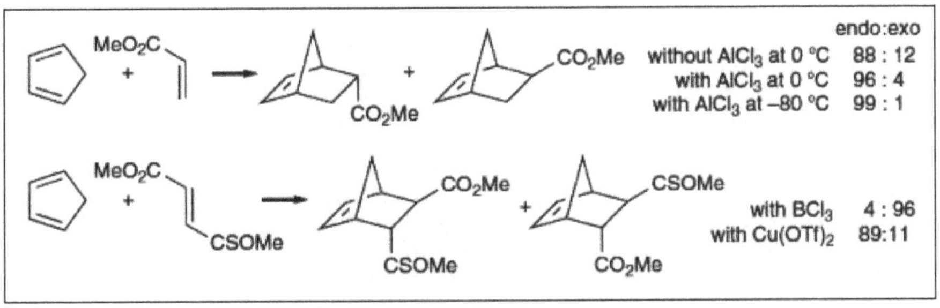

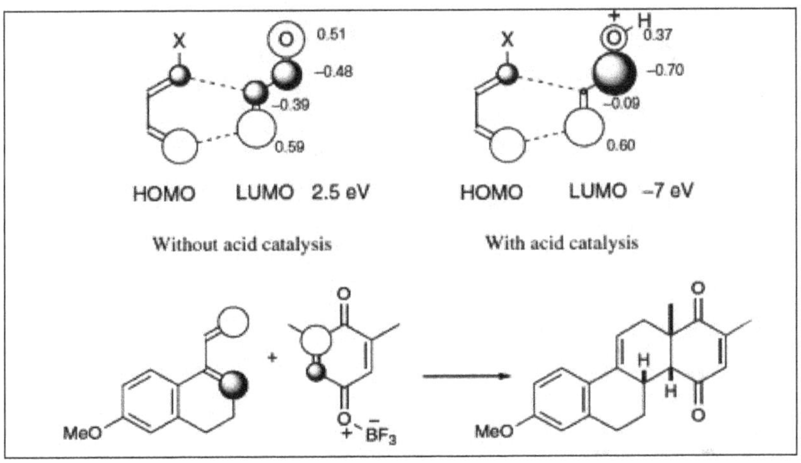

Site Selectivity of Diels-Alder Reactions

In cycloadditions, site selectivity is a pair of regioselectivity of a polyfunctional molecule having multiple sites in which the counterpart of another molecule reacts. The adduct obtained from anthracene arises due to product stability by achieving two isolated benzene ring and the largest coefficients in the HOMO at the 9,10-positions.

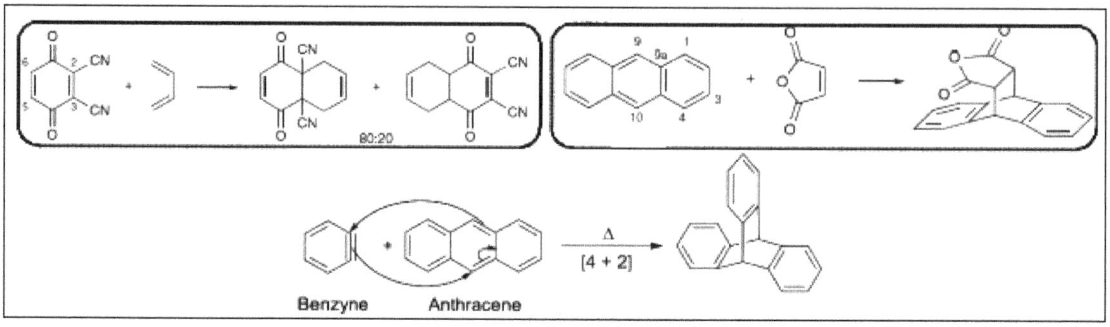

The dimerisation of hexatriene is another example of site selectivity. The former retains a conjugated diene system and is likely to be lower in energy than the latter. The former product is thermodynamically favoured.

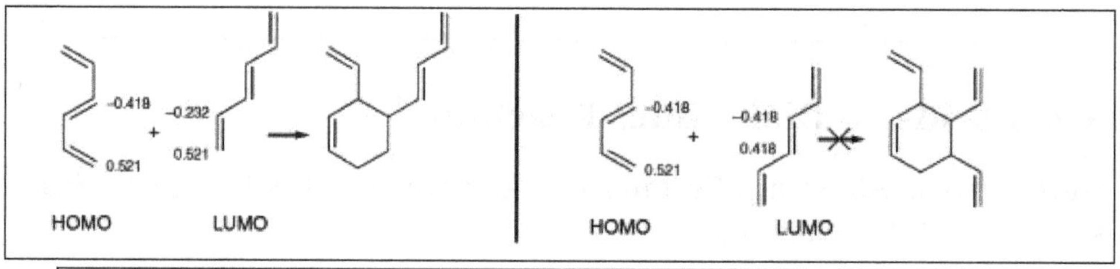

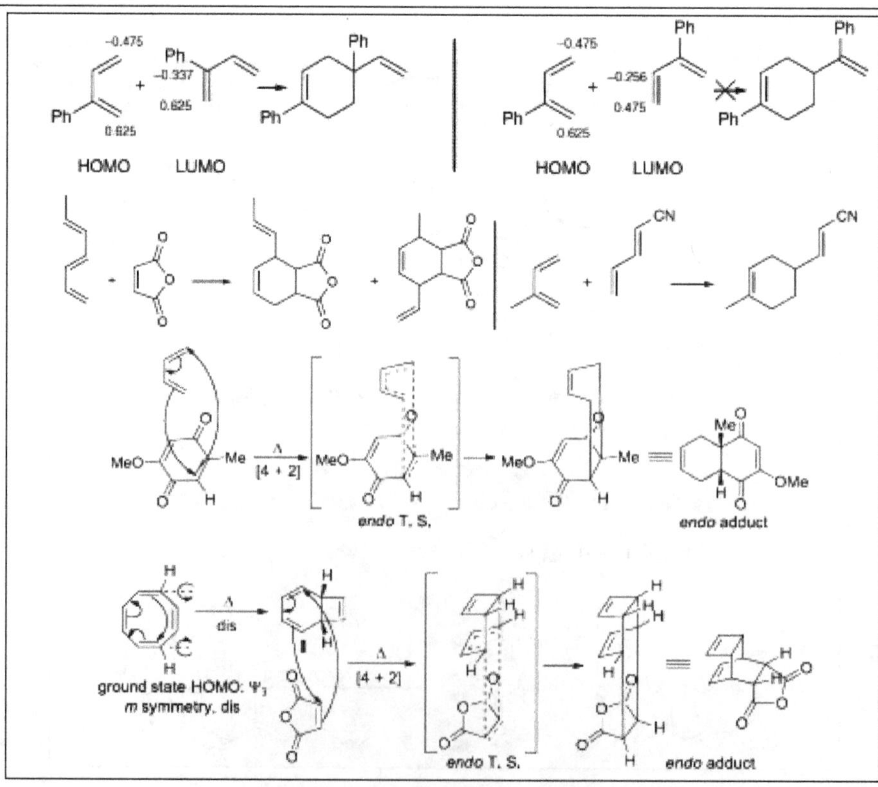

Among the four isomers (A-D), isomer (A) is more reactive for the highest HOMO energy and it remains in equilibrium in very low extent. The smallest energy is between the diene-HOMO of (C) and the dienophile-LUMO of (D). The reverse variation suffers from steric hindrance. The product has always endo selectivity.

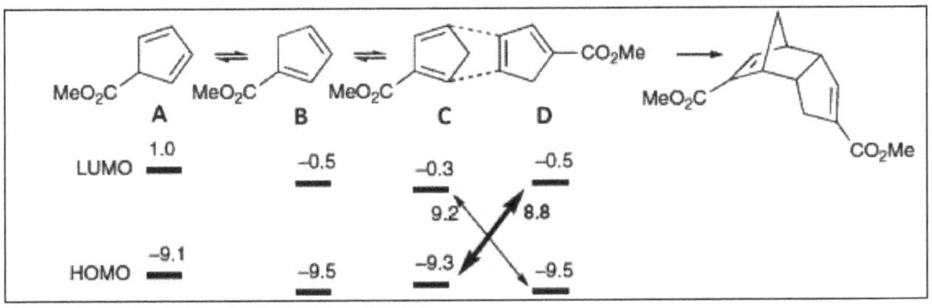

Periselectivity

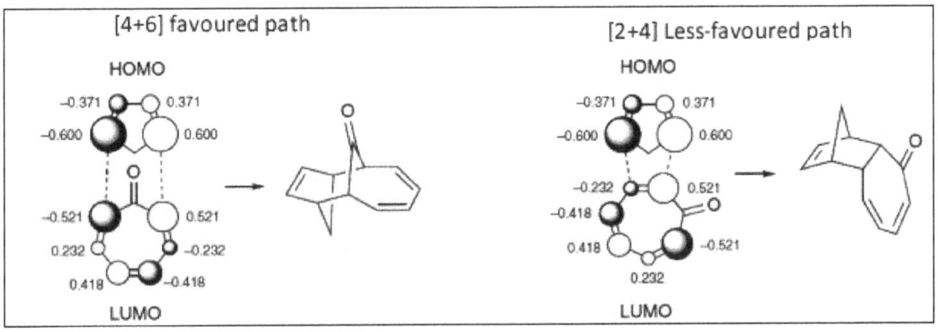

Periselectivity is a special case of site-selectivity which considers whether whole of the conjugated array of electrons or a large part of it, or only a small part of it is involved in pericyclic reaction. Periselectivity decides [4+6] cycloaddition adduct between cyclopentadiene and tropone prefers over [2+4] cycloaddition although both are thermodynamically much different in energy. Rather, the frontier orbitals indicate that the longer conjugated system of the tropone is more reactive than the shorter. The largest coefficients of the LUMO of tropone are at C-2 and C-7. It results the bonding to these sites with lower energy more than bonding to C-2 and C-3.

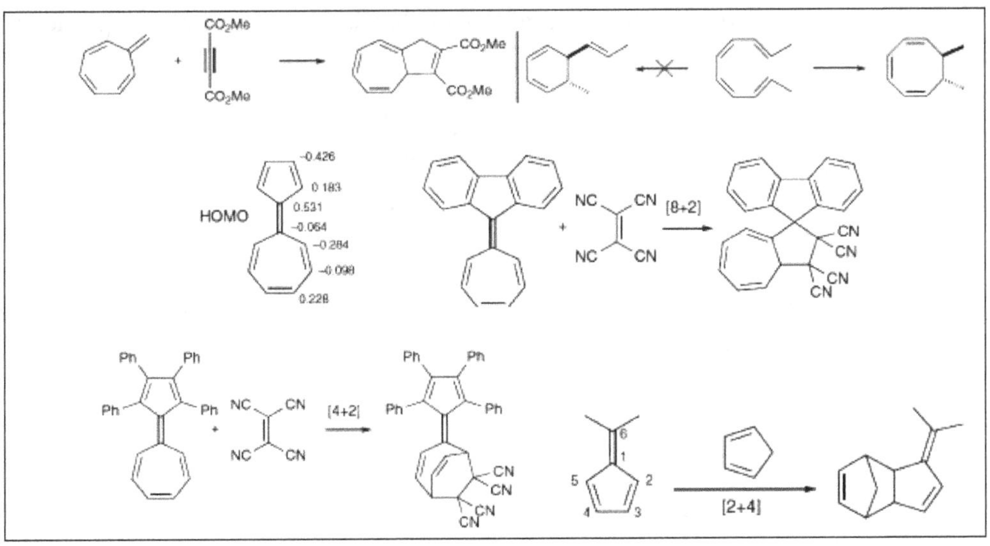

Rates, Regioselectivity and Site Selectivity of 1,3-Dipolar Cycloadditions

1,3-dipolar cycloaddtions is faster rate of concerted cycloaddition reaction between 1,3-dipole (like diazomethane, nitrone, ozone) with high-energy HOMOs and low energy–LUMO alkenes (dipolarophiles) carrying electron-withdrawing substituents. Frontier orbital interactions resemble this reaction with normal Diels-Alder reactions. The reaction of diazomethane or phenyl azide with methyl acrylate shows the regioselectivity.

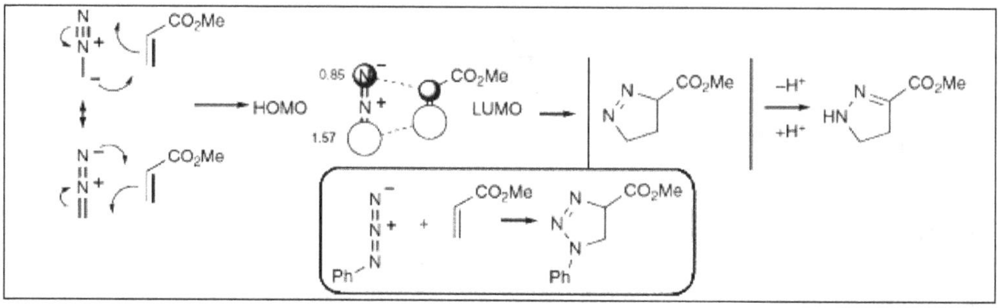

A smaller separation between the HOMO of the dipole and the LUMO of the dipolarophile is called dipole-HO controlled and between the LUMO of the dipole and the HOMO of the dipolarophile is called dipole-LU controlled. The regioselectivity of of 1,3-dipolar cycloadditions are shown below. The smallest energy separation (7.8 eV) is with X-substituted dipolarophiles implies a dipole-LU controlled reactions. Conversely, regioselectivity for phenyl azide reacting with a Z-substituted alkene is dipole-HO-controlled regiochemistry with $E_{LUMO} - E_{HOMO} = 9.5$ eV (compared to dipole-LU-controlled regiochemistry $E_{LUMO} - E_{HOMO} = 10.7$ eV).

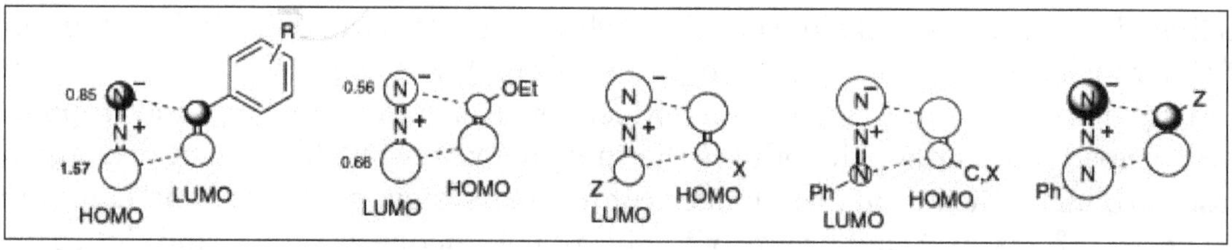

Overall ozonolysis reaction exhibits regioselectivity. The first step has no regiochemistry; the second step is a cycloreversion, but the third stage the cycloaddition of a ketone oxide to a carbonyl group, is highly regioselective.

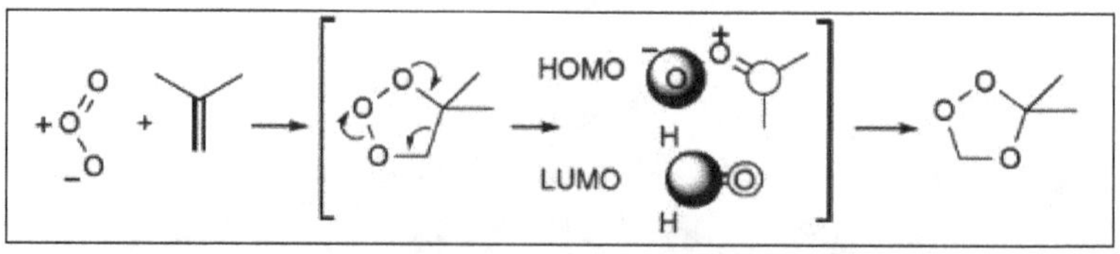

The site selectivity of 1,3-dipolar cycloadditions is analogous to the Diels-Alder reactions.

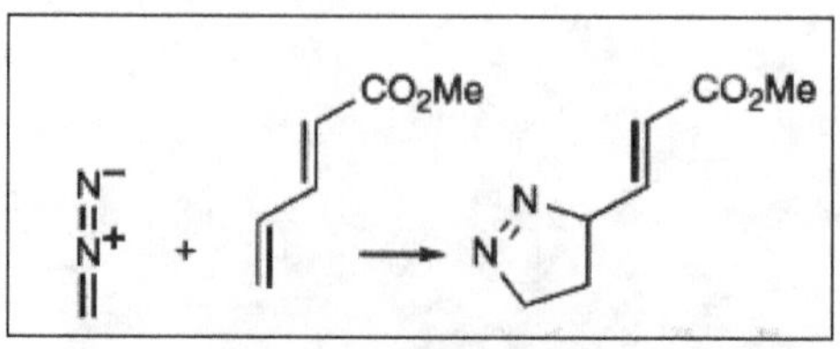

[4+2] Cycloaddition of Cations

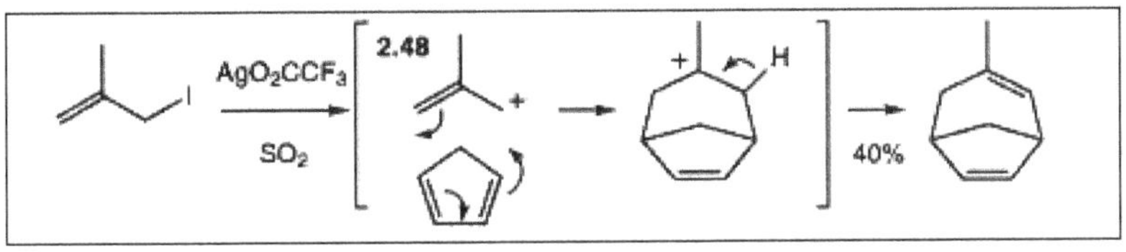

[4+6] Cycloaddition

The reaction between cyclopentadiene and tropone produces thermodynamically favoured exo adduct with favourable exo transition structure. Frontier orbitals indicate a repulsive interaction (wavy lines) between C-3, C-4 on the tropone and C-2 on the diene (and between C-5 and C-6 on the tropone and C-3 on the diene) in the endo transition structure.

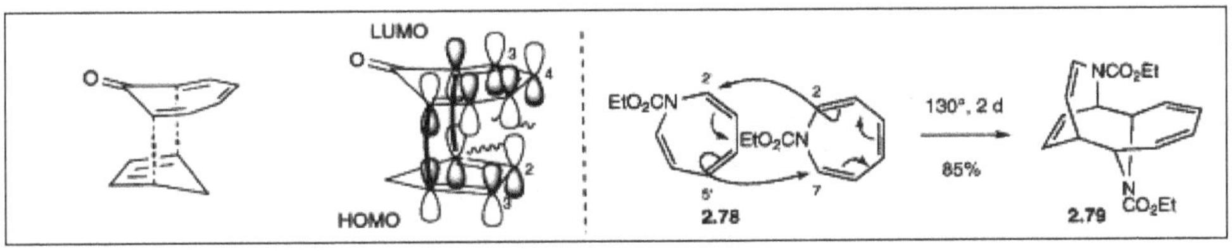

[8+2] Cycloaddition

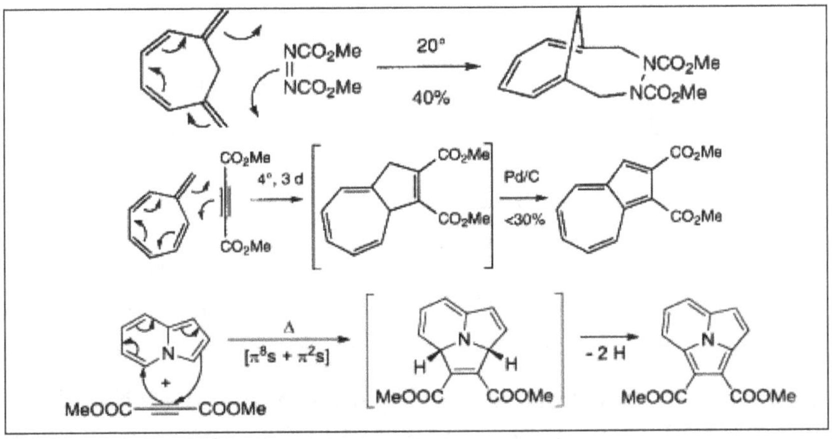

[14+2] Cycloaddition

[14+2] Cycloaddition of heptafulvalene (HOMO) with tetracyanoethylene (LUMO) involves an antiaromatic transition state. The reaction is termed as $[\pi_{14}{}^{a} + \pi_{2}{}^{s}]$ and is thermally allowed process. This is possible because heptafulvalene is flexible enough not to lose the conjugation through seven double bonds.

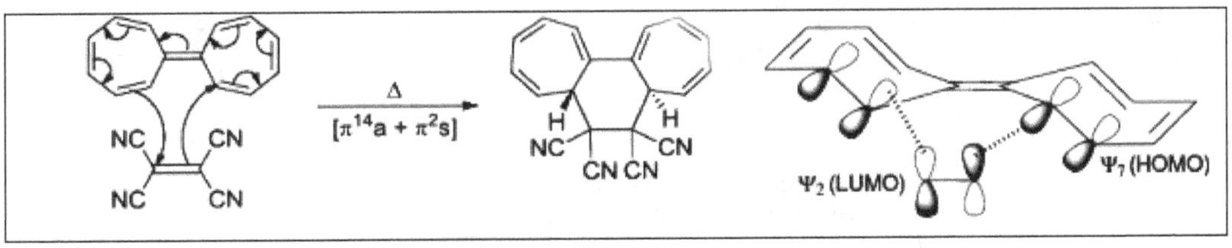

Cycloaddition with Multiple Components

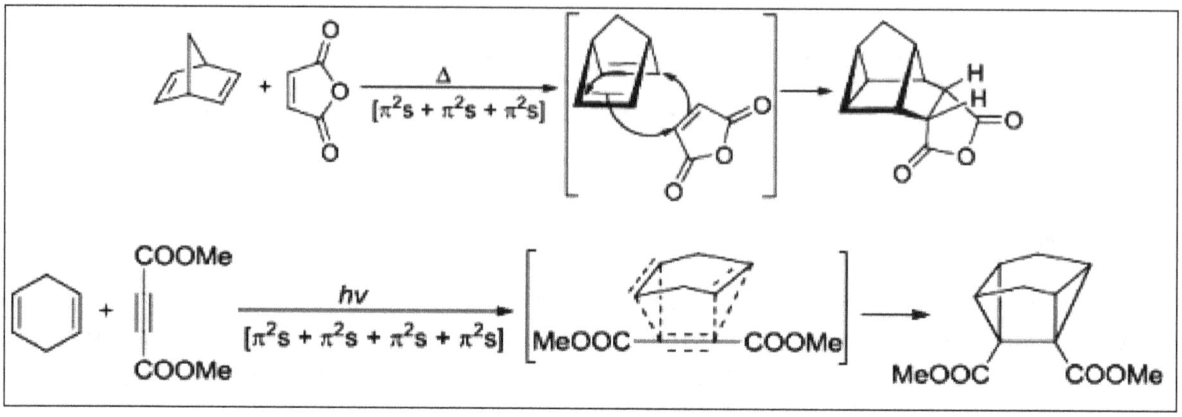

Ketene Cycloaddition

Ketenes undergo faster cycloadditions with double bonds to give cyclobutanones. It uses the interaction between the LUMO of the ketene and the HOMO of the ketenophile. The conjugation of the C—Cl bonds with the carbonyl group of the ketene will lower, by negative hyperconjugation, the energy of the LUMO, which is more or less π^*_{CO}. A C or X substituent on the alkene raises the energy of its HOMO, and the energy separation between the frontier orbitals is reduced.

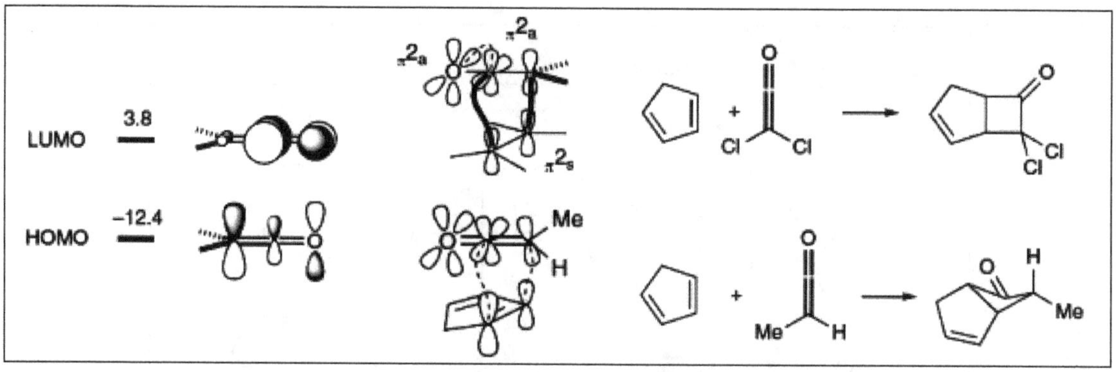

The regioselectivity in the reaction between cyclopentadiene and dichloroketene giving the cyclobutanone is explained by the overlap from the large LUMO coefficient on the central atom of the ketene and the larger coefficient in the HOMO of the diene at C-1.

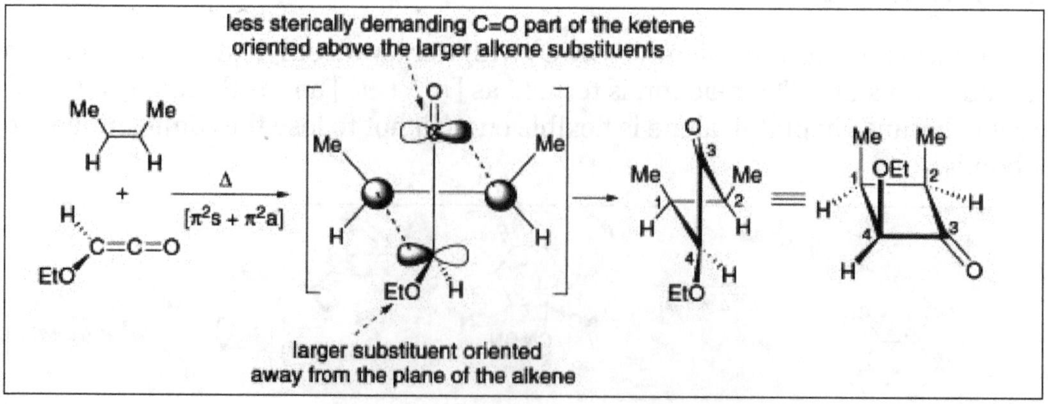

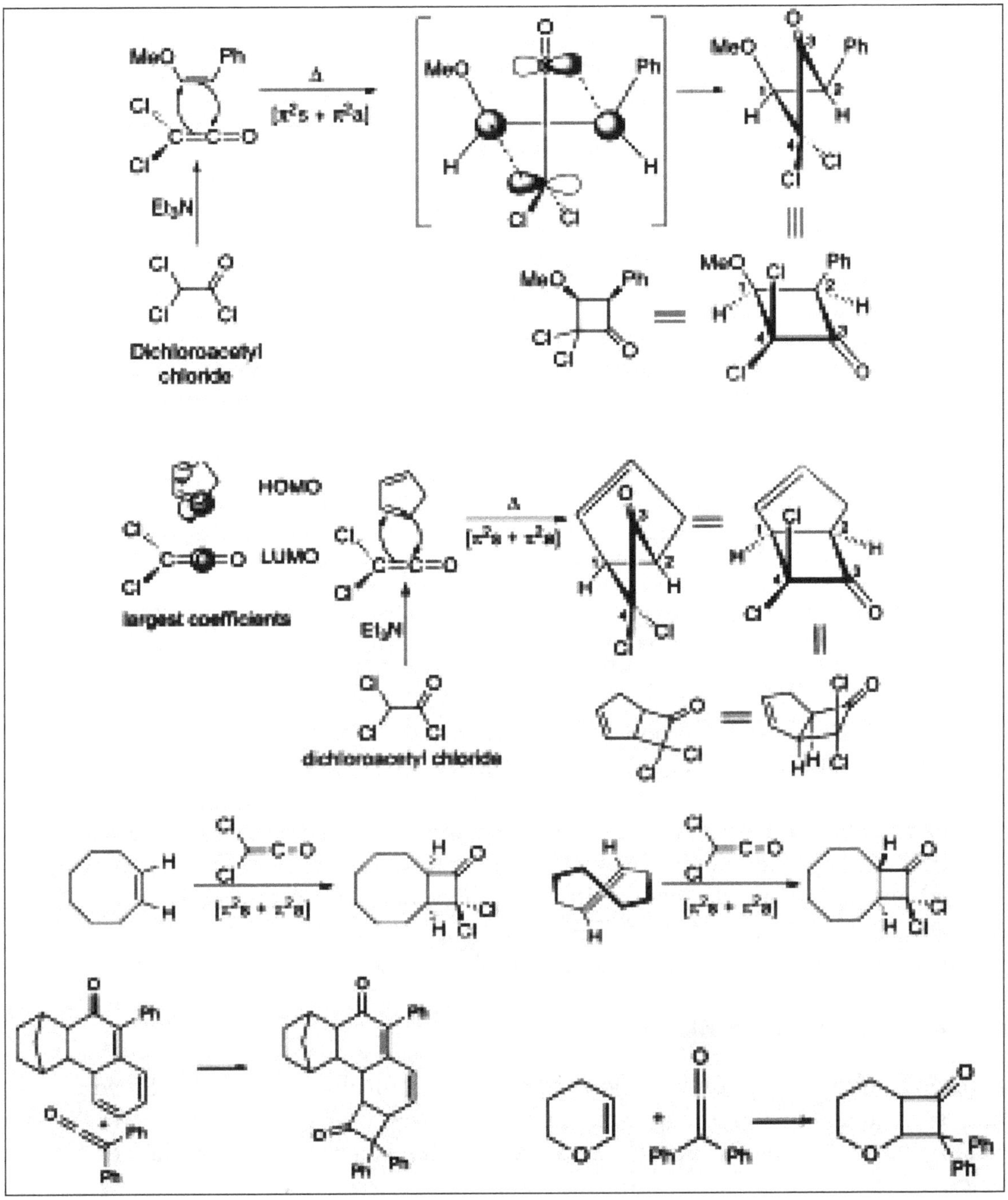

Both HOMO(ketone)/LUMO(ketene) and LUMO(ketone)/HOMO(ketene) interactions lead to the same conclusions about regioselectivity.

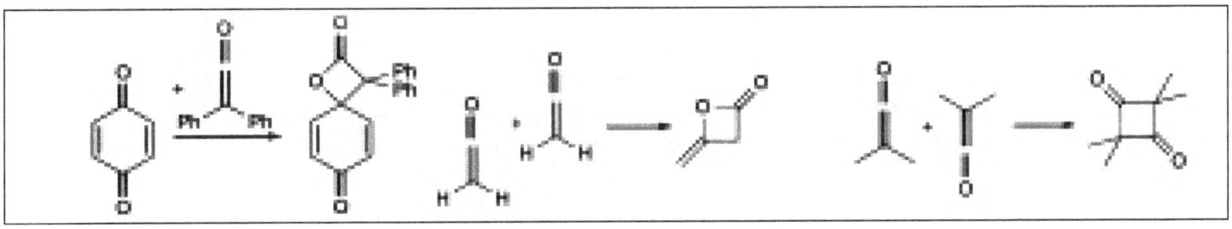

Retro Diels Alder Reaction

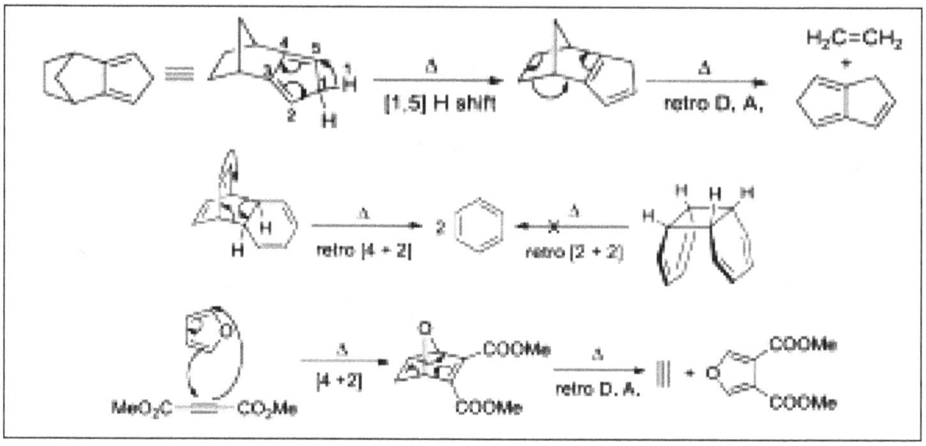

Forbidden Reactions

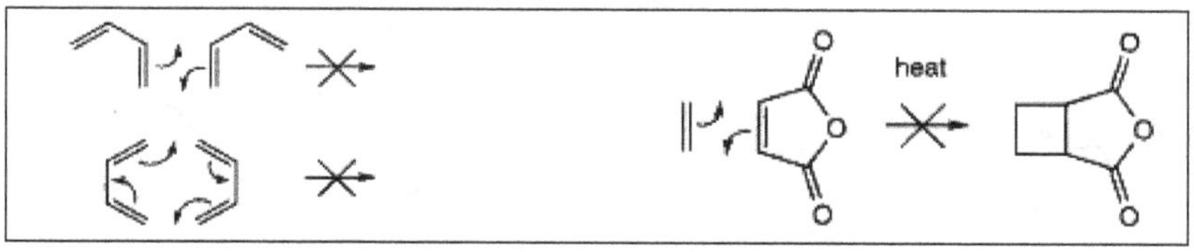

[4+2], [4+6] and [8+2] cycloaddition are commonplace. Burtadiene dimerizes following [4+2] cycloaddition reactions, not by [2+2] or [4+4] pathway. The simplest answer is 6-membered ring is easiest to form energetically than that of 4 or 8 membered ring due to ring strain or little probability to get two ends close. This argument does not apply to the reaction between ethylene and maleic anhydride. To get 4-membered ring with lower energy, a high kinetic barrier has to overcome. Thus, a total number of [4n+2] electron involved in cycloaddition reaction is thermally allowed; but not 4n number of electron system. Thus, the selection rule of orbital symmetry proposed by Woodward Hoffmann is justified. They are accomplished by photochemical means.

Photochemical Cycloadditions

Cycloaddtion by photoactivation considers the first electronically excited state. Since photoactivation puts so much energy into the molecule, many pathways including self and crossdimerization are available in the first electronically excited state. This is accentuated by the observation that photo induced Diels-alder reactions are very rare.

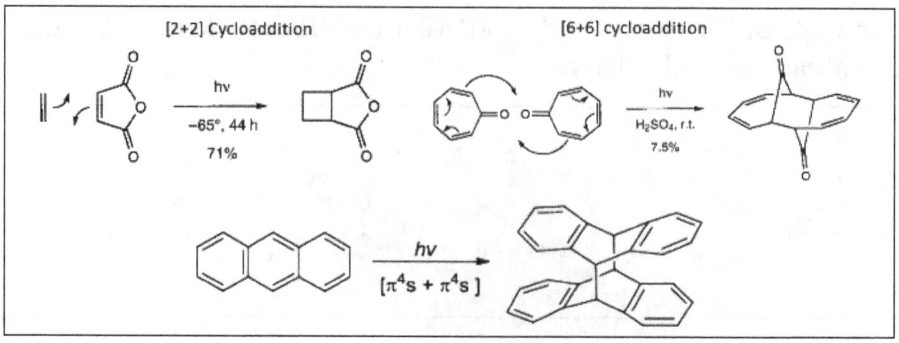

All Cycloadditions are not Pericyclic

The reaction of enamine with methyl vinyl ketone provides hetero Diels Alder adduct. However, the product is cycloadduct, the reaction is not included in pericyclic because of isolation of zwitterionic intermediate due to stepwise reaction. Even, the nucleophilic enamine reacts readily with either electrophilic methyl vinyl ketone or methyl acrylate gives different types of adducts.

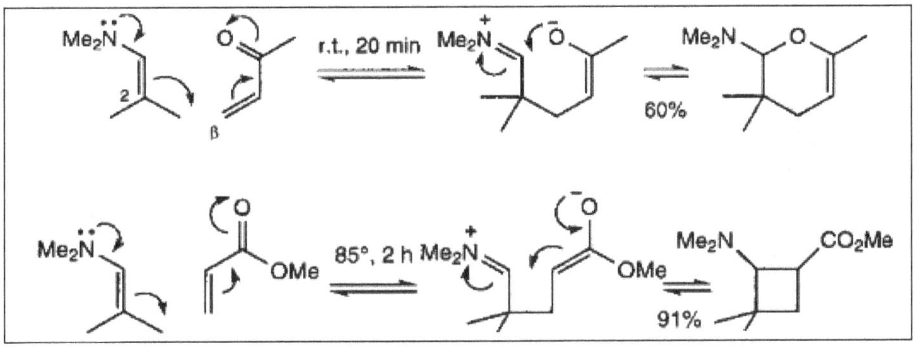

Cheletropic Reactions

The stereospecific and suprafacial insertion of carbine into double bond (alkene) is known as Cheletropic reaction.

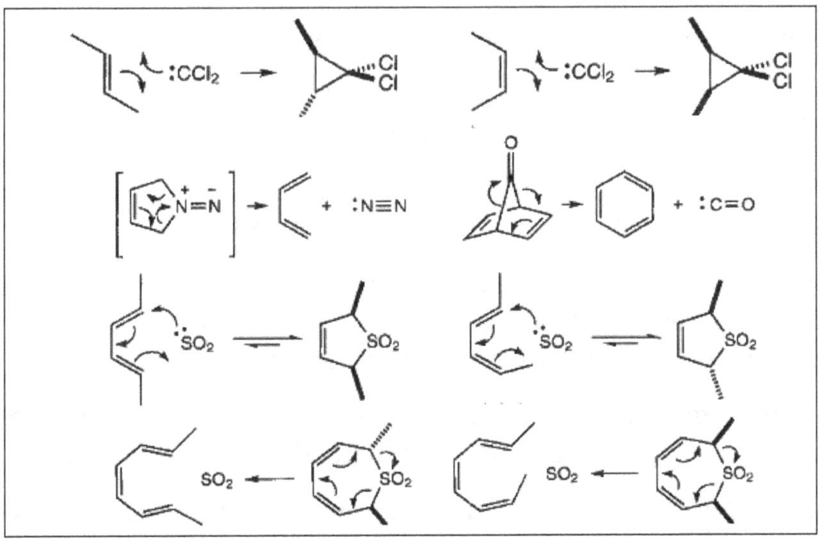

Intramolecular Cycloaddition Reaction

When diene and the dienophile are part of the same molecule, such type of molecule gives Diels-Alder reaction known as intramolecular Diels-Alder reaction.

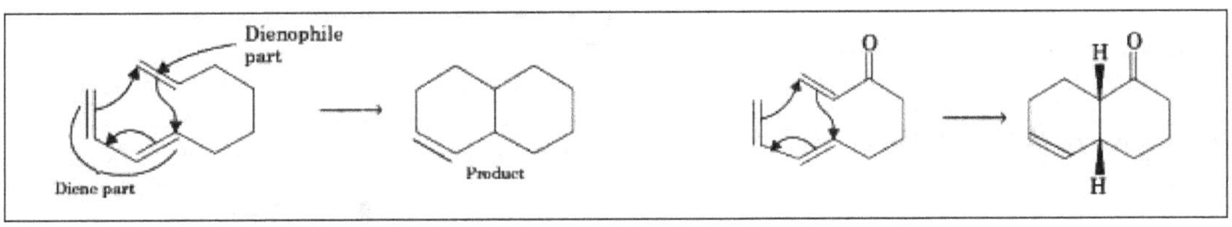

Sigmatropic Rearrangement

Sigmatropic rearrangement is defined as the intramolecular migration of an allylic σ-bond or atom or group or ligand along with an σ-bond from a migration origin to the terminus of a polyene system through reorganization of π-bonds. This process is concerted involving simultaneous breaking of one σ-bond with the formation of new σ-bond and rearrangements of π-bonds. The numbers of π-bonds remain unchanged.

Classification of Sigmatropic Rearrangement

The sigmatropic rearrangement can be represented by [i,j] shift where i and j are defined as the centre of migration origin and centre of migration terminus respectively. The difference of sigmatropic rearrangement from electrocyclic and cycloaddition reaction is that the later are classified by the number of π electrons involved in the cyclic transition state without migration of bonds.

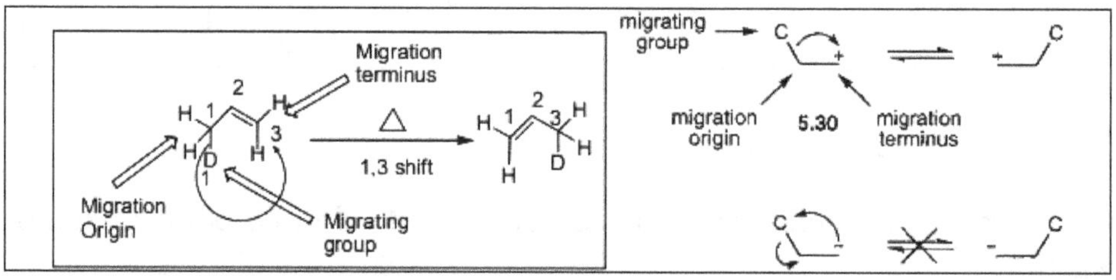

Basic Features and Conditions for Sigmatropic Rearrangement

- The symmetry of migrating orbital which migrates from migratory origin to migratory terminus must be in same phase and suprafacial.
- Suprafacial in phase interaction of lobes must take place by maintaining a reliable orientation and certain distance between migratory origin and terminus.
- Depending on the way of migration, stereochemical features would be resulted. If the migratory group is chiral, the configuration of the chiral centre will depend on the total number of electron involves during the process [4n or (4n+2)].
- HOMO of either migrating group and LUMO rest of the residue or vice-versa has to be considered in order to get symmetry allowed kinetically and thermodynamically more stable product.
- That sigmatropic rearrangement will be favoured which follows the Woodward –Hoffmann principle of orbital symmetry. The involvement of any steric crowding or generation of strain in the product molecule will avoid the subsequent mechanism.

Source/Influencing Factor/Mode of Activation

- Heat or Thermal (Δ).
- Light or Photochemical (hγ).

Criteria for Sigmatropic Reaction

- Mode of activation (Δ/hγ).

- Total number of electron participates. [4n or (4n+2)]π.
- Stereochemical outcome (stereospecific, stereoselective).
- The nature of orbitals interacting (σ and π).

Representative Examples of Various Types of Sigmatropic Rearrangement

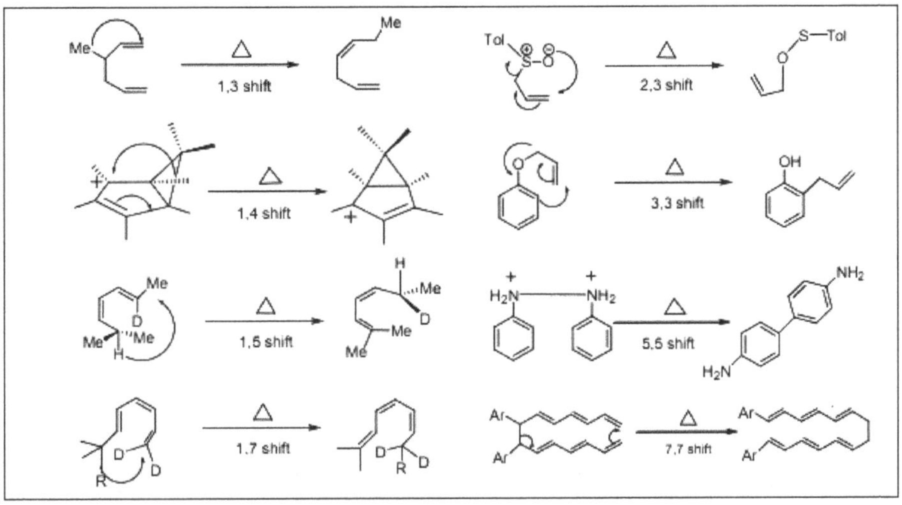

Case 1: [1,3] Sigmatropic Shift

A thermal [1,3] Sigmatropic shift is illustrated below:

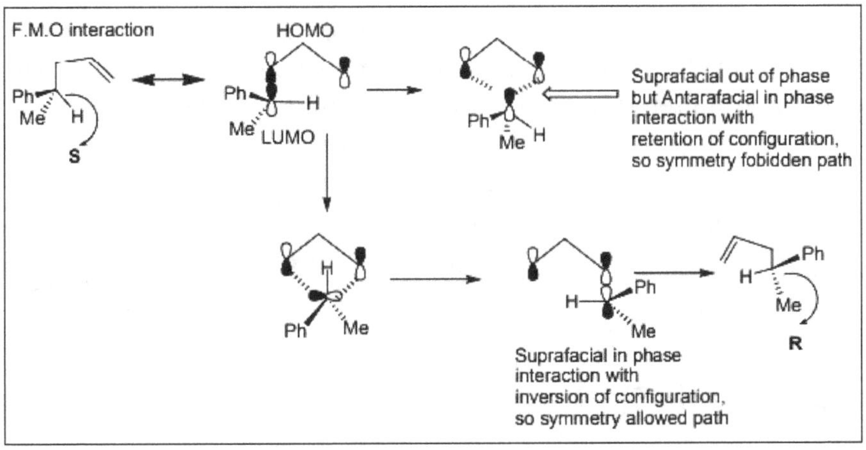

Orbital symmetry considers that the suprafacial path is thermally symmetry forbidden while antarafacial path might be followed; but it is geometrically impossible. Therefore, thermal [1,3] shift is very unlike. The process should be symmetry allowed under photochemical condition. However, the examples for 1,3 shift in photochemical process are rare. Because, when photon is absorbed by the molecule, the LUMO of the ground state will become HOMO of the excited state known as photochemical HOMO. Such situation is very unlikely.

1. A thermal ring expansion (left) with antarafacial [1,3] shift while a reversible suprafacial [1,3] shift (right) with retention of configuration in the migrating group are observed.

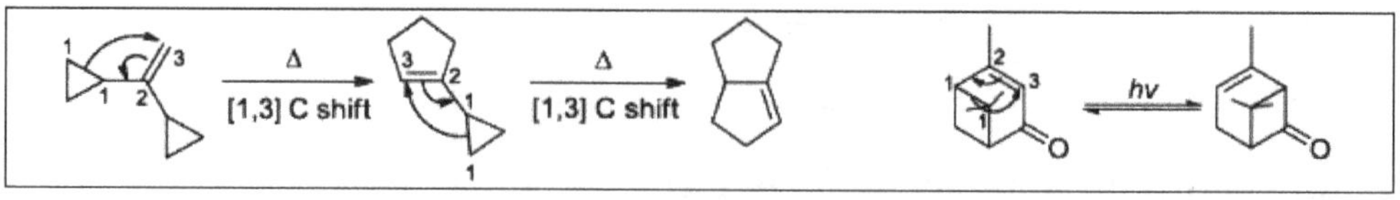

2. The reaction involves [1,3] sigmatropic shift under thermal conditions which is characterized by suprafacial with inversion at the migrating apical carbon.

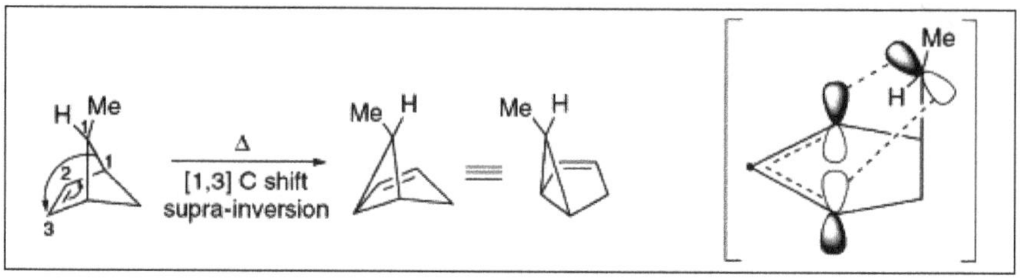

3. [1,3] suprafacial shift of an alkyl group proceeds with inversion at the migrating center at high temperature (300 °C).

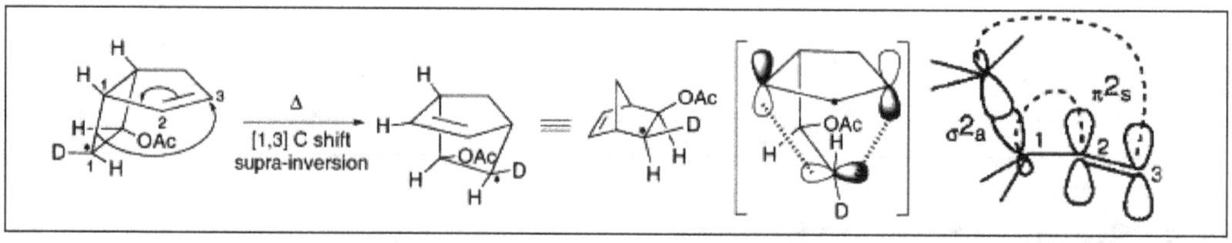

[1,4] Sigmatropic Shift

[1,4] Sigmatropic shift involves allylic cation. The example is the degenerate rearrangement of bicyclic cation in which the bridge moves around perimeter of the five membered ring. The endo and exo —Me-groups remain unaltered from spectroscopic point of view.

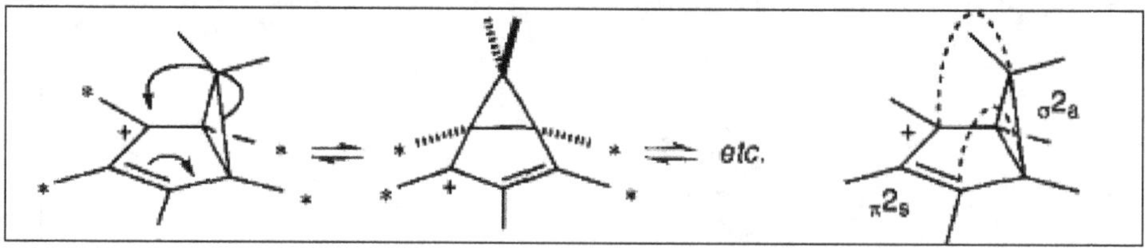

Case 2: [1,5] Sigmatropic Shift

A thermal [1,5] Sigmatropic shift is illustrated below:

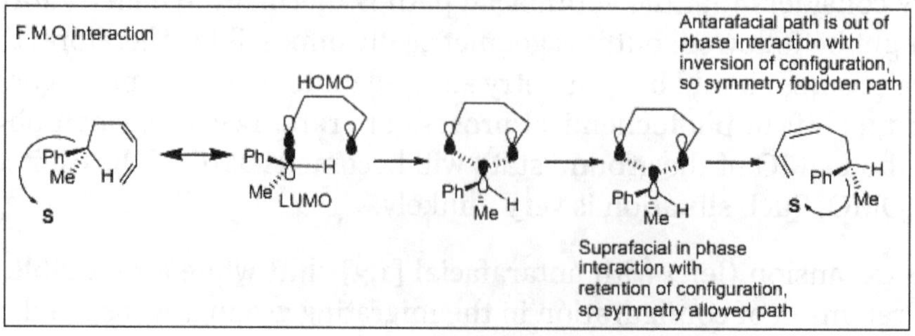

Among all possibilities of sigmatropic migration, always 1,5 shift occurs due both symmetrically and geometrically favourable migration. H's and alkyl groups migration favours suprafacial [1,5] shift. The [1,5] sigmatropic shift results symmetry allowed suprafacial migration with retention of configuration.

Stereochemistry of Migration

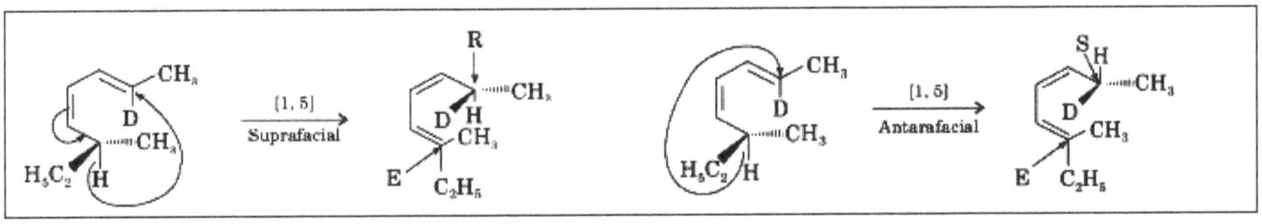

Suprafacial migration renders R-configuration at C-5 and E-configuration at C-1 (left) whereas antarafacial migration yields S-configuration at C-5 and Econfiguration at C-2 (right).

1. The structures of the dienes present in the reaction mixture via sequential thermal [1,5] suprafacial sigmatropic H and D—shifts.

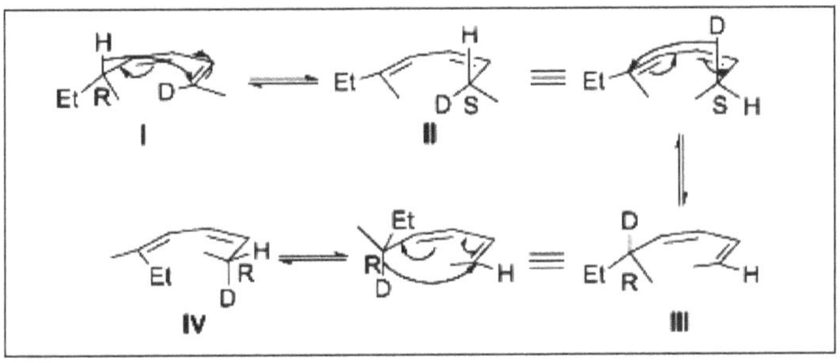

2. The isomers of monosubstituted cycloheptatriene are interconvertible following successive suprafacial [1,5] 'H shifts. This reaction involving [1,5] 'H shifts is reasonably slower than that of cyclopentadienes. In contrary, photochemically [1,5] 'H shifts occurs only in open chain system where the migration can be antarafacial.

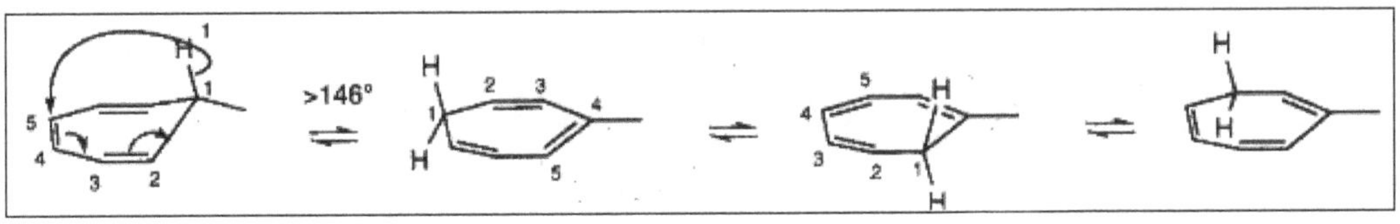

3. Isomeric methylcycloheptadienes (1—4) are in a thermal equilibration as a result of successive [1,5] 'H shifts.

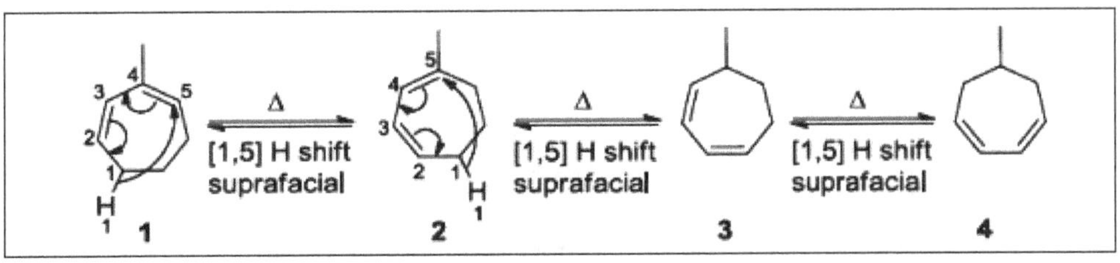

4. Both thermally allowed [1,5] H and D—shifts shows the scrambling of deuterium label of 1-deuteroindene.

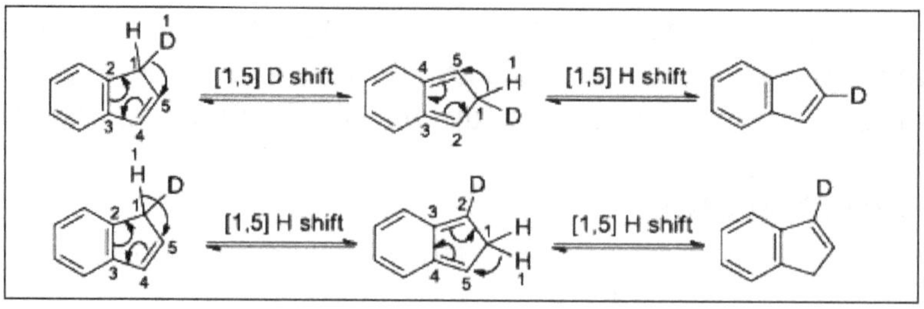

5. The deuterium labeling is found consistent with thermal suprafacial [1,5] deuterium shift for optically active 1-deuterio-1-methyl-3-tert-butylindene. Three products were obtained as optical active.

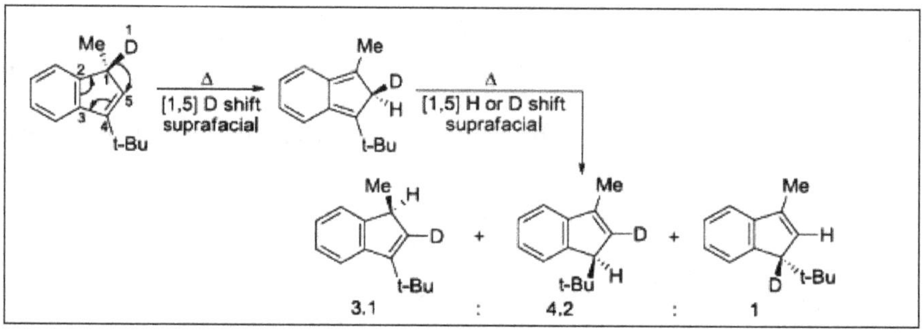

6. The given molecule represents fluxtionality. The fluxionality of the σ (sigma) bonded atom in cyclopentadienide involves migration.

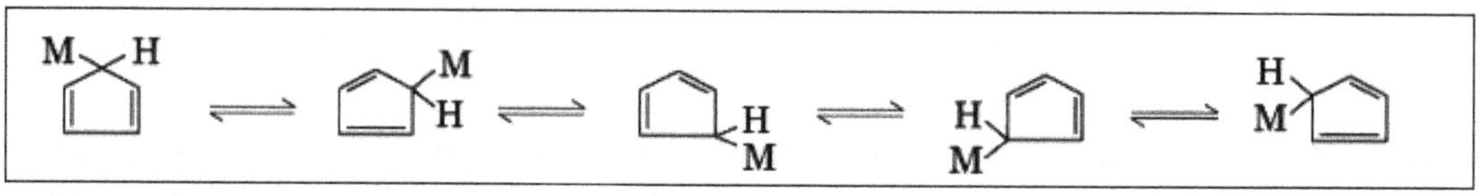

7. Both reactions involve a suprafacial [1,5] deuterium shift under thermal condition.

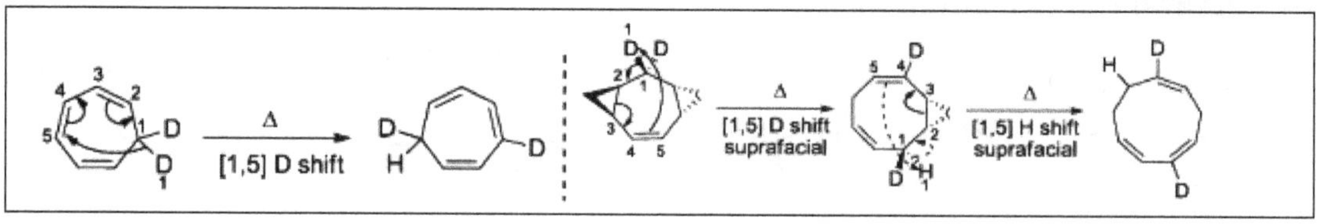

8. A relatively less stable diene with exocyclic double bonds is switched to a stable diene containing endocyclic double bond by [1,5] H-shift.

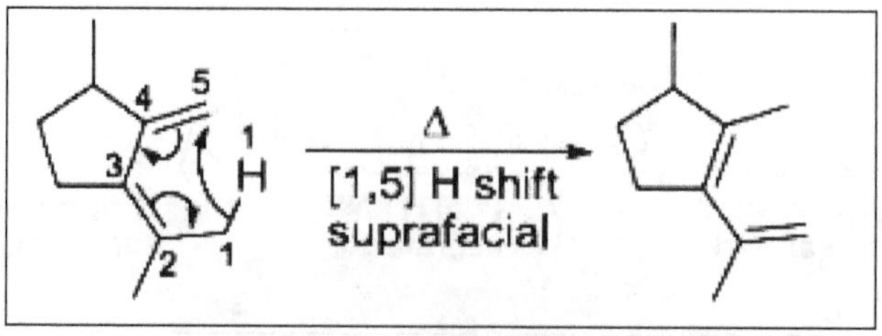

9. The isomers of 1,4-di(cycloheptatrienyl)benzene are interconverted by thermal [1,5] and photochemical [1,7] suprafacial sigmatropic hydrogen shifts.

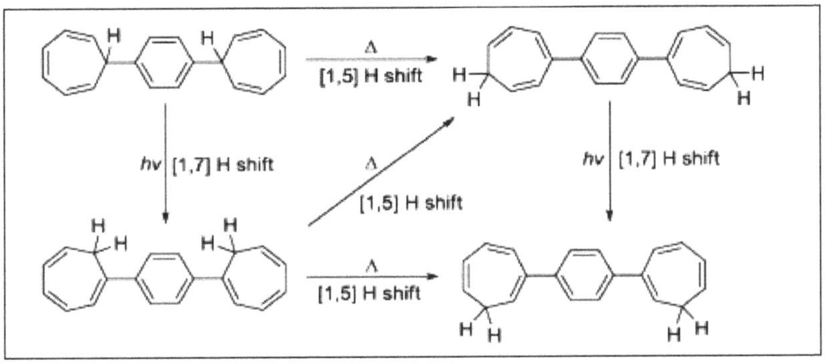

10. Only suprafacial [1,5] hydrogen shift is thermally allowed for the conversion into toluene.

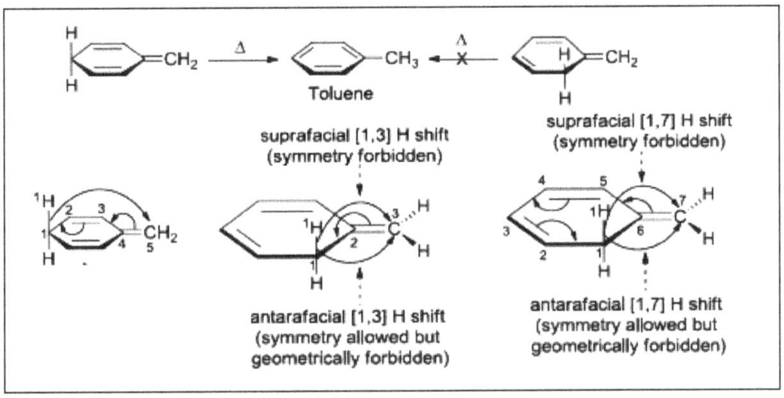

11. Suprafacial [1,5] H shift of Z-cyclodeca1,2,4 triene followed by an 6π electrocyclic ring closure in disrotatory manner yield bicycle[4,4,0] deca2,4-diene.

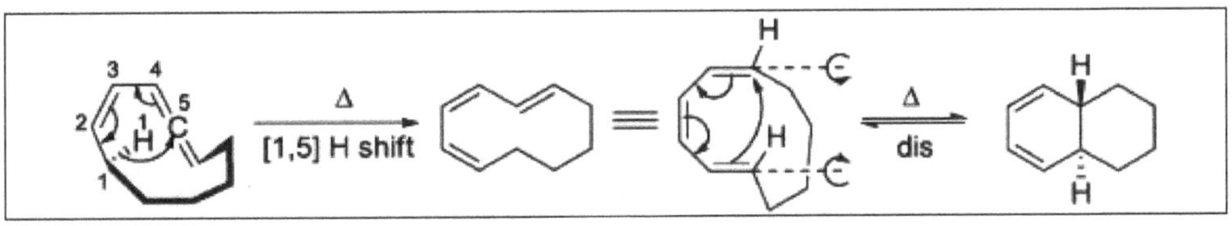

12. S-(+)- α-phellandrene loses its optical activity on distillation owing to spontaneous and geometrically favourable [1,5] H shift of chiral hydrogen within the molecule.

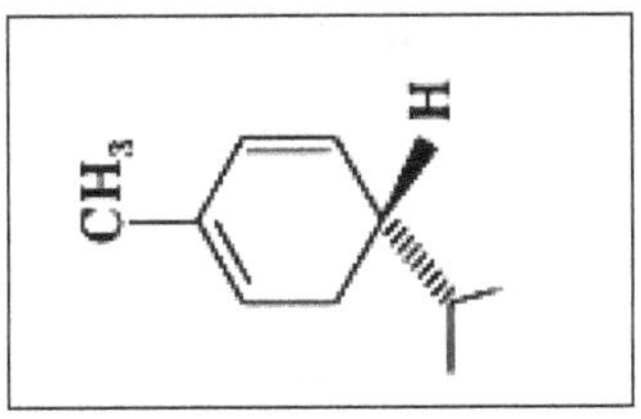

13. An 6π electrocyclic disrotatory ring closure followed by suprafacial [1,5] H shift results 7-fused-4 to stable 6-fused-5 membered ring.

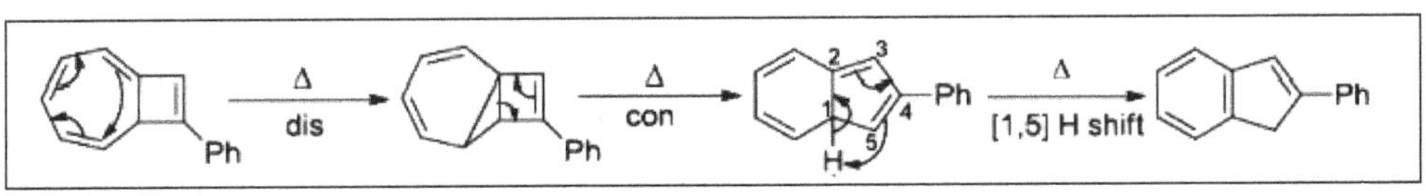

14. Three successive thermal suprafacial [1,5] sigmatropic shifts including one σ—bond and two hydrogen occurs with retention of configuration.

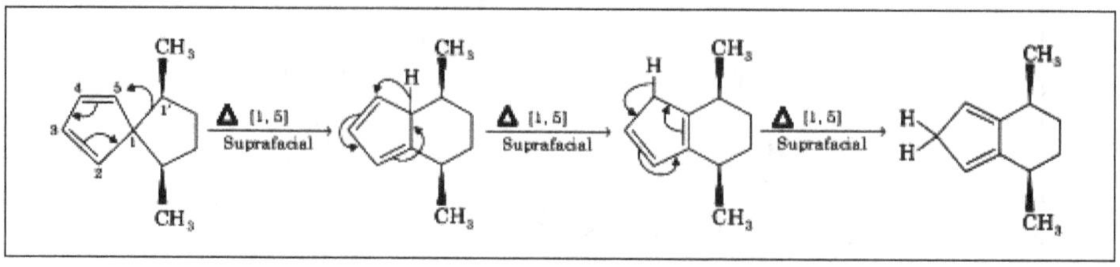

15. The given compound shows only a singlet for the methyl group in 1H NMR. Because the [1, 5] migration of hydrogen takes place in the preference to the methyl group and the migration is so fast at room temperature that only a singlet is observed for the methyl group.

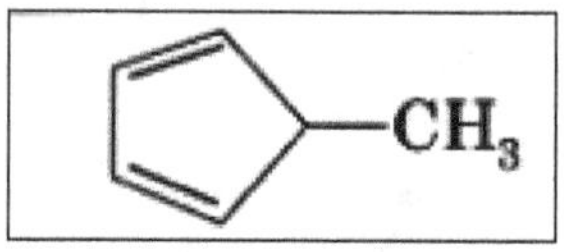

16. Walk rearrangement of norcaradiene with [1,5] shifts.

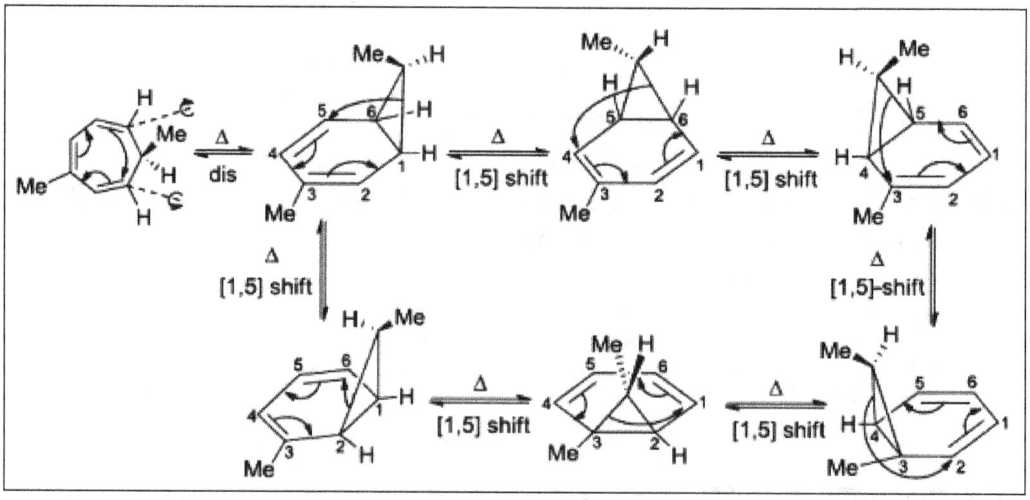

17. Isomerization of substituted cycloheptatrienes takes place through sequences of [1,5] sigmatropic rearrangements and electrocyclic reactions.

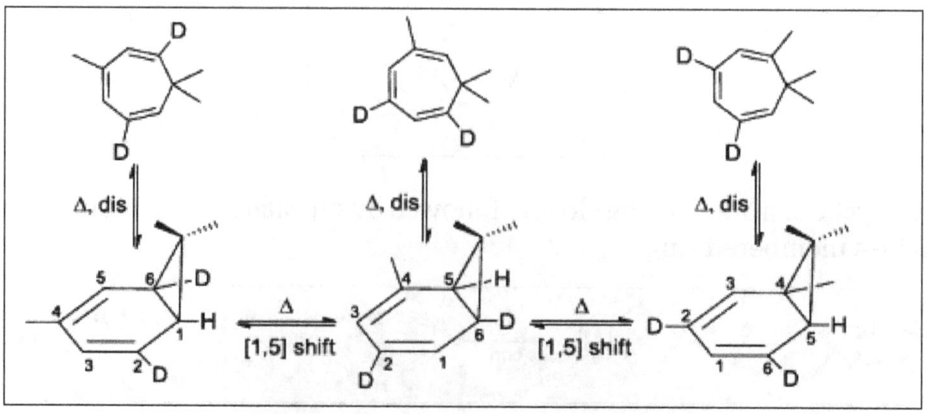

Case 3: [1,7] Sigmatropic Shift

A thermal [1,7] Sigmatropic shift is illustrated below:

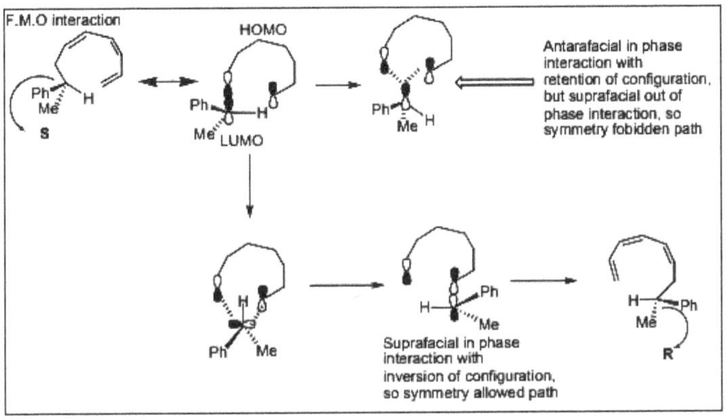

Thermally [1,7] shift although is suprafacial symmetry forbidden process, it occurs due to achievement of favourable geometry.

1. Unlike [1,3] shift, antarafacial migration is allowed for [1,7] hydrogen shift since the geometrical restrictions on antarafacial T.S of [1,7] hydrogen shift is not as severe as in the [1,3] shift case due to flexibility of a π-system of seven carbon atoms.

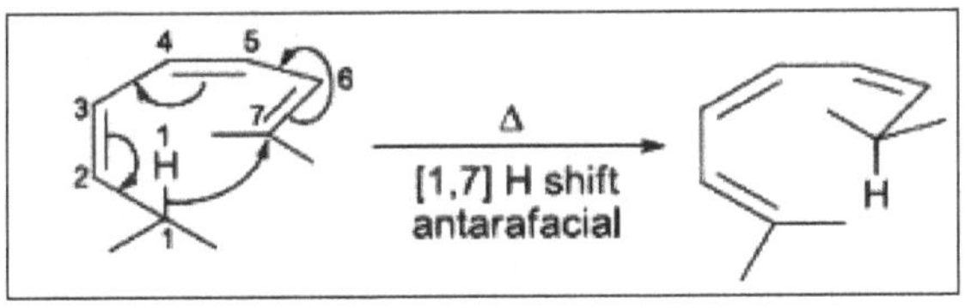

2. The thermal equilibrium between precalciferol (pre-vitamin D2) and calciferol (vitamin D_2) is illustrated by [1,7] Hydrogen shift.

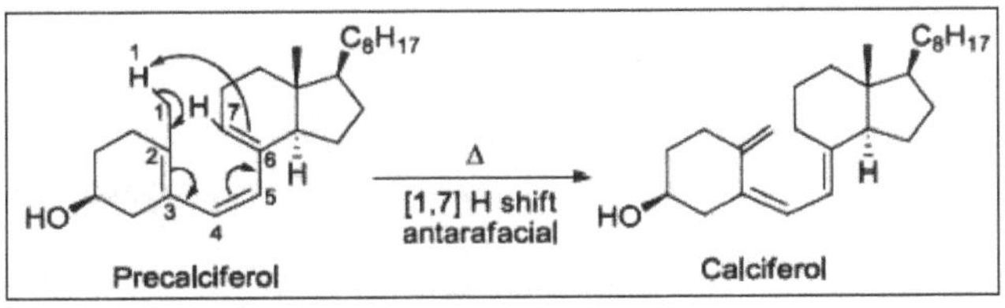

[2,3] Sigmatropic Shift: Wittig Rearrangement [$\sigma^2_s+\omega^2_s+\pi^2_s$]

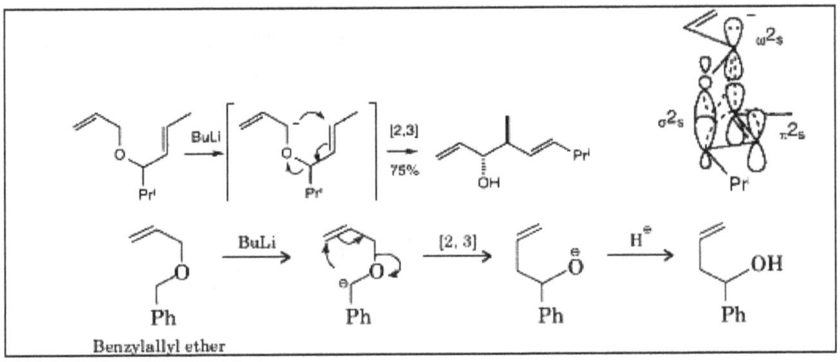

[3,3] Sigmatropic shift

Cope Rearrangements

A [3,3] sigmatropic rearrangement of 1,5-dienes under thermal condition is called the Cope rearrangement. The reaction proceeds through a 6-membered cyclic T.S. The Cope rearrangement is a reversible process; the equilibrium mixture has a great proportion of the thermodynamically stable isomer. It is an intramolecular process because rearrangement of a mixture of two different hexa-1,5-dienes does not lead to the formation of cross products.

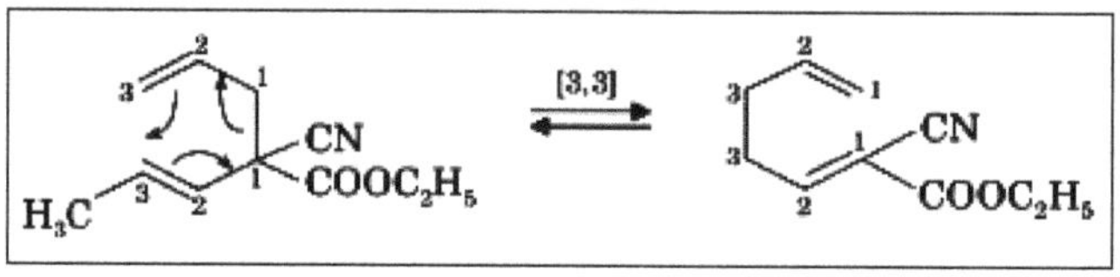

1. In [3,3] sigmatropic shifts, a chair-like transition state is energetically preferred to a boat-like structure. Rearrangement of the meso diene through such a transition state then would give the cisetrans isomer while in the case of the rearrangement of the racemic mixture the trans-trans isomer is the major product.

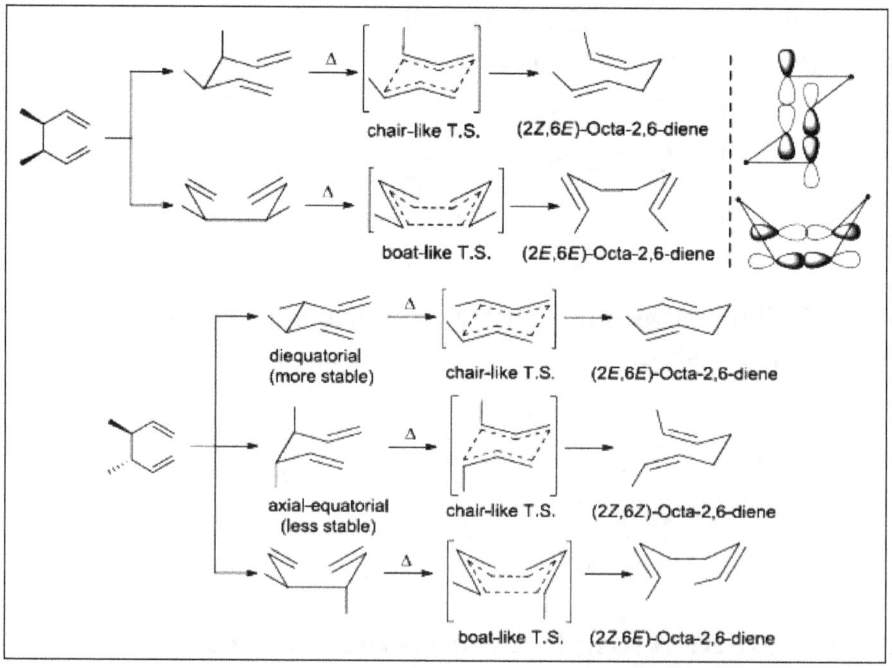

2. The following examples represent [3,3] shift mostly via a suitable chair conformation.

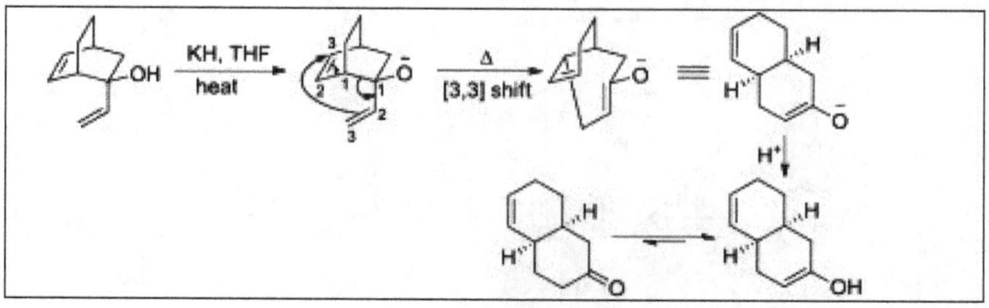

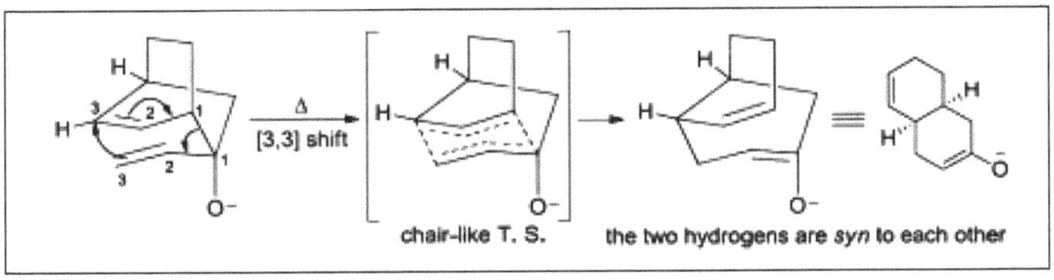

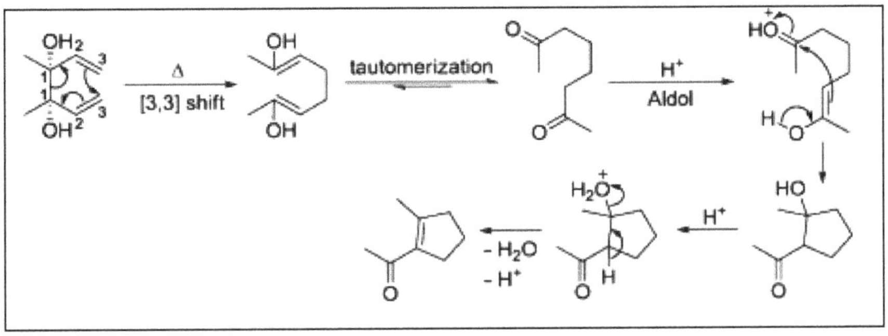

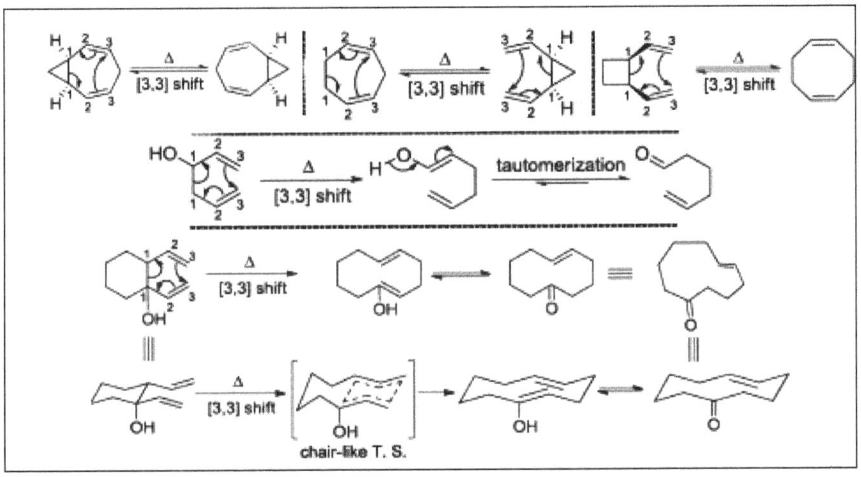

3. Diyne system can be thermally activated to suprafacial [3,3] sigmatropic shift via reactive allene intermediate which allows thermal electrocyclic ring closing reaction through conrotatory mode to furnish conjugated exocyclic double bond.

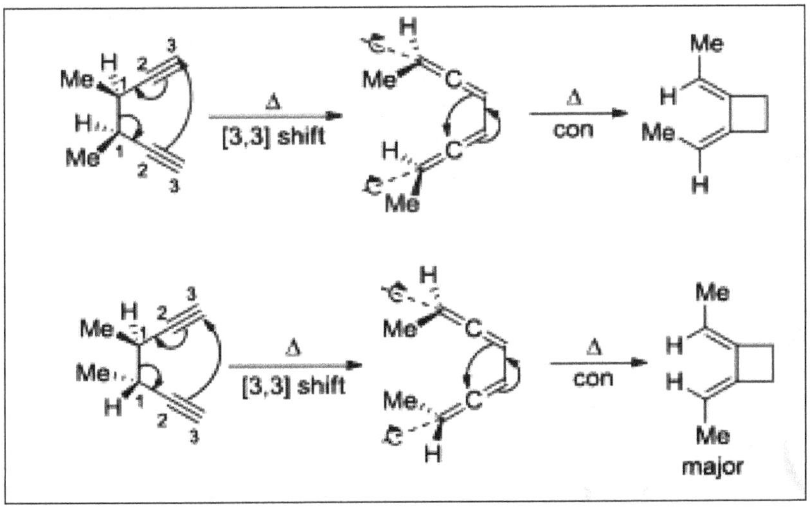

Claisen Rearrangement

The Claisen rearrangement is an typical example of [3,3] sigmatropic rearrangement; for instance, the conversion of allyl vinyl ether to pent-4-enal or an allyl phenyl ether to an ortho-substituted phenol on heating.

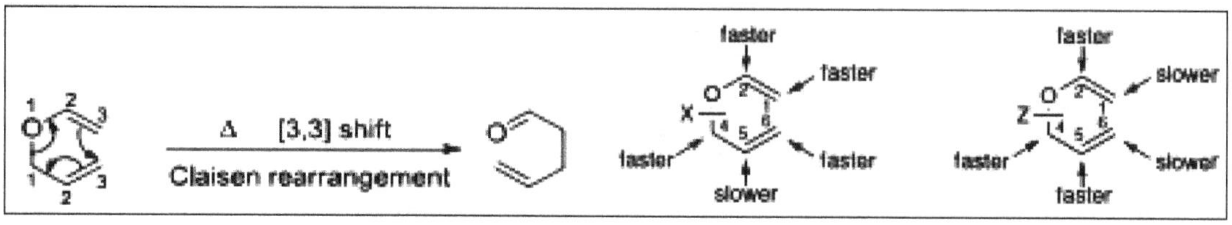

1. The mechanism of Claisen rearrangement for various isomers is shown in chair like T.S. Most favorable geometry arises in case of two larger methyl substituents in the pseudoequatorial position (E,E) in the transition state. The order of increasing rates toward the Claisen rearrangement is E,E > Z,E ≈ E,Z > Z,Z.

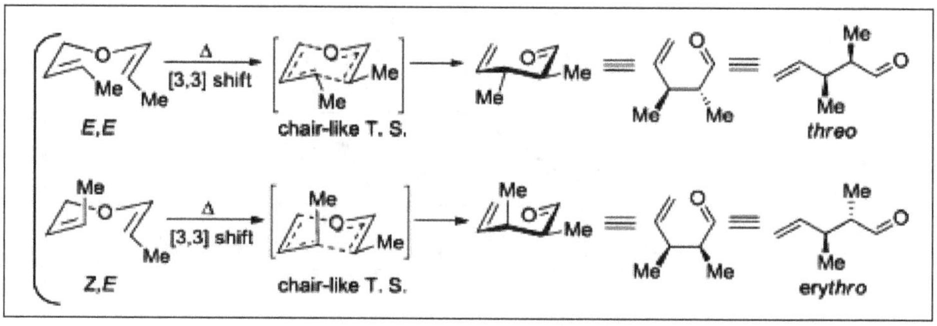

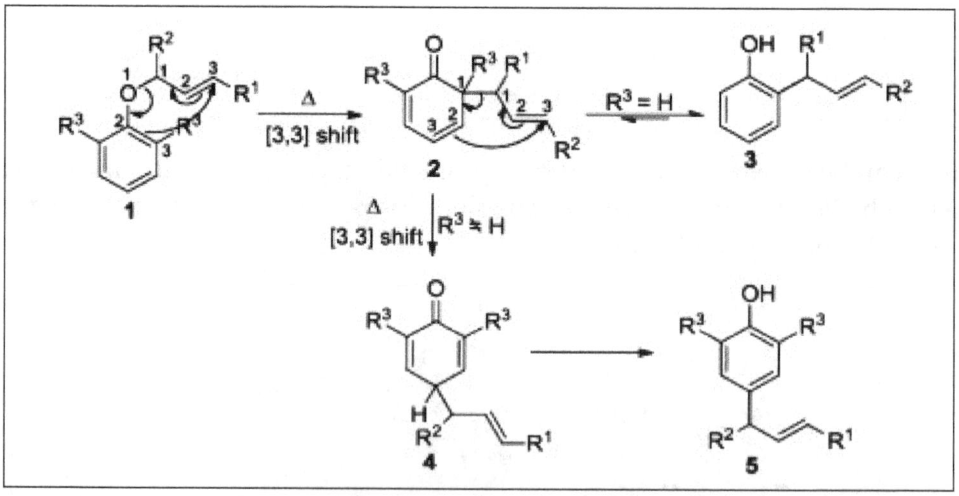

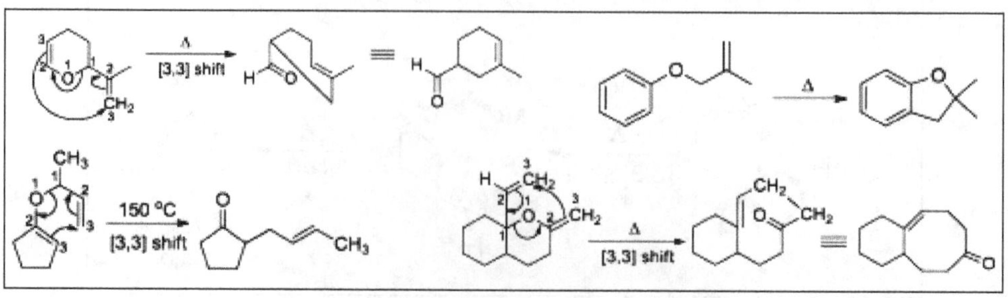

2. Citral is formed by following [3,3] sigmatropic rearrangement.

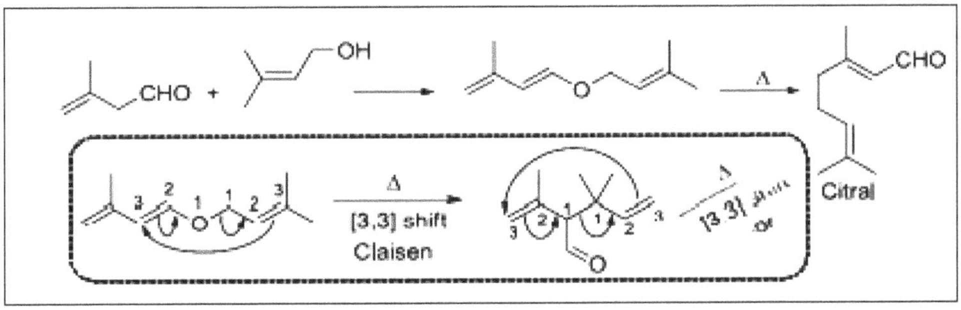

3. Ireland-Claisen rearrangement.

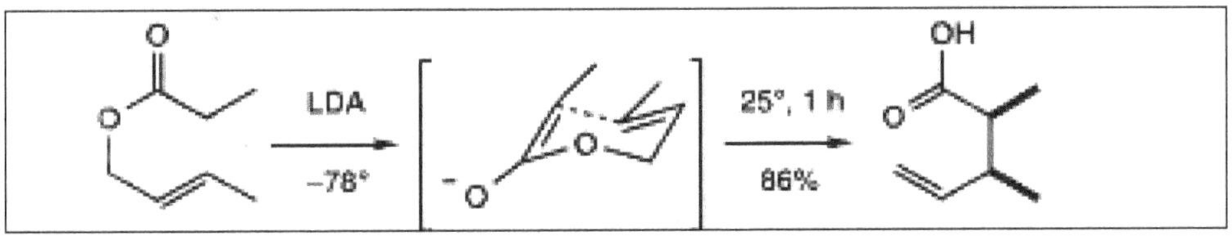

The β-ketoesters can rearrange to β-ketoacids following a [3,3] shift which is decarboxylated to give γ, δ-unsaturated ketone.

4. Kimel-Cope rearrangement.

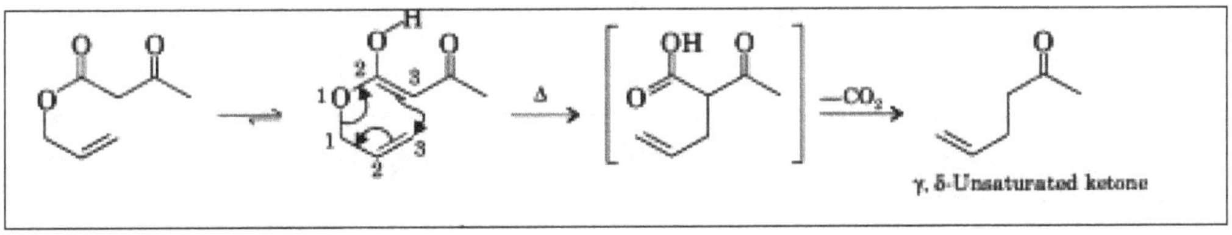

5. A typical example for anionic oxy-Cope rearrangement is depicted by converting allyl alcohol to anion and then [3,3] shift occurs.

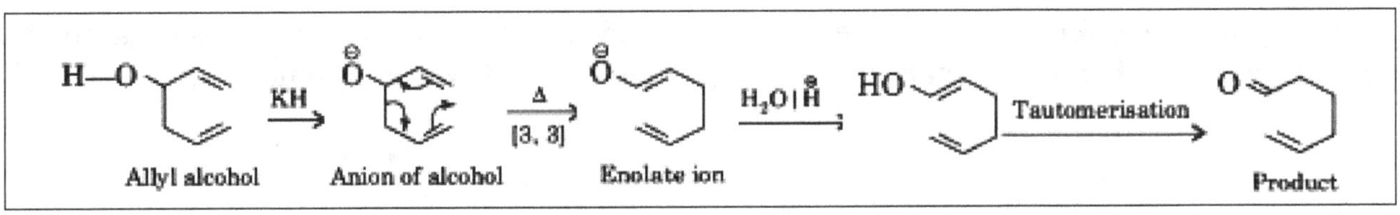

6. The following molecules do not exhibit Cope rearrangement.

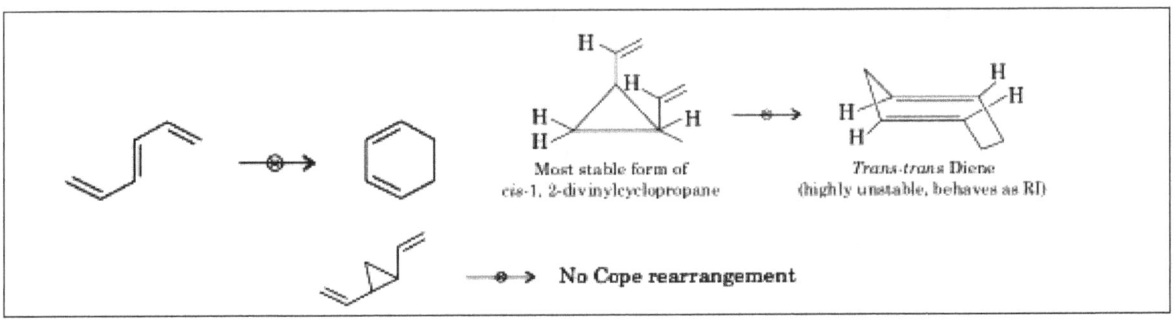

[5,5] Sigmatropic Shift

Example:

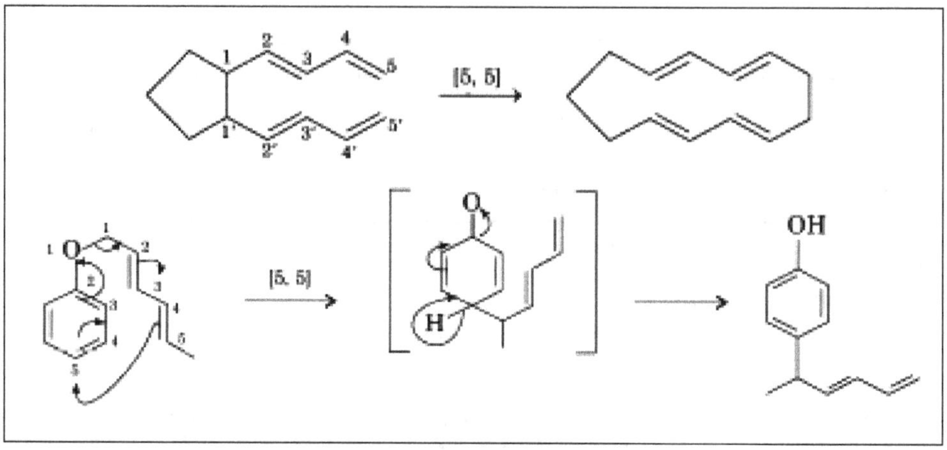

Example: Benzidine rearrangement. An acid-catalyzed rearrangement of hydrazobenzenes involving [5,5] sigmatropic shifts proceeds suprafacially through 10-membered T.S.

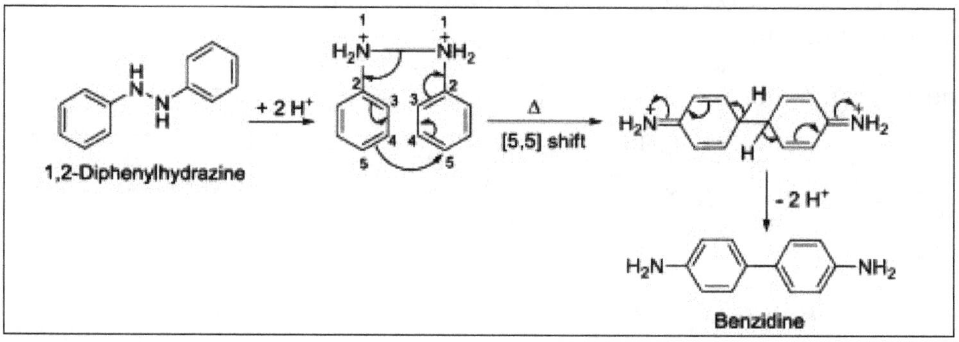

Hückel-Mobius Rule in Sigmatropic Rearrangements

According to Hückel-Mobius rule, a thermal sigmatropic rearrangement take places via aromatic T.S whereas photochemical sigmatropic rearrangement proceeds via antiaromatic T.S.

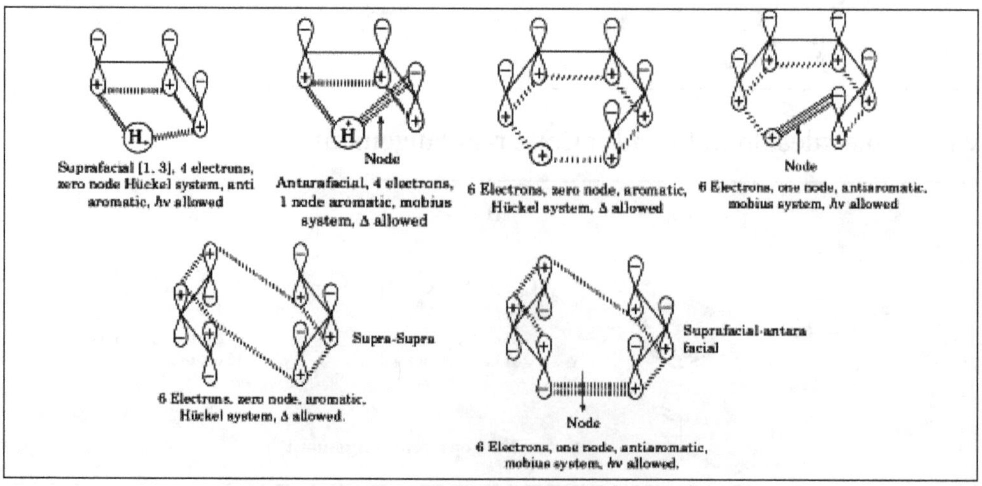

No. of allylic πe⁻	Mode of activation	Interaction	Feasibility	Product Stereo-chemistry
4nπ	Thermal (Δ)	Suprafacial	Symmetry allowed	Retention
		Antarafacial	Symmetry forbidden	Inversion
4nπ	Photochemic al(hγ)	Antarafacial	Symmetry allowed	Inversion
		Suprafacial	Symmetry forbidden	Retention
(4n+2)π	Thermal (Δ)	Antarafacial	Symmetry allowed	Inversion
		Suprafacial	Symmetry forbidden	Retention
(4n+2)π	Photochemic al(hγ)	Suprafacial	Symmetry allowed	Retention
		Antarafacial	Symmetry forbidden	Inversion

Typical Examples on Cycloadditions

Examples:

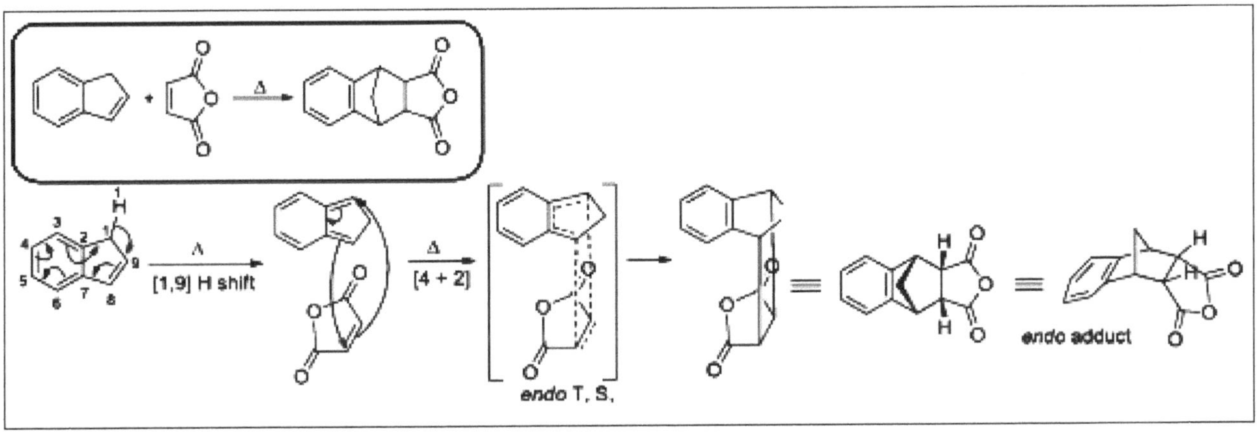

Examples:

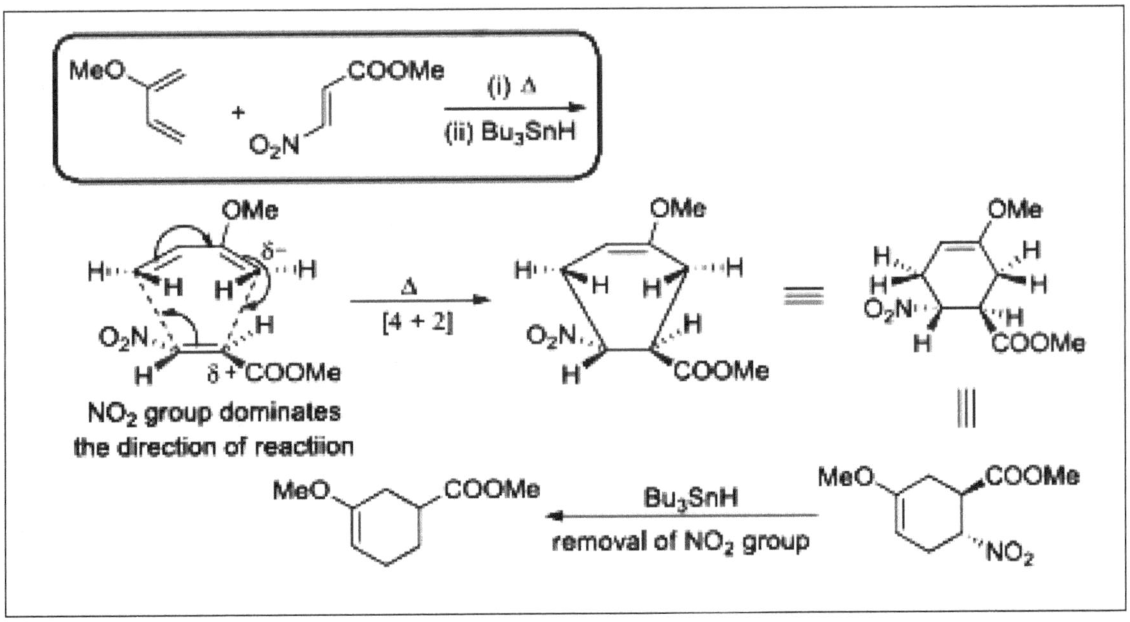

Table: A correlation among the entire factors associated with Cycloaddition Reaction.

No. of total π e^- involved	Symmetry	Mode of activation	Interaction	Feasibility	Reactants stereo-chemistry	Product Stereochemistry
$4n\pi$	m(s)	Thermal (Δ)	Supra-supra	Symmetry allowed	trans- trans cis-cis	trans
			Supra-antara	Symmetry forbidden	trans-cis	cis
$4n\pi$	C_2(s)	Photochemical(hγ)	Supra-supra	Symmetry allowed	trans- trans cis-cis	cis
			Supra-antara	Symmetry forbidden	trans-cis	trans
$(4n+2)\pi$	C_2(s)	Thermal (Δ)	Supra-supra	Symmetry allowed	trans- trans cis-cis	cis
			Supra-antara	Symmetry forbidden	trans-cis	trans
$(4n+2)\pi$	m(s)	Photochemical(hγ)	Supra-supra	Symmetry allowed	trans- trans cis-cis	trans
			Supra-antara	Symmetry forbidden	trans-cis	cis

References

- Symmetry-rules-for-chemical-reactions: sciencedirect.com, Retrieved 15, January 2020
- A-Brief-Description-on-Pericyclic-Reactions: midnaporecollege.ac.in, Retrieved 24, April 2020

Group Theory and Vibrational Spectroscopy

5

The spectroscopy which focuses on the infrared region of the electromagnetic spectrum is termed as vibrational spectroscopy. It is based on absorption spectroscopy. This chapter closely examines the key concepts of group theory and vibrational spectroscopy such as molecular vibrations and vibration modes to provide an extensive understanding of the subject.

Molecular Vibrations

A diatomic molecule has only a single bond that can vibrate; we say it has a single vibrational mode. As you may expect, the vibrational motions of polyatomic molecules are much more complicated than those in a diatomic. Firstly, there are more bonds that can vibrate; and secondly, in addition to stretching vibrations, the only type of vibration possible in a diatomic, we can also have bending and torsional vibrational modes. Since changing one bond length in a polyatomic will often affect the length of nearby bonds, we cannot consider the vibrational motion of each bond in isolation; instead we talk of normal modes involving the concerted motion of groups of bonds. As a simple example, the normal modes of a linear triatomic molecule are shown below:

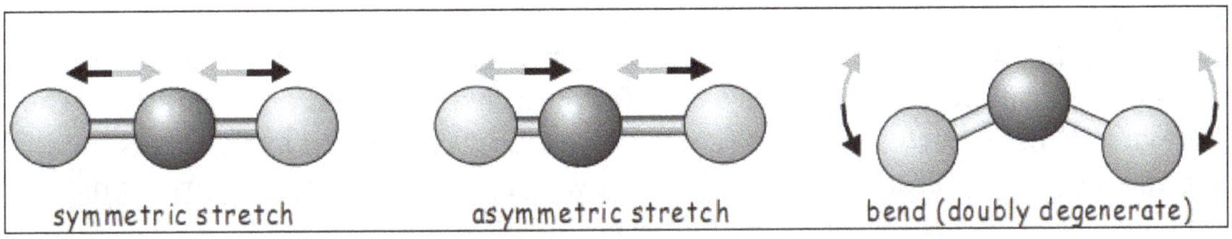

Once we know the symmetry of a molecule at its equilibrium structure, group theory allows us to predict the vibrational motions it will undergo using exactly the same tools we used above to investigate molecular orbitals. Each vibrational mode transforms as one of the irreps of the molecule's point group.

Molecular Degrees of Freedom – Determining the Number of Normal Vibrational Modes

An atom can undergo only translational motion, and therefore has three degrees of freedom corresponding to motion along the x, y, and z Cartesian axes. Translational motion in any arbitrary direction can always be expressed in terms of components along these three axes. When atoms combine to form molecules, each atom still has three degrees of freedom, so the molecule as a whole has 3N degrees of freedom, where N is the number of atoms in the molecule. However, the fact that each atom in a molecule is bonded to one or more neighbouring atoms severely hinders its translational motion, and also ties its motion to that of the atoms to which it is attached. For these reasons, while it is entirely possible to describe molecular motions in terms of the translational motions of individual atoms (we will come back to this in the next section), we are often more interested in the motions of the molecule as a whole. These may be divided into three types: translational; rotational and vibrational.

Just as for an individual atom, the molecule as a whole has three degrees of translational freedom, leaving 3N-3 degrees of freedom in rotation and vibration.

The number of rotational degrees of freedom depends on the structure of the molecule. In general, there are three possible rotational degrees of freedom, corresponding to rotation about the x, y and z Cartesian axes. A non-linear polyatomic molecule does indeed have three rotational degrees of freedom, leaving 3N-6 degrees of freedom in vibration (i.e3N-6 vibrational modes). In a linear molecule, the situation is a little different. It is generally accepted that to be classified as a true rotation, a motion must change the position of one or more of the atoms. If we define the z axis as the molecular axis, we see that spinning the molecule about the axis does not move any of the atoms from their original position, so this motion is not truly a rotation. Consequently, a linear molecule has only two degrees of rotational freedom, corresponding to rotations about the x and y axis. This type of molecule has 3N-5 degrees of freedom left for vibration, or 3N-5 vibrational modes.

Determining the Symmetries of Molecular Motions

The procedure for determining the normal vibrational modes of a polyatomic molecule is very similar to that used in previous sections to construct molecular orbitals. In fact, virtually the only difference between these two applications of group theory is the choice of basis set. The motions of a molecule may be described in terms of the motions of each atom along the x, y and z axis. Consequently, it probably won't come as too much of a surprise to discover that a very useful basis for describing molecular motions comprises a set of (x, y, z) axes centred on each atom. This basis is usually known as the 3N Cartesian basis (since there are 3N Cartesian axes, 3 axes for each of the N atoms in the molecule). Note that each molecule will have a different 3N Cartesian basis, just as every molecule has a different atomic orbital basis.

Our first task in investigating motions of a particular molecule is to determine the characters of the matrix representatives for the 3N Cartesian basis under each of the symmetry operations in the molecular point group. We will use the H_2O molecule, which has C_{2v} symmetry, as an example.

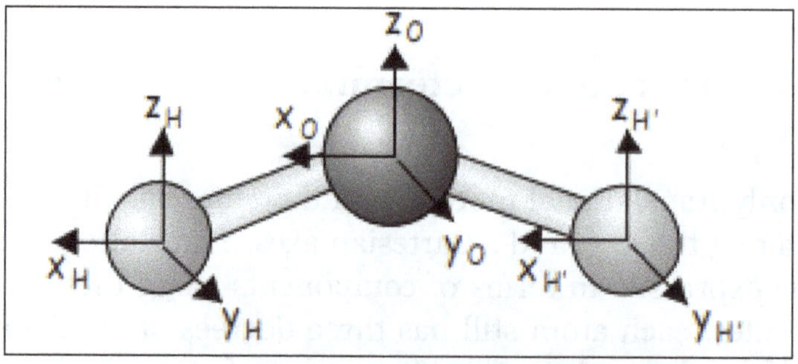

One way of determining the characters would be to construct all of the matrix representatives and take their traces. we can also determine the character of a matrix representative under a particular symmetry operation by stepping through the basis functions and applying the following rules:

- Add 1 to the character if the basis function is unchanged by the symmetry operation.
- Add −1 to the character if the basis function changes sign under the symmetry operation.
- Add 0 to the character if the basis function moves when the symmetry operation is applied.

For H_2O, this gives us the following characters for the 3N Cartesian basis (check that you can obtain this result using the rules above and the basis vectors as drawn in the figure):

Operation	E	C_2	$\sigma_v(xz)$	$\sigma_v'(yz)$
χ_{3N}	9	-1	3	1

There is an even quicker way to work out the characters of the 3N Cartesian basis if you have a character table in front of you. The character for the Cartesian basis is simply the sum of the characters for the x, y and z (or T_x, T_y, and T_z) functions listed in the character table. To get the character for the 3N Cartesian basis, simply multiply this by the number of atoms in the molecule that are unshifted by the symmetry operation.

The C_{2v} character table is shown below:

C2v	E	C_2	σv	σ_v'	h = 4
A_1	1	1	1	1	z, x^2, y^2, z^2
A_2	1	1	-1	-1	xy, Rz
B_1	1	-1	1	-1	x, xz, Ry
B_2	1	-1	-1	1	y, yz, Rx

x transforms as B_1, y as B_2, and z as A_1, so the characters for the Cartesian basis are,

Operation	E	C_2	$\sigma_v(xz)$	$\sigma_v'(yz)$
χCart	3	−1	1	1

We multiply each of these by the number of unshifted atoms (3 for the identity operation, 1 for C_2, 3 for σv and 1 for σ_v') to obtain the characters for the 3N Cartesian basis.

χ_{3N}	9	−1	3	1

Reassuringly, we obtain the same characters as we did previously. Which of the three methods you use to get to this point is up to you.

We now have the characters for the molecular motions (described by the 3N Cartesian basis) under each symmetry operation. At this point, we want to separate these characters into contributions from translation, rotation, and vibration. This turns out to be a very straightforward task. We can read the characters for the translational and rotational modes directly from the character table, and we obtain the characters for the vibrations simply by subtracting these from the 3N Cartesian characters we've just determined. The characters for the translations are the same as those for χCart. We find the characters for the rotations by adding together the characters for R_x, R_y and R_z from the character table (or just Rx and Ry if the molecule is linear). For H_2O, we have:

Operation	E	C_2	$\sigma_v(xz)$	$\sigma_v'(yz)$
χ_{3N}	9	-1	3	1
χ_{Trans}	3	-1	1	1
χ_{Rot}	3	-1	-1	-1
$\chi_{Vib}=\chi_{3N}-\chi_{Trans}-\chi_{Rot}$	3	1	3	1

The characters in the final row are the sums of the characters for all of the molecular vibrations. We can find out the symmetries of the individual vibrations by using the reduction equation to determine the contribution from each irrep.

In many cases you won't even need to use the equation, and can work out which irreps are contributing just by inspection of the character table. In the present case, the only combination of irreps that can give the required values for χ_{Vib} is $2A_1 + B_1$.

So far this may all seem a little abstract, and you probably want to know is what the vibrations of H_2O actually look like. For a molecule with only three atoms, it is fairly easy to identify the possible vibrational modes and to assign them to the appropriate irrep.

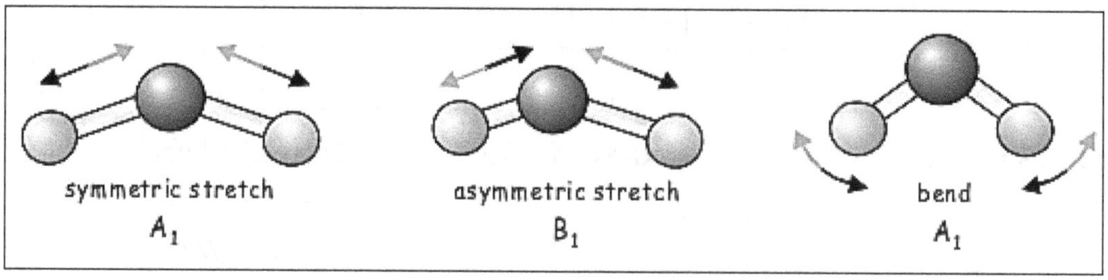

For a larger molecule, the problem may become much more complex, and in that case we can generate the SALCs of the 3N Cartesian basis, which will tell us the atomic displacements associated with each vibrational mode. We will do this now for H_2O.

We generate the SALCs of each symmetry by applying the appropriate projection operator to each of the basis functions (or in this case, basis vectors) f_i in turn,

$$\phi_i = \sum_g \chi_k(g) g f_i$$

In this case we have 9 basis vectors, which we will label x_H, y_H, z_H, x_O, y_O, z_O, $x_{H'}$, $y_{H'}$, $z_{H'}$, describing the displacements of the two H atoms and the O atom along Cartesian axes. For the SALCs of A_1

symmetry, applying the projection operator to each basis vector in turn gives (check that you can obtain this result):

$$\phi_1(x_H)=x_H-x_{H'}+x_H-x_{H'}=2x_H-2x_{H'}$$
$$\phi_2(y_H)=y_H-y_{H'}-y_H+y_{H'}=0$$
$$\phi_3(z_H)=z_H+z_{H'}+z_H+z_{H'}=2z_H+2z_{H'}$$
$$\phi_4(x_O)=x_O-x_O-x_O-x_O=0$$
$$\phi_5(y_O)=y_O-y_O-y_O+y_O=0$$
$$\phi_6(z_O)=z_O+z_O+z_O+z_O=4z_O$$
$$\phi_7(x_{H'})=x_{H'}-x_H+x_{H'}-x_H=2x_{H'}-2x_H$$
$$\phi_8(y_{H'})=y_{H'}-y_H-y_{H'}+y_H=0$$
$$\phi_9(z_{H'})=z_{H'}+z_H+z_{H'}+z_H=2z_{H'}+2z_H$$

We see that the motion characteristic of an A1 vibration (which we have identified as the symmetric stretch and the bending vibration) may be summarised as follows:

- $2(x_H-x_{H'})$: The two hydrogen atoms move in opposite directions along the x axis.
- $2(z_H+z_{H'})$: The two hydrogen atoms move in the same direction along the z axis.
- $4z_O$: The oxygen atom moves along the z axis.
- There is no motion of any of the atoms in the y direction.

The asymmetric stretch has B_1 symmetry, and applying the projection operator in this case gives:

$$\phi_1(x_H)=x_H+x_{H'}+x_H+x_{H'}=2x_H+2x_{H'}$$
$$\phi_2(y_H)=y_H-y_{H'}-y_H-y_{H'}=0$$
$$\phi_3(z_H)=z_H-z_{H'}+z_H-z_{H'}=2z_H-2z_{H'}$$
$$\phi_4(x_O)=x_O+x_O+x_O+x_O=4x_o$$
$$\phi_5(y_O)=y_O+y_O-y_O+y_O=0$$
$$\phi_6(z_O)=z_O-z_O+z_O-z_O=0$$
$$\phi_7(x_{H'})=x_{H'}+x_H+x_{H'}+x_H=2x_{H'}+2x_H$$
$$\phi_8(y_{H'})=y_{H'}+y_H-y_{H'}-y_H=0$$
$$\phi_9(z_{H'})=z_{H'}-z_H+z_{H'}-z_H=2z_{H'}-2z_H$$

In this vibrational mode, the two H atoms move in the same direction along the x axis and in opposite directions along the z axis. We have now shown how group theory may be used together with the 3N Cartesian basis to identify the symmetries of the translational, rotational and vibrational modes of motion of a molecule, and also to determine the atomic displacements associated with each vibrational mode.

Molecular Vibrations using Internal Coordinates

While it was fairly straightforward to investigate the atomic displacements associated with each vibrational mode of H_2O using the 3N Cartesian basis, this procedure becomes more complicated for larger molecules. Also, we are often more interested in how bond lengths and angles change in a vibration, rather than in the Cartesian displacements of the individual atoms. If we are only interested in looking at molecular vibrations, we can use a different procedure from that described above, and start from a basis of internal coordinates. Internal coordinates are simply a set of bond lengths and bond angles, which we can use as a basis for generating representations and, eventually, SALCs. Since bond lengths and angles do not change during translational or rotational motion, no information will be obtained on these types of motion.

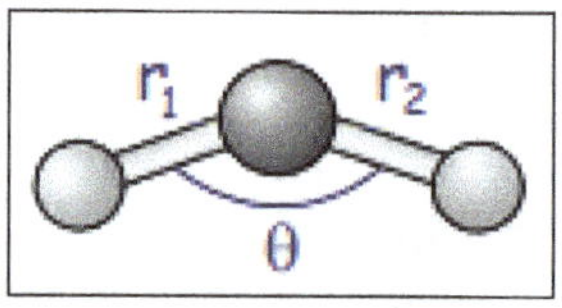

For H_2O, the three internal coordinates of interest are the two OH bond lengths, which we will label r and r', and the HOH bond angle, which we will label θ. If we wanted to, we could separate our basis into two different bases, one consisting only of bond lengths, to describe stretching vibrations, and one consisting of only bond angles, to describe bending vibrations. However, the current example is simple enough to treat all the basis functions together.

As usual, our first step is to work out the characters of the matrix representatives for this basis under each symmetry operation. The effects of the various transformations on our chosen basis, and the characters of the corresponding representatives, are:

$$E(r,r',\theta)=(r,r',\theta) \qquad \chi(E)=3$$

$$C_2(r,r',\theta)=(r',r,\theta) \qquad \chi(C_2)=1$$

$$\sigma_v(xz)(r,r',\theta)=(r,r',\theta) \qquad \chi(\sigma_v)=3$$

$$\sigma_v'(yz)(r,r'',\theta)=(r',r,\theta) \qquad \chi(\sigma_v')=1$$

These are the same characters as we found before using the 3N Cartesian basis, and as before, we can see by inspection of the character table that the representation may be reduced down to the sum of irreps $2A_1 + B_1$. We can now work out the symmetry adapted linear combinations of our new basis set to see how the bond lengths and angle change as H_2O vibrates in each of the three vibrational modes.

Again, we will use the projection operator $\phi_i = \sum_g \chi_k(g) g f_i$ applied to each basis function in turn. Firstly the A_1 vibrations:

$$\phi_1(r)=r+r'+r+r'=2(r+r')$$

$$\phi_2(r')=r'+r+r'+r=2(r+r')$$

$$\phi_3(\theta)=\theta+\theta+\theta+\theta=4\theta$$

From these SALCs, we can identify ϕ_1 (and ϕ_2, which is identical) with the symmetric stretch, in which both bond lengths change in phase with each other, and ϕ_3 with the bend.

Now for the B_1 vibration:

$$\phi_4(r) = r - r' + r - r' = 2(r + r')$$
$$\phi_5(r') = r' - r + r' - r = 2(r' + r)$$
$$\phi_6(\theta) = \theta - \theta + \theta - \theta = 0$$

ϕ_4 and ϕ_5 are not linearly independent, and either one may be chosen to describe the asymmetric stretch, in which one bond lengthens as the other shortens.

Group Theory in Spectroscopy

Symmetry can help to solve many of the issues encountered in chemistry, and group theory is the primary tool that is utilized to identify symmetry. If we know how to determine the symmetry of small molecules, we can determine the symmetry of other targets. This is not only limited to the symmetry of molecules but also to the symmetries of local atoms, molecular orbitals, rotations, and vibrations of bonds. A typical example is the knowledge of the symmetries of molecular orbital wave functions allowing the identification of the nature and characteristics of the binding. Also, the particular methods associated with certain symmetries allow us to decide if the transition is prohibited and to understand the bands observed in infrared or Raman spectrum. A *symmetry operation* to a molecule is an operation that leaves the physical proprieties of the molecule unchanged. This is equivalent to having the molecule unchanged before and after the symmetry operation is performed. In other words, when we do a symmetry operation on a molecule, every point of the molecule will be in an equivalent position.

The application of group theory in spectroscopy intends to investigate the way in which symmetry considerations influence the interaction of light with matter. Group theory can be used to understand the molecular orbitals in a molecule and to determine the possible electronic states accessible by absorption of a photon. Another important function of group theory is the investigation of the light that excites different vibrational modes of a polyatomic molecule. A photon of the appropriate energy is able to excite an electronic transition in an atom, subject to the following selection rules:

$$\Delta n = \text{Integer}$$

$$\Delta l = \pm 1$$

$$\Delta L = 0, \pm 1$$

$$\Delta S = 0$$

$$\Delta J = 0, \pm 1; J = 0$$

In general, different types of spectroscopic transitions obey different selection rules. The common transitions involve changing the electronic state of an atom and involve absorption of a photon in the UV or visible part of the electromagnetic spectrum. There are analogous electronic transitions in molecules, which we will consider here. The absorption of photons in the infrared region of the spectrum controls the vibrational excitation in molecules and the absorption of photons in the microwave region commands rotational excitation. Typically, each excitation executes its own selection rules, but the general methodology for establishing the selection rules is identical in all cases. The determination of the conditions under which the probability of transition is not zero is a simple process. Therefore, the first step in understanding the origins of selection rules is to learn how transition probabilities are computed, and this requires some quantum mechanics concepts. Overall, group theory plays a very important role in spectroscopy, which we can see from various applications of group theory in spectroscopy such as infrared spectrum, Raman spectrum, electronic spectrum, and so on. Typically, the change in electronic energy is greater than in vibrational energy, which is also greater than in rotational energy. Figure illustrates the different energy levels in a molecule.

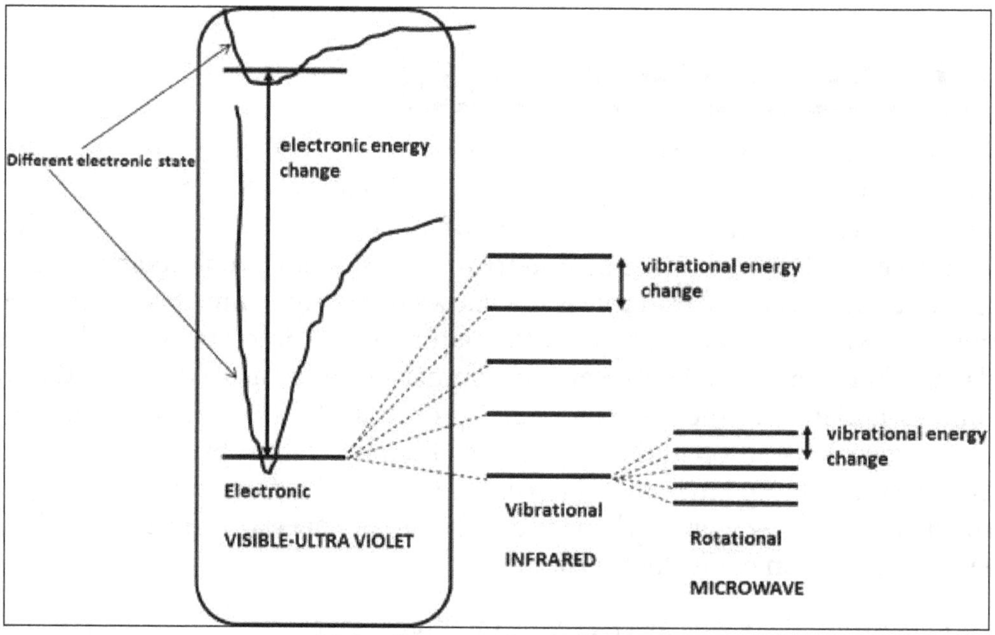

Molecular energy levels diagram.

Electronic Transitions in Molecules

When an electron is excited from one electronic state to another, this is what is called an electronic transition. The selection rules for electronic transitions are governed by the transition moment integral. Due to the fact that the electrons are coupled between two vibrational states that are between two electronic states, it is important to consider both the electronic state symmetries and the vibration state symmetries. This modification of the transition moment integral produces the symmetry of the initial electronic and vibrational states called "bra" and the final electronic and vibrational states named "ket."

This appears to be a modified version of the transition moment integral. If we assume that we have a molecule in an initial state, we can determine which final states can be accessed by the absorption of a photon. So, we need to determine the symmetry of an electronic state. The symmetry of an electronic state is obtained by identifying any unpaired electrons and taking the direct product of the unrepresentative of the molecular orbitals in which they are detected. The total symmetric unrepresentative always holds the ground state of a closed-shell molecule in which all electrons are paired. The determination of the unrepresentative electric dipole operator allows obtaining the electronic states accessible by absorption of photons. Light that is linearly polarized along the x, y, and z axes transforms in the same way as the functions x, y, and z in the appropriate character table. From the C_{3v} character table, we see that x- and y-polarized light transforms as E, while z-polarized light transforms as in the appropriate character table.

The excitation from one energy level to a higher energy level happens during the electronic transitions in a molecule. The change of energy associated with these transitions gives structural information of the molecule and determines many other molecular properties such as color. Planck's relation provides the relationship between the energy involved in the electronic transition and the frequency of radiation. Planck's equation is sometimes termed the Planck-Einstein:

$$E = h\gamma$$

where $h = 6.55 \times 10^{-34}$ J.s is a Planck constant. Electronic transitions in molecules occur between orbitals and they must cohere to angular momentum selection rules. Figure shows possible electronic transitions of p, s, and n electrons. In the process of transition $\sigma \rightarrow \sigma^*$, electrons occupying a "HOMO" of a "sigma bond" can get excited to the "LUMO" of that bond. Similarly, in the process of transition $\pi \rightarrow \pi^*$, electrons from a "pi-bonding orbital" can get excited to the "antibonding-pi orbital" of that bond. Auxochromes with free electron pairs denoted as n have their own transitions. The following molecular electronic transitions exist:

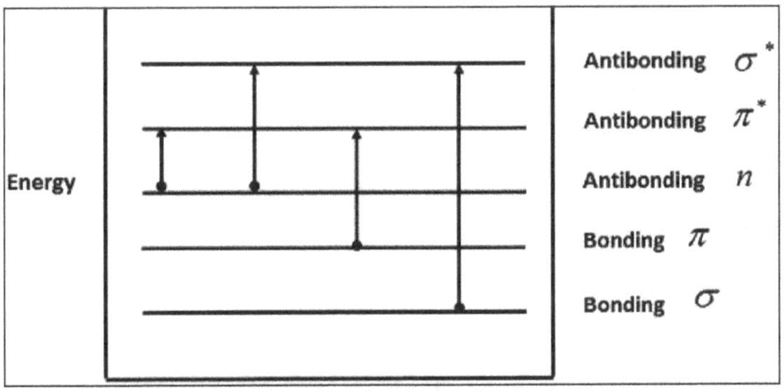

Absorbing species containing p, s, and n electrons.

Vibrational Transitions in Molecules

All molecules vibrate. While these vibrations can originate from several events, the most basic of these occurs when an electron is excited within the electronic state from one eigenstate to another. When an electron is excited from one eigenstate to another within the electronic state, there is a change in interatomic distance, which results in a vibration occurring. A vibration occurs when an electron remains within the electronic state but changes from one eigenstate to another. Just as in electronic transitions, the selection rules for Vibrational transitions are dictated by the transition moment integral. Light polarized along the x, y, and z axes of the molecule may be used to excite vibrations with the same symmetry as the x, y, and z functions listed in the character table. For example, in the C_{2v} point group, x-polarized light may be used to excite vibrations of B_1 symmetry, y-polarized light to excite vibrations of B_2 symmetry, and z-polarized light to excite vibrations of A_1 symmetry. In H_2O, we would use z-polarized light to excite the symmetric stretch and bending modes, and x-polarized light to excite the asymmetric stretch. Shining y-polarized light onto a molecule of H_2O would not excite any vibrational motion. For instance, let us consider a simple case of a vibrating diatomic molecule where the restoring force is proportional to the displacement,

$$F = -kx.$$

The potential energy is:

$$V = \frac{1}{2}kx^2$$

and the allowed energy can be obtained from Schrodinger equation,

$$E_v = \left(v + \frac{1}{2}\right)\hbar\omega$$

where,

$$\omega = \left(\frac{k}{\mu}\right)^{1/2}, v = 0,1,2,3,4...,$$

and, $$\mu = \frac{m_1 m_2}{m_1 + m_2}.$$

The vibrational terms of the molecule can therefore be given by,

$$G_v = \left(v + \frac{1}{2}\right)\frac{1}{2\pi c}\left(\frac{k}{\mu}\right)^{1/2}.$$

Selection Rules for IR and Raman Spectroscopy

A vibrating bond in a molecule behaves like oscillating electric field and there is a change electric dipole moment. Electromagnet waves in the region of infrared also have electric field component with it. When internal electric field component due to vibrating molecule interacts with the external electric field component of electromagnet waves, the transitions from ground vibrational state to higher vibrational excited state may occur or not i.e. this transition is allowed or not. In this section we learn selection rules about these types of transitions.

According to quantum mechanical rule $\Delta\upsilon = \pm 1$ i.e. transition $\upsilon_0 \rightarrow \upsilon_{+1}$ is allowed. In quantum mechanics the basis for infrared spectroscopic selection rule is the value of the transition moment integral of the type:

$$\int \psi_g{}^* \mu\, \psi_e dT$$

Transition occurs from vibrational ground state (initial) ψ_g to vibrationally excited state ψ_e by absorbing one quanta of energy. This is called fundamental transition. The μ is molecular dipole moment vector. The probability of such transition from ψ_g ground vibrational state to ψ_e first excited vibrational state is directly proportional to transition moment given by the integral of the type given in $\int \psi_g{}^* \mu \, \psi_e dT$ .Since ψ_g ground state wave function is always real we have $\psi_g{}^* = \psi_g$.

Thus equation can be written as:

$$\int \psi_g \mu \psi_e dT$$

dT Means integration is carried over all variables of the wave function.

The spectroscopic activity of a vibrational mode depends upon the value of transition moment integral $\int \psi_g \mu \psi_e dT$. If this integral has zero value for a particular transition, then the probability of occurring this transition will be zero and no absorption of IR radiations will occur. Thus the transition is forbidden and vibration will vibrationally inactive. If the integral is non-zero, then there will be absorption of IR radiations by the molecule and the transition will be allowed .Hence it will be IR active.

In practice, the integral itself does not need to be calculated to determine a selection rule. It is sufficient to determine the symmetry of transition moment function, $\int \psi_g \mu \psi_e dT$. The idea of symmetry is important when considering the integral of odd functions. Since ψ_g ground state wave function is always real and symmetric. It follows that, for a vibrational transition to be allowed, the symmetry of the excited state wave function must be the same as the symmetry of the transition moment operator. Solving of integrals of the type $\int \psi_g \mu \psi_e dT$ is a cumbersome process. Group theory helps and simplifies the solutions of such integrals. According to group theory and symmetry considerations," if the direct product in this integral contains totally symmetrical representation (an irreducible representation for which χ =+1 for all symmetry operations of the point group) then the integral will be non-zero and there will be interaction between electric field component of electromagnetic radiation(in IR region)and electric field component of oscillating electric dipole moment of the vibrating molecule and hence absorption of IR radiation will occur and vibration will be IR active".

Thus only difficulty left, in assigning the spectroscopic activity to vibrational modes, is to solvethis integral. The transition moment integral,

$$\int \psi_g \mu \psi_e dT$$

Can be resolved into three components along X, Y, Z axes and these resolved integrals can be written as:

$$\int \psi_g \mu_x \psi_e dT$$

$$\int \psi_g \mu_y \psi_e dT$$

$$\int \psi_g \mu_z \psi_e dT$$

If any of these integrals given by equations 2, 3, 4 is nonzero then according to symmetry point of view this transition will be allowed. So problem now is how to check the symmetries of direct product in these three component integrals? We know that vibrational ground state $\upsilon_o = 0$, i.e. ψ_g has mathematical similarity to that of s-orbital. Since a sphere is totally symmetrical in a point group, the ground vibrational state ψ_g will be totally symmetrical for all molecules.

Symmetries of components of transition moment vectors, μx, μy, μz is the same as that of X, Y, Z axes or that of translational vectors T_x, T_y, T_z. Their symmetries can be directly obtained from the character table of the point group to which the molecule belongs. The symmetry properties of vibrationally excited states ψ_e are the same as that of the vector which describes/forms the basis, for the particular vibrational mode (Γ_{vib}).

So we have to evaluate the direct product of ψ_g (totally symmetric), the irreducible representation for which X,Y,Z axes or T_x, T_y, T_z form the basis for representations and ψ_e, irreducible representation of the vibrational mode to be considered for IR activity, spectroscopic activity of which we want to determine. In the integral $\int \psi_g \mu \psi_e dT$ the direct product of $\psi_g . \psi_e$ will give the product as ψ_e. The direct product of a totally symmetric irreducible representation with any non-symmetrical irreducible representation equals to non-symmetric irreducible representation. For example in C_2v point group $A_1 \times B_1 = B_1$. So direct productμ ψ_e must contain totally symmetric irreducible representation in it. So we have to check the symmetries of μ or μ_x, μ_y, μ_z. The symmetry of ψ_e is the same as that of irreducible representation the activity of which we want to know and it is given. A normal mode belonging to any symmetry species as that of X,Y,Z axes or T_x, T_y, T_z translational vectors will be IR active. Symmetry characteristics of transition moment operator are summarized as:

Symmetry characteristics of transition moment operator		
Type of Transition	μ transforms as	
Electric dipole	x, y, z	Infrared spectra
Electric polarizability	x^2, y^2, z^2, xy, xz, yz	Raman spectra

To explain the points made so far about vibrational transition let us take few examples.

Examples of H_2O/SO_2 (C_{2v}) Point Group Molecules

In module -30 we have obtained Γ_{3N} and then Γ_{vib} as $2A_1 + B_2$ for water molecule. Now we will find the activity of A_1 and B_2 modes i.e. that of ψ_e which belongs to either A_1 or B_2 symmetries. In C_{2v} point group the totally symmetric representation is A_1. We have to find the irreducible representations for which X,Y, Z axes or T_x, T_y, T_z form the basis for representations i.e. symmetry species of μ_x, μ_y, μ_z components. Using character table of $C_2{}^v$ point group the symmetries of these vectors are:

C_{2v}	E	$C_2(z)$	σ	σ	Linear functions, rotations
			$_v$(xz)	$_v$(yz)	
A_1	+1	+1	+1	+1	z
A_2	+1	+1	-1	-1	R_z
B_1	-1	-1	+1	-1	x, R_y
B_2	-1	-1	-1	+1	y, R_x

- $\mu_x = T_x$ or x belongs to → B_1 symmetry.
- $\mu_y = T_y$ or y belongs to → B_2 symmetry.
- $\mu_z = T_z$ or z belongs to → A_1 symmetry.

In order to check the activity of $2A_1$ and B_2 modes we have to evaluate the following direct products. Check whether the direct products equal to or contain totally symmetric irreducible representation in it.

$$\int \psi_g \mu_x \psi_e dT \quad \int A_1 \otimes B_1 \otimes A_1 \equiv B_1$$

$$\int \psi_g \mu_y \psi_e dT \quad \int A_1 \otimes B_2 \otimes A_1 \equiv B_2$$

$$\int \psi_g \mu_z \psi_e dT \quad \int A_1 \otimes A_1 \otimes A_1 \equiv A_1$$

Here ψ_g has A1 symmetry; μ_x, μ_y μ_z have B_1, B_2, A_1 symmetries respectively and ψ_e has the symmetries of the type A_1, B_2. Out of the three integrals given by equations if any of these contain totally symmetric representation then the vibrational mode will be IR active. Out of these three equations the direct product integral in equation contains a totally symmetric representation A_1. Thus the overall integral $\int \psi_g \mu \psi_e dT$ has A_1 representation in it and thus integral is nonzero. Hence the vibrational mode is IR active i.e. A_1 mode is IR active. Let us check the activity of B_2 mode for this we have to consider the following direct products and evaluate these:

$$\int \psi_g \mu_x \psi_e dT \quad \int A_1 \otimes B_1 \otimes B_2 \equiv A_2$$

$$\int \psi_g \mu_y \psi_e dT \quad \int A_1 \otimes B_2 \otimes B_2 \equiv A_1$$

$$\int \psi_g \mu_z \psi_e dT \quad \int A_1 \otimes A_1 \otimes B_2 \equiv B_2$$

Out of the equations above; equations contains totally symmetric representations. Hence over all intergral contains A_1 mode and thus it is non zero and the B_2 mode is also IR active. Thus in $\Gamma_{vib} = 2A_1 + B_2$ both A_1 and B_2 modes are IR active for H_2O/SO_2 (C_{2v}) point group molecules.

Examples of $POCl_3$/ $CHCl_3$ (C_{3v}) Point Group Molecules

We have obtained Γ_{3N} and then Γ_{vib} as $\Gamma_{vib} = 3A_1 + 3E$ for $POCl_3$ molecule. Now we will find the activity of A_1 and E modes i.e. ψ_e belongs to either A_1 or E symmetries. In C_{3v} point group the totally symmetric representation is A_1 .We have to find the irreducible representations for which X,Y,Z axes or T_x,T_y,T_z form the basis for representations i.e. symmetry species of μ_x, μ_y, μ_z components. Using character table of C_{3v} point group symmetries of these vectors i.e., of these X,Y,Z axes or T_x,T_y,T_z are:

C_{3v}	E	$2C_3$ (z)	$3\sigma_v$	Linear functions, rotations
A_1	+1	+1	+1	Z
A_2	+1	+1	-1	R_z
E_1	+2	-1	0	(x,y) (R_x,R_y)

- $\mu_x = T_x$ or x belongs to→ E.
- $\mu_y = T_y$ or y belongs to → E.
- $\mu_z = T_z$ or z belongs to → A_1.

In order to find the activity of A1 and E modes we have to evaluate the following direct product integrals i.e. have to check the activity when ψ_e has the symmetry as $3A_1$ and 3E .Whether these are IR active or not.

For this we have to consider the integrals of the type given below for A1 mode and solve them:

$$\int \psi_g \mu_x \psi_e dT \quad \int A_1 \otimes E \otimes A_1 \equiv E$$

$$\int \psi_g \mu_y \psi_e dT \quad \int A_1 \otimes E \otimes A_1 \equiv E$$

$$\int \psi_g \mu_z \psi_e dT \quad \int A_1 \otimes A_1 \otimes A_1 \equiv A_1$$

Out of the equations above the last equation contains totally symmetric representation. Overall integral is non-zero. Hence A_1 mode is IR active. Now let us take E mode and check its activity. For this we have to consider the following direct products and evaluate these:

$$\int \psi_g \mu_x \psi_e dT \quad \int A_1 \otimes E \otimes E \equiv A_1$$

$$\int \psi_g \mu_y \psi_e dT \quad \int A_1 \otimes E \otimes E \equiv A_1$$

$$\int \psi_g \mu_z \psi_e dT \quad \int A_1 \otimes A_1 \otimes E \equiv E$$

Out of the integral in equations above, the first two equations contain the totally symmetric representation. So over all integral is non-zero. So the vibrational mode E is also IR active.

Thus in $\Gamma_{vib} = 3A_1 + 3E$, both A_1 and E modes are IR active for $POCl_3$/$CHCl_3$/(C_{3v}) point group molecules.

Symmetry Selection Rule for Raman Spectroscopy

In Raman spectroscopy transition between vibrational states can be described in similar manner as described in case of IR spectroscopy, by the integral of the type:

$$\int \psi_g \alpha \psi_e dT$$

Where ψ_g and ψ_e represent the ground and first excited vibrational states respectively. The α describes the polarizability of molecule under investigation. Raman effect depends upon the extent of induced dipole by laser beam. Induced dipole is directly proportional to polarizability of the molecule,

$$P = \alpha E$$

Polarizability means the extent to which electron distribution in the molecule can be distorted by the applied laser beam. As mentioned earlier α is a tensor with a 3 x 3 array of component terms such as; α_{xx} α_{xy}, α_{xz}, α_{yx}, α_{yy}, α_{yz} α_{zx}, α_{zy} α_{zz}.

$$\begin{bmatrix} P_x \\ P_y \\ P_z \end{bmatrix} = \begin{bmatrix} \alpha_{xx} & \alpha_{xy} & \alpha_{xz} \\ \alpha_{yx} & \alpha_{yy} & \alpha_{yz} \\ \alpha_{zx} & \alpha_{zy} & \alpha_{zz} \end{bmatrix} \begin{bmatrix} E_x \\ E_y \\ E_z \end{bmatrix}$$

For Vibrational Transition in Raman Spectrum

$\alpha_{jk} = \alpha_{kj}$(where j,k = x,y,or z). Therefore, out of nine terms only six terms are to be considered. $\alpha_{xy} = \alpha_{yx}$, $\alpha_{zx} = \alpha_{xz}$, $\alpha_{yz} = \alpha_{zy}$, $\alpha_{xx}(\alpha_x^2)$, $\alpha^{yy}(\alpha_y^2)$, α_{zz}. (α_z^2). Thus there are six distinct components of the integral $\int \psi_g \alpha \psi_e dT$. These integrals are:

1. $\int \psi_g \left(\alpha_{xx=x^2}\right) \psi_e dT$
2. $\int \psi_g \left(\alpha_{yy=y^2}\right) \psi_e dT$
3. $\int \psi_g \left(\alpha_{zz=z^2}\right) \psi_e dT$
4. $\int \psi_g \left(\alpha_{xy=yx}\right) \psi_e dT$
5. $\int \psi_g \left(\alpha_{yz=zy}\right) \psi_e dT$
6. $\int \psi_g \left(\alpha_{zx=xz}\right) \psi_e dT$

For a vibration to be Raman active any one of the above six integral should be non-zero. We need to find the symmetry of direct product of these integrals. ψ_g is totally symmetrical and symmetry of $\alpha_{xy} = \alpha_{yx}$; $\alpha_{yz} = \alpha_{zy}$; $\alpha_{zx} = \alpha_{xz}$; $\alpha_{xx}(\alpha_x^2)$; $\alpha_{yy}(\alpha_y^2)$; $\alpha_{zz}(\alpha_z^2)$ is the same as the symmetry of binary functions ; xy =yx ; yz=zy ; zx=xz ; xx(x^2); yy(y^2) ; zz(z^2). Symmetries of these binary functions can be noted down from column (which has binary functions as basis) of character table of that point group.

Example of C_{2v} Point Group Molecule H_2O/SO_2

In earlier modules we have obtained Γ3N and then Γ_{vib} as $2A_1+B_2$ for water molecule. Now we will find the Raman activity of A_1 and B_2 modes i.e. that of ψ_e which belongs to either A_1 or B_2 symmetries. In C_{2v} point group the totally symmetric representation is A_1. For Raman activity we have to find the irreducible representations for which xy =yx ; yz=zy ; zx=xz ; xx(x^2); yy($^y2$) ; zz(z^2) binary functions form the basis for irreducible representations i.e. symmetry species of $\alpha_{xy} = \alpha_{yx}$, $\alpha_{zx} = \alpha_{xz}$, $\alpha_{yz} = \alpha_{zy}$, $\alpha_{xx}(\alpha_x^2)$, $\alpha_{yy}(\alpha_y^2)$, α_{zz}. (α_z^2) components. Using character table of C_{2v} point group the symmetries of these vectors can be obtained easily.

C_{2v}	E	C_2 (z)	σ_v(yz)	Quadric functions
A_1	+1	+1	+1	x^2, y^2, z^2
A_2	+1	+1	-1	Xy
B_1	+1	-1	+1	Xz
B_2	+1	-1	-1	yz

Symmetries of $\alpha_{xy} = \alpha_{yx} = A_2$; $\alpha_{yz} = \alpha_{zy} = B_2$; $\alpha_{zx}, = \alpha_{xz}, = B_1$; $\alpha_{xx}(\alpha_x^2) = A_1$; $\alpha_{yy}(\alpha_y^2) = A_1$; $\alpha_{zz}.\ (\alpha_z^2) = A_1$. For checking the Raman activity of A_1 mode above six direct product integrals can be rewritten as:

(1a) $A_1{}^{\otimes}A_1{}^{\otimes}A_1 \equiv A_1$ (4d) $A_1{}^{\otimes}A_2{}^{\otimes}A_1 \equiv A_2$

(2b) $A_1{}^{\otimes}A_1{}^{\otimes}A_1 \equiv A_1$ (5e) $A_1{}^{\otimes}B_2{}^{\otimes}A_1 \equiv B_2$

(3c) $A_1{}^{\otimes}A_1{}^{\otimes}A_1 \equiv A_1$ (6f) $A_1{}^{\otimes}B_1{}^{\otimes}A_1 \equiv B_1$

Integrals 1a, 2b, 3c all contain A_1 representation and thus the whole integral $\int \psi_g \alpha \psi_e dT$ is non-zero or A_1 mode is Raman active. Symmetric modes are Raman active and always polarized.

For checking the Raman activity of B_2 mode above six direct product integrals can be rewritten as:

$(7) A_1{}^{\otimes}A_1{}^{\otimes}\ B_2 \equiv B_2$ $(10) A_1{}^{\otimes}A_2{}^{\otimes}\ B_2 \equiv B_1$
$(8) A_1{}^{\otimes}A_1{}^{\otimes}\ B_2 \equiv B_2$ $(11) A_1{}^{\otimes}B_2{}^{\otimes}\ B_2 \equiv A_1$
$(9) A_1{}^{\otimes}A_1{}^{\otimes}\ B_2 \equiv B_2$ $(12) A_1{}^{\otimes}B_1{}^{\otimes}\ B_2 \equiv A_2$

Of these six integrals (11) contains totally symmetric representation. So the whole integral is non zero. Hence B_2 mode is also Raman active.

So for H_2O molecule both vibrational modes $2A_1$ and B_2 are Raman Active.

Example of C_{3v} Point Group Molecule $POCl_3/CHCl_3$

For C_{3v} point group $\Gamma_{vib} = 3A_1 + 3E$. Using character table of C_{3v} point group symmetries of components of α are noted as: $\alpha_{xy} = \alpha_{yx} = E$; $\alpha_{yz} = \alpha_{zy} = E$; $\alpha_{zx}, = \alpha_{xz} = E$; $\alpha_{xx}\left(\alpha_x{}^2\right) = A_1$; $\alpha_{yy}\left(\alpha_y{}^2\right) = A_1$; $\alpha_{zz.}\left(\alpha_z{}^2\right) = A_1$ and ψ_e e is either of A_1 or E symmetry i.e. we have to check the Raman activity of A_1 and E modes.

C_{3V}	E	$2C_3$ (z)	$3\sigma_v$	Quadratic functions
A_1	+1	+1	+1	$x^2 + y^2, z^2$
A_2	+1	+1	−1	–
E	+2	−1	0	$(x^2 - y^2, xy)\ (xz, yz)$

For this we have to consider the following six direct products integrals of $\int \psi_g \alpha\ \psi_e dT$. For checking the activity of A_1 mode these can be written in direct product form as:

$(13) A_1{}^{\otimes}A_1{}^{\otimes}A_1 \equiv A_1$ $(16) A_1{}^{\otimes}E^{\otimes}\ A_1 \equiv E$
$(14) A_1{}^{\otimes}A_1{}^{\otimes}A_1 \equiv A_1$ $(17) A_1{}^{\otimes}E^{\otimes}\ A_1 \equiv E$
$(15) A_1{}^{\otimes}A_1{}^{\otimes}A_1 \equiv A_1$ $(18) A_1{}^{\otimes}E^{\otimes}\ A_1 \equiv E$

The integrals in equations 13, 14, 15 contain totally symmetric representation A1. So the whole integral is non- zero. Hence A_1 mode is Raman active.

For checking the activity of E mode 6 similar integrals can be written in direct product form as:

$$(19) A_1 \otimes A_1 \otimes E \equiv E \quad (22) A_1 \otimes E \otimes E \equiv A_1$$
$$(20) A_1 \otimes A_1 \otimes E \equiv E \quad (23) A_1 \otimes E \otimes E \equiv A_1$$
$$(21) A_1 \otimes A_1 \otimes E \equiv E \quad (24) A_1 \otimes E \otimes E \equiv A_1$$

Integrals in 22, 23, 24 contain totally symmetric representation. So whole integral is non zero. Hence E mode is also Raman active. So for $POCl_3/CHCl_3$ molecule all vibrational modes $3A_1$ and 3E are Raman Active.

For C_{2v} and C_{3v} point group molecules we have noticed that all modes are active in both IR and Raman. For centrosymmetric molecules i.e. which have centre of inversion 'i' the, "mutual Exclusion Principle applies". According to this rule, there exists no common vibrational mode which is active in both IR and Raman. A vibration which is active in IR will be Raman inactive and vice-versa. The reason for this is that, T_x, T_y, T_z, which form the basis for vibrational modes, are antisymmetric with repect to 'i ' i.e. these are ungerade, whereas the product x^2, y^2, z^2 etc. are always symmetric with respect to 'i' and cannot show IR active bands.

Potential Energy Diagrams

The energy of a molecule can be approximated as $E = E_{rot} + E_{vib} + E_{elec} + E_{tran} + E_{spin} + E_{nucl}$. If we make the approximation that the wavefunction is separable, then the problem can be reduced to several smaller tasks. This amounts to the assumption of the Born-Oppenheimer approximation, which permits the nuclear and electronic motions to be analyzed separately. In fact, this approximation breaks down, and leads to the Jahn-Teller phenomenon and other vibronic effects. Let us now address the pure vibrational problem. Exchange of energy between a molecule and the electromagnetic field occurs when hn = ΔE, where ΔE is the difference between initial and final quantized states. In terms of energies,

$$\nu = \frac{c \text{ cm/sec}}{\lambda \quad \text{cm}} = \text{Hz} \quad \bar{\nu} = \frac{1}{\lambda} = \frac{\nu}{c} = \text{cm}^{-1}$$

Infrared absorption spectra usually cover the range 200-4000 cm-1 or 50-2.5 micrometers (microns). The conversion factor 1 e.v. = 8066 cm^{-1} or 23 kcal/mole is also useful to remember. Translational energies are about 200 cm^{-1} at room temperature and rotational energies are 1-100 cm^{-1}. Most infrared spectrometers provide the spectrum in the form of % transmittance vs. wavenumber. Transmittance spectra tend to emphasize weak absorptions in the spectrum. Spectra of the same sample recorded at different concentrations will have different relative peak heights, when displayed as % transmittance. Conversion to absorbance spectra (absorbance = -log10 (transmittance) is an option available on most spectrometers. Absorbance spectra should be used whenever peak ratios or concentration information is desired (e.g., in kinetics, where the decrease or increase of concentration must be monitored). The other important aspect to IR and Raman spectroscopy is that the time scale of the measurement amounts to the time it takes for a vibration

(~ 0.1 psec). Even rapidly isomerizing species show distinct vibrational spectra in contrast to slower techniques, such as NMR spectroscopy. Figure shows the vibrational potential energy surface for a harmonic oscillator (A).

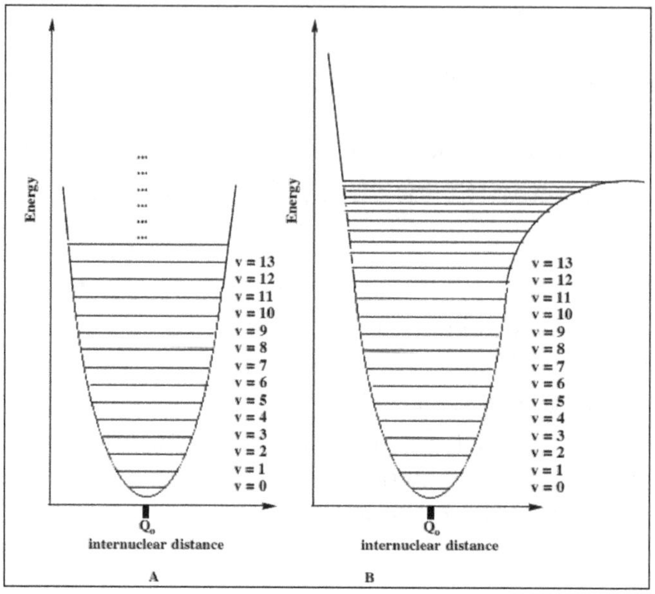

Figure shows potential energy vs. internuclear distance diagrams for a diatomic molecule which behaves like an ideal harmonic oscillator (A) and that observed for a real molecule (B). Q_o denotes the equilibrium internuclear distance and v the vibrational quantum numbers for the stretching vibration.

And for a nonideal diatomic molecule (B). These plots depict how the total electronic energy of the molecule varies as the internuclear distance varies away from its internuclear values Q_o. In the simplest approximation the bond joining the atoms is assumed to behave as a spring. In polyatomics, where the nuclear motion may be complex, it is still assumed that the overall vibrational motion along the coordinate Q can be described with a spring-like force. Then the restoring force, F, is given by Hooke's Law:

$$F = k(Q_o - Q)$$

And by integration the energy E is given by:

$$E = 1/2k\ (Q_o - Q)^2$$

This equation generates the parabolic surface of Figure A. The nuclei are constrained to move on this potential surface, and solution of the Schrödinger wave equation yields the quantized vibrational energy levels with energies.

$$E_v = h\nu\ (v + 1/2)$$

Here v is vibrational quantum number 0,1,2 .... The frequency of vibration ν is given by,

$$\nu = 1/2\pi\sqrt{k/\mu},$$

Where k is the bond force constant and m is the reduced mass for two nuclei of masses m_1 and m_2.

$$\frac{1}{\mu}=\frac{1}{m_1}+\frac{1}{m_2}$$

This yields the quantized vibrational level scheme shown in Figure A. Because transitions between the v = 0 and v = 1 levels dominate in infrared or Raman spectroscopy, the harmonic oscillator description provides a useful approximation for real molecules, 5.1 B, near the bottom of the potential well.

There are several predictions of the harmonic oscillator model. First, the frequency n becomes smaller as m increases (heavier atoms), or as k decreases. The force constant k, which represents the second derivative or curvature of the potential surface of Q_0, often parallels bond strengths, D_0; however, there's not necessarily a correlation between the two. These facts are illustrated by the data in Table for the halogens.

Table: Vibrational frequencies $\bar{v}$, Force Constants, k, and Dissociation Energies, D_o for the Halogens.

	$\bar{v}$, cm^{-1}	k (mdyn/Å)	D_o kcal/mole
F_2	892	4.45	37
Cl_2	546	3.19	58
Br_2	319	2.46	46
I_2	215	1.76	36

Normal Modes in Polyatomic Molecules

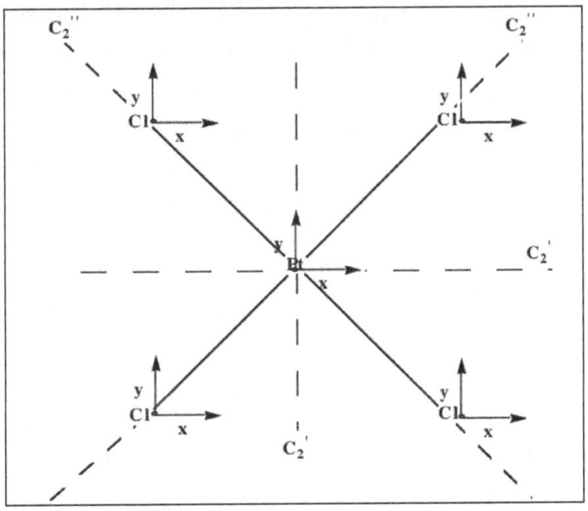

Choice of Cartesian coordinate system for $PtCl_4^{2-}$ – σ_v is in the x_z and y_z planes with 2 σ_d between these planes.

Consider a molecule containing N atoms. Because most of the mass resides in the nuclei, the translational, rotational and vibrational motions of the molecule can be approximated by considering only nuclear motions. This assumption of separability of nuclear motion from that of the electrons amounts to the Born-Oppenheimer approximation. If there are N nuclei, then each nucleus possesses 3N degrees of freedom in three dimensional spaces. Three of these coordinates describe translational motion of the center of mass and three more degrees of freedom describe rotations around the center of mass. For linear molecules only two degrees of freedom are needed to describe rotations. The remaining 3N-6 degrees of freedom (3N-5 if linear) describe vibrations. Symmetry species of translations, rotations, and vibrations can be determined by considering the character of the representation spanned by Cartesian vectors localized on each atom. For example, consider the $PtCl_4^{2-}$ ion of Figure, with the z axis assumed to emerge from the plane of the page for each atom.

Although the choice of coordinate system is arbitrary, we choose it so the operations of the D4h group transform the x, y, z vectors into ± themselves, or else into one another. Recall, that to compute the character for the representation spanned by the 3N coordinates, denoted X(3N), you need only sum the diagonal elements of the 15-dimensional transformation matrix. Under a $\hat{C}_4$ rotation, for example, the chlorine atoms interchange. No vector from these atoms can go into ± itself or some fraction thereof. They contribute zero to the character. Similarly, the vectors x and y on Pt go to +y and -x, respectively, which together contribute zero to the character. Only z on Pt, which goes into z on a $\hat{C}_4$ rotation (i.e., is unchanged), contributes a +1 to the diagonal of the transformation matrix. So the character for $\hat{C}_4$ is +1. Characters for the other operations appear below:

	$\hat{E}$	$2\hat{C}_4$	$\hat{C}_2$	$2\hat{C}_{2'}$	$2\hat{C}_{2''}$	$\hat{i}$	$2\hat{S}_4$	$\hat{\sigma}_h$	$2\hat{\sigma}_v$	$2\hat{\sigma}_d$
$\chi(3N)$	15	1	−1	−1	−3	−3	−1	5	1	3

$$= A_{1g} + A_{2g} + B_{1g} + B_{2g} + E_g + 2_{A2u} + B_{1u} + 3_{Eu}$$

Now remove from this 15-dimensional reduced representation the translations, which transform like x, y, z = $A_{2u} + E_u$ and the rotations R_x, R_y, $R_z = E_g + A_{2g}$ to yield the χ (3N-6) vibrations of the molecule = $A_{1g} + B_{1g} + B_{2g} + A_{2u} + B_{1u} + 2_{Eu}$. This list comprises the normal modes of vibration. They describe the symmetry representations for all possible vibrational motions or vibrational wavefunctions of the PtCl42- ion. Any motion of the molecule that leaves the center of mass fixed, and which does not rotate the molecule, can be represented as a superposition of the normal modes of vibration. It is informative to consider the symmetries of the vibrational coordinates and use the projection operator to obtain a visual representation. In the case of $PtCl_4^{2-}$, there will be an ambiguity in the nature of the E_u vibrations obtained. There are two modes of this symmetry in the list of possible normal modes and the exact nature of each can only be determined by solving the vibrational Hamiltonian. Mixing may occur between the symmetry adapted vibrational coordinates of the same symmetry. This problem goes beyond what simple group theory can determine.

We could solve the symmetry coordinate problem with Cartesian displacements (and subtract out rotations and translations); however, it is customary to use "internal coordinates" that correspond to bond stretches, bends, and torsions. Vibrational motions depicted in the internal coordinate system correspond better to intuitive notions of bond stretching and bending. Unlike the cartesian coordinates, problems arise because the internal coordinates are not always orthogonal and linearly independent. One must be careful to eliminate redundancies that arise in the internal coordinates, and to modify the internal coordinate motions to ensure the center of mass remains fixed. If the center of mass shifts then the motion would not be a pure vibration, but contains some translational motion as well. For example, the four Cl-Pt-Cl angles are not linearly independent.

Any three fix the value of the remaining angle in the plane, because all four must sum to 360°. Thus, only three of the angles are linearly independent. If we attempt to construct four linear combinations from these four angles with the projection operator, then one will be specious.

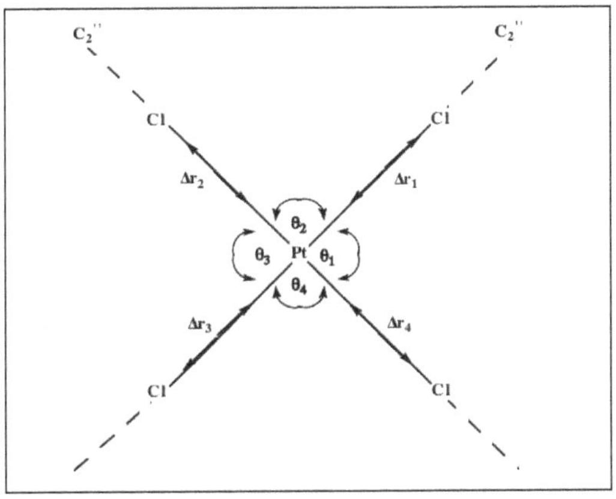

Choice of internal coordinate system for $PtCl_4^{2-}$.

First, consider $PtCl_4^{2-}$ and the set of coordinates for Pt-Cl bond stretching Δr_1, Δr_2, Δr_3, and Δr_4 shown in Figure. The character of the Pt-Cl bond stretches is:

	E	$2C_4$	C_2	$2C_{2'}$	$2C_{2''}$	i	$2S_4$	σ_h	$2\sigma_v$	$2\sigma_d$
$\chi(pt-Cl)$	4	0	0	0	2	0	0	4	0	2

$$= A_{1g} + B_{2g} + E_u$$

Using the projection operator, the totally symmetric stretch is found to be:

$$P_{a1g}(\Delta r_1) = N(\Delta r_1 + \Delta r_2 + \Delta r_3 + \Delta r_4)$$

At present we won't concern ourselves with normalization conditions.

The B_{2g} coordinate is found by applying $P^{B_1g}(\Delta r_1)$ to yield:

	E	C_4	C_4^3	C_2	C_2'	C_2'	C_2''	C_2''	i	S_4	S_4^3
$\hat{R}(\Delta r_1)$	Δr_1	Δr_2	Δr_4	Δr_3	Δr_2	Δr_4	Δr_1	Δr_3	Δr_3	Δr_2	Δr_4
$\chi(B_{2g})$	1	−1	−1	1	−1	−1	1	1	1	−1	−1
	σ_h	σ_v	σ_v	σ_d	σ_d						
$\hat{R}(\Delta r_1)$	Δr_1	Δr_2	Δr_4	Δr_1	Δr_3						
$\chi(B_{2g})$	1	−1	−1	+1	+1						

$$P^{B_2g}(\Delta r_1) = N(\Delta r_1 + \Delta r_3 - \Delta r_2 - \Delta r_4)$$

Also for the degenerate stretching vibrations,

$$\rho^{Eu}(\Delta r_1) = N(\Delta r_1 - \Delta r_3)$$

The other partner may be generated by application of $\hat{C}_4$.

$$N\hat{C}_4(\Delta r_1 - \Delta r_3) = N(\Delta r_2 - \Delta r_4)$$

The orthogonal linear combination derived from the sum and difference of the preceding pair is normally used.

$$N'[\Delta r_1 + \Delta r_2 - \Delta r_3 - \Delta r_4]$$
$$N'[\Delta r_1 + \Delta r_4 - \Delta r_2 - \Delta r_3]$$

These stretching vibrations are depicted in Figure. It is important to reemphasize that the center of mass must not move during a normal vibration. Otherwise the motion would include some translational motion and not be a pure vibration. For the degenerate e_u stretching vibration, some motion of the central Pt must be added, as shown in the figure, so the center of mass remains fixed. It is always required that no center of mass motion (a translation) occur in a pure vibration, and this condition must always be applied to the SALC one obtains.

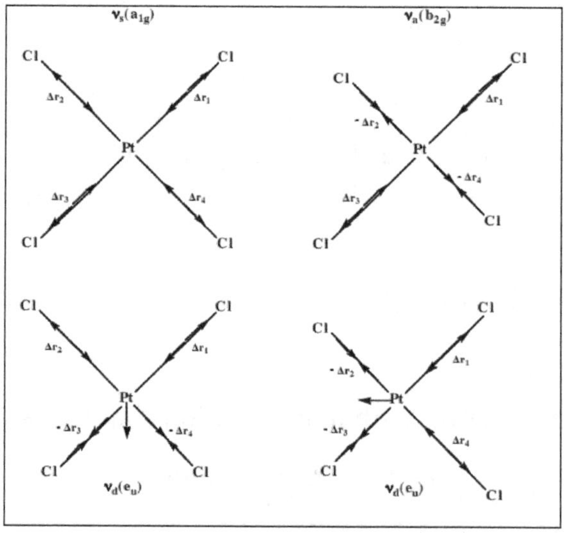

Symmetry adapted linear combinations of the a1g, b2g, and eu Pt-Cl stretching vibrations in $PtCl_4^{2-}$.

The ν_{a1g} and ν_{b2g} vibrations are often called the symmetric and antisymmetric stretches, ν_s and ν_a, in MX_4 complexes. They usually occur at similar energies with $\nu_s > \nu_a$. The Eu mode is called ν_d and can mix with the E_u angle deformation (vide infra). In these complexes, $\nu_d > \nu_s > \nu_a$. As is always the case, the frequencies shift to lower energies as the mass of the halide increases. Metalligand stretching vibrations in transition metal complexes usually occur below 450 cm^{-1}. The motions that involve bending or deformation of metal-ligand bond angles fall at even lower frequencies.

	$\nu_s(a_{1g})$	$\nu_a(b_{2g})$	$\nu_d(e_u)$
$PtCl_4^{2-}$	330 cm^{-1}	312 cm^{-1}	313 cm^{-1}
$PtBr_4^{2-}$	208	194	227
PtI_4^{2-}	155	142	180

Bending and Angle Deformation Vibrations

Next consider the bending vibrations θ_1 - θ_4 of $PtCl_4^{2-}$ shown in Figure. The characters for the representation are:

	E	$2C_4$	C_2	$2C_{2'}$	$2C_{2''}$	i	$2S_4$	σ_h	$2\sigma_v$	$2\sigma_d$
$\chi(\theta)$	4	0	0	2	0	0	0	4	2	0

$= a_{1g} + b_{1g} + e_u$

Recall that one representation will prove to be redundant, because if θ_1, θ_2, and θ_3 are known then θ_4 is fixed by the requirement that their sum totals 360°. Application of the projection operator for the a_{1g} vibration yields:

$$a_{1g}(q) = N(\theta_1 + \theta_2 + \theta_3 + \theta_4)$$

However, it is impossible for all four angles to increase simultaneously, and this a_{1g} "bending vibration" is a nonsensical consequence the internal coordinate redundancy. This arises because $\theta_1 \rightarrow \theta_4$ were not linearly independent.

The b_{1g} and e_u SALC's are genuine symmetry coordinates for vibrations. Application of the projection operator yields:

	E	C_4	C_4^3	C_2	$C_{2'}$	$C_{2'}$	$C_{2''}$	$C_{2''}$	i	S_4	S_4^3	σ_h	σ_v	σ_v	σ_d	σ_d
$R(\theta_1)$	θ_1	θ_2	θ_4	θ_3	θ_1	θ_3	θ_4	θ_2	θ_3	θ_2	θ_4	θ_1	θ_1	θ_3	θ_4	θ_2

$$P^{b_1g}(\theta_1) = N(\theta_1 + \theta_3 - \theta_2 - \theta_4)$$

For the e_u mode we obtain $(\theta_1 - \theta_3)$ and its partner $(\theta_2 - \theta_4)$. These bending vibrations are depicted in Figure. Note the added motion of Pt necessary to leave the center of mass fixed in the e_u mode.

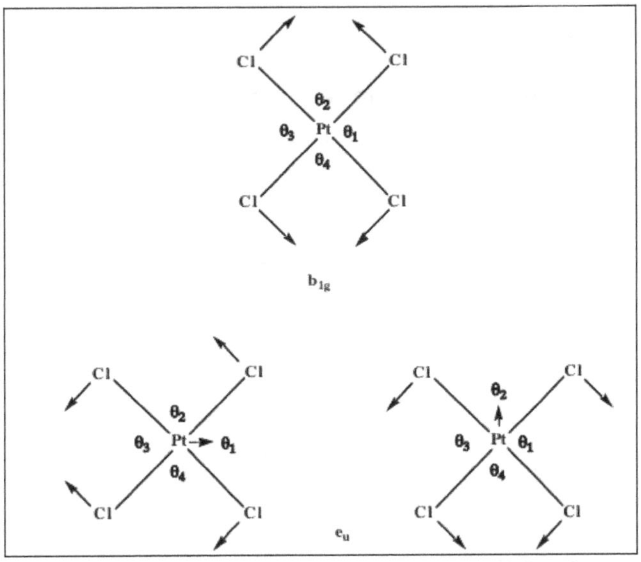

Sketch of the b1g and eu bending vibrations of $PtCl_4^{2-}$.

Finally, consider the out-of-plane bending vibrations. As yet, the a_{2u} and b_{1u} vibrational modes in the complete list of 3N-6 vibrations are not accounted for by the in plane bends and stretches. We anticipate that the out-of-plane motions should account for the missing vibrations. The out of plane bends can be depicted by a vector displacement perpendicular to the plane of the molecule for each of the four chloride ligands. A + to denotes displacement above the plane for each chlorine atom, and a - denotes displacement below the plane. Since this spans a four dimensional representation, and since only a_{2u} and b_{1u} are missing, it is clear that the two extra dimensions must arise from redundancies in the coordinates.

	4	0	0	0	2	0	0	4	0	2
	E	$2C_4$	C_2	$2C_{2'}$	$2C_{2''}$	i	$2S_4$	σ_h	$2\sigma_v$	$2\sigma_d$
$\chi(+)$	4	0	0	0	−2	0	0	−4	0	2

$= a_{2u} + b_{1u} + e_g$

The e_g representation must be redundant, because only the a_{2u} and b_{1u} vibrations are left. Application of the projection operator for them yields:

	E	C_4	C_4^3	C_2	$C_{2'}$	$C_{2'}$	$C_{2''}$	$C_{2''}$	i	S_4	S_4	σ_h	σ_v	σ_v	σ_d	σ_d
$\hat{R}_{(1+)}$	+1	+2	+4	+3	−4	−2	−1	−3	−3	−2	−4	−1	+4	+2	+1	+3
$\chi(a_2u)$	1	1	1	1	−1	−1	−1	−1	−1	−1	–	−1	1	1	1	1
$\chi(b_1u)$	1	−1	−1	1	1	1	−1	−1	−1	1	1	−1	−1	−1	1	1

$$P^a 2u\,(+1) = 2(+1) - 2(-1) + 2(+2) - 2(-2) + 2(+3) - 2(-3)$$

$$+\ 2(+4) - 2(4)$$

$$= 4\{(+1) + (+2) + (+3) + (+4)\}$$

$$P^b 1u\,(+1) = 4\{(+1) + (-2) + (+3) + (-4)\}$$

Calculating these linear combinations uses the relation that $-(+1) = +(-1)$. In other words, minus a positive displacement of atom #1 is the same as a negative displacement of the same atom. The out of plane vibrations are sketched in Figure.

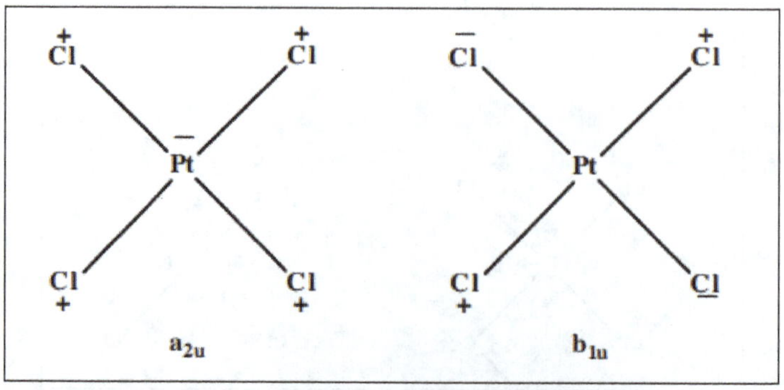

Sketch of the a_{2u} and b_{1u} bending vibrations of $PtCl_4^{2-}$.

For $PtCl_4^{2-}$, the complete set of observeable vibrations are:

	Stretches		In-plane bends		Out-of-plane bends
$\nu_6(e_u)$	$\nu_1(a_{1g})$	$\nu_4(b_{2g})$	$\nu_2(b_{1g})$	$\nu_7(e_u)$	$\nu_3(a_{2u})$
$\nu_d(pt-Cl)$	$\nu_s(Pt-Cl)$	$\nu_a(Pt-Cl)$	$\delta_d(PtCl_2)$	$\delta_d(PtCl_2)$	π
313 cm^{-1}	330 cm^{-1}	312^{-1}	171 cm^{-1}	165 cm^{-1}	147 cm^{-1}

The $\nu_s(b_{1u})$ mode is not IR or Raman active, as will be discussed and cannot be observed.

Vibration Modes

It is useful to discuss analysis of vibrational modes in terms of stretching of bonds in between atoms and angle deformations. The internal coordinates can be of the type:

- Bond stretchings, i.e., symmetric or anti symmetric stretches.
- Angle deformation.
- Rocking.
- Wagging.
- Scissoring.
- Twisting.

In rocking, wagging, or twisting, the coordinates, the angles and bond lengths within the groups involved do not change. Rocking may be distinguished from wagging by the fact that the atoms in the group stay in the same plane. Some of these modes of vibrations are shown in figure.

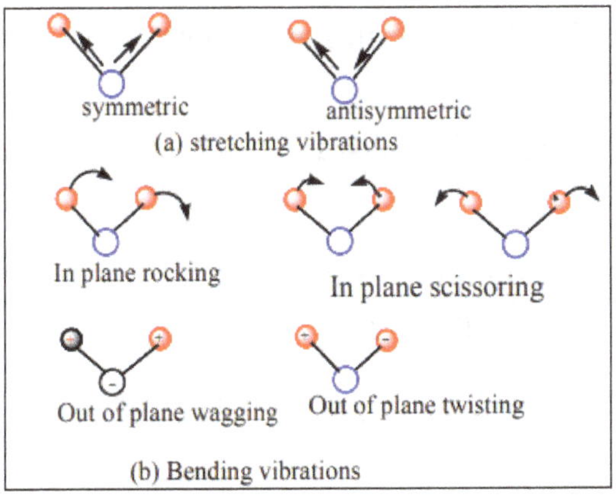

Some modes of vibrations. ⊕ indicates out of plane motion and ⊖ indicates back of plane motion.

The most important internal coordinates are:

- Change in bond length (stretching).
- Angle deformation (bending).

Only these will be used as basis for representations to get Γ_{str} (stretchings) and Γ_{bend} (angle deformation). To find symmetry species of Γ_{str} and Γ_{bend} reducible representations, these will be reduced to irreducible representations. These results will then be correlated with those that were obtained using Γ_{3N} method i.e., $\Gamma_{vib} = \Gamma_{3N} - \Gamma_{T+R}$ method.

To explain this method let us take few examples:

Example of H_2O Molecule Belonging to C_{2v} Point Group

Various symmetry elements present in water molecules together with (i) Two O-H stretches are shown by arrows pointing from O to H and (ii) double headed arrow shows:

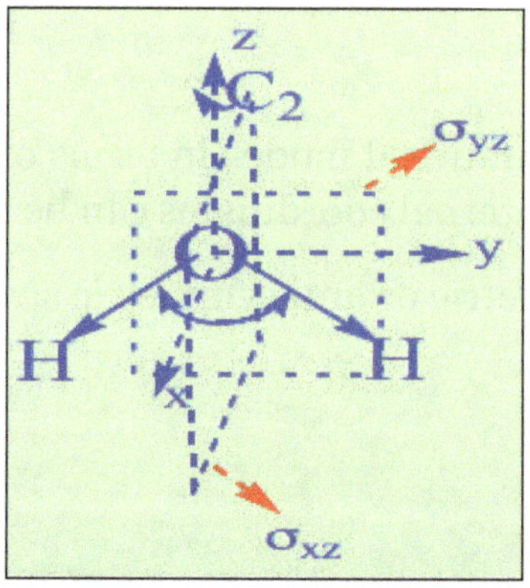

Symmetry elements and bond vectors in H2O molecule.

The steps involved are:

1. Find reducible representation Γ_{bond} or Γ_{str} using two bond vectors as basis.

- Find the number of unshifted bond vectors, then follow the following rules for finding χ which contributed towards Γ_{bond}.
- A vector which is unshifted by the symmetry operations of the point group contributes $\chi(R) = +1$.
- A vector which is shifted to another position by the symmetry operations of the point group contributes zero to $\chi(R) = 0$.
- A vector which is unshifted but only its direction is changed by the symmetry operations of the point group contributes $\chi(R) = -1$.

Using character table of C_{2v} point group let us get reducible representation Γ_{bond} or Γ_{str} (i) Find Γ_{str} as follows:

C_{2v}	E	C_2	σ_{xz}	σ_{yz}
Unshifted O-H stretches	2	0	0	2
Multiplication factor	x1	x1	x1	x1
Γ_{str}	2	0	0	2

The symmetry operations C_2 and σ_{xz} interchange both O-H vectors. Therefore, $\chi(R) = 0$. Under E and σ_{yz} symmetry operations both vectors are not shifted $\chi(R) = +1$.

Let us now reduce Γ_{str}.

C_{2v}	E	C_2 (z)	σ_v (xz)	σ_v (yz)	Linear functions, rotations	Quadratic functions
A_1	+1	+1	+1	+1	z	x^2, y^2, z^2
A_2	+1	+1	−1	−1	R_z	xy
B_1	+1	−1	+1	−1	x, R_y	xz
B_2	+1	−1	−1	+1	y, R_x	yz
Γ_{str}	2	0	0	2		

Follow the standard reduction formula to reduce Γ_{str}. Since there are four operations in C_{2v} point group, the order of the group is four. The equations for calculations are:

- $A_1 = 1/4[1.1.2 + 1.1.0 + 1.1.0 + 1.1.2] = 1/4[4] = 1$. It means A_1 is present only once.
- $B_1 = 1/4[1.1.2 + 1.-1.0 + 1.1.0 + 1.-1.2] = 1/4[0] = 0$. It means B_1 is not present.
- $A_2 = 1/4[1.1.2 + 1.1.0 + 1.-1.0 + 1.-1.2] = 1/4[0] = 0$. It means A_2 is not present.
- $B_2 = 1/4[1.1.2 + 1.-1.0 + 1.-1.0 + 1.1.2] = 1/4[0] = 1/4[4] = 1$. It means B_2 is present only once.

Thus $\Gamma_{str} = A_1 + B_2$.

If we compare the result with $\Gamma_{vib} = \Gamma_{3N} - \Gamma_{T+R}$ method,

$$\Gamma_{vib} = 2A_1 + B_2$$

Thus we are short by A_1 mode. How to account this mode? For this we take double headed arrow <HOH bond angle as basis for getting Γ_{bend}.

2. Find Γ_{bend} as follows:

Now let us examine the effect of symmetry operations of C_{2v} point group on double headed arrow.

E → + 1; C_2 = +1(because one half of the angle is taken into another half, final angle is not changed); σ_{xz} = +1 (because one half of the angle is taken into another half, final angle is not changed); σ_{yz} = + 1(double headed arrow lies in this plane).

C_{2v}	E	C_2	σ_{xz}	σ_{yz}
Unshifted double headed arrow	1	1*	1*	1
Multiplication factor	x1	x1	x1	x1
Γ_{bend}	1	1	1	1

* half angle goes into other half only.

Γ_{bend} can be reduced as:

A_1 = 1/4[1.1+1.1 + 1.1 + 1.1] = 1/4[4] = 1. It means A_1 is present only once.

Try for other irreducible representations A_2, B_1 and B_2, you will get zero results for this.

Therefore, $\Gamma_{\text{internal coordinates}} = \Gamma_{str} + \Gamma_{bend} = [A_1 + B_2] + [A_1] = 2A_1 + B_2$.

The result is same as that obtained earlier using Γ_{3N} as the basis for representation. Out of these three modes of H2O two modes, A_1 + B_2, are stretchings and one A_1 mode is angle deformation. These modes of vibration are shown in figure.

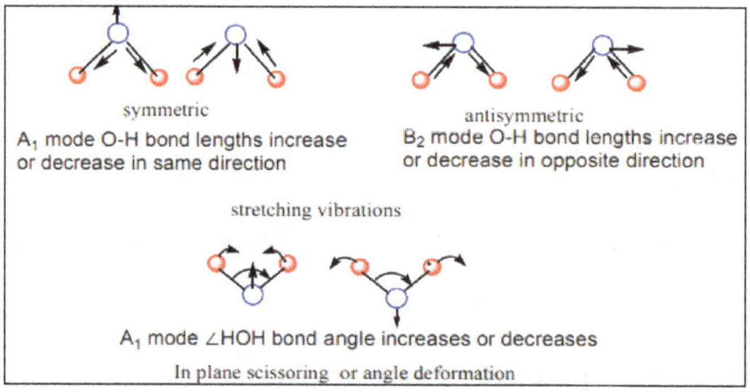

Symmetric and anti-symmetric stretchins and angle deformation in H_2O.

Assigning Vibrational Frequencies to Various Mode

Let us take actual IR and Raman spectra H_2O. The tentative vibrational spectrum of H_2O molecule is given in table.

Table: Tentative vibrational spectrum of H_2O molecule.

Vibrational Spectra of H_2O molecule	
IR, cm^{-1}	Raman ,cm^{-1}
3657	3657(Pol.)
1595	1595(pol.)
3756	3756(depol.)

Before assigning these frequencies to the observed symmetry species of vibrational modes, following general observations are to be noted. These are based upon large number of experimental data on large number of molecules and from chemical intuitions:

- Stretching frequencies are higher than the bending frequencies involving similar atoms.
- Symmetric frequencies are always Raman polarized.

Since $\Gamma_{str} = A_1 + B_2$, and $\Gamma_{bend} = A_1$.

Bands at 3657(Pol.) cm^{-1} and 1595(pol.) cm^{-1} [polarized in Raman bands are of A_1 type always] Further higher frequency 3657(Pol.) cm^{-1} is due to symmetric O-H stretch and lower frequency 1595(pol.) cm^{-1} is due to < HOH angle deformation. The third frequency at 3756(depol) cm^{-1} has B_2 symmetry i.e., associated with antisymmetric O-H stretch. This O-H stretch is near to symmetric

O-H stretch. Now we can assign H_2O vibrations as follows:

- Symmetric O-H stretch ----- 3657(Pol.) cm^{-1}, υ_1, (polarized, A_1).
- < HOH angle deformation ----1595(pol.) cm^{-1}, υ_2, (polarized, A_1).
- Anti symmetric O-H stretch --3756(depol) cm^{-1}, υ_3 (depolarized, B_2).

Vibrational frequencies are generally written in decreasing order within each symmetry species and working further according to order of species given in character table. Here there are two symmetric frequencies of A_1 type. Totally symmetric species are considered first .In this spectrum A_1 mode at 3657(Pol.) cm^{-1} is higher. Therefore it is assigned as υ_1 and next lower frequency belonging to A_1 type is at 1595(pol.) cm^{-1}, so it is assigned as υ_2. Next symmetry species is B_2 which is not symmetric. Here there is only one frequency which is at 3756(depol) cm^{-1}. So it is assigned as υ_3. This convention is generally followed all through.

Overtones, Hot Bands and Combination Bands

Due to anharmonicity it is possible to excite two or more vibrational modes simultaneously to give rise to overtones, combination bands etc i.e. excitation is from υ_0 to υ_2 or υ_0 to $(\upsilon_1+\upsilon_2)$ or υ_0 to $(\upsilon_1-\upsilon_2)$ or υ_0 to $(2\upsilon_1+\upsilon_2)$. A fundamental that is forbidden in IR may be possibly be allowed as an overtone or combination band. Appearance of overtones or combination band is determined by symmetry selection rules i.e. we have to consider the direct product and vanishing integrals involving ground state symmetry and excited state symmetry$(\upsilon_2, \upsilon_3 \text{ -----})$. The origin of overtones, hot bands, combination bands is shown in the Figure.

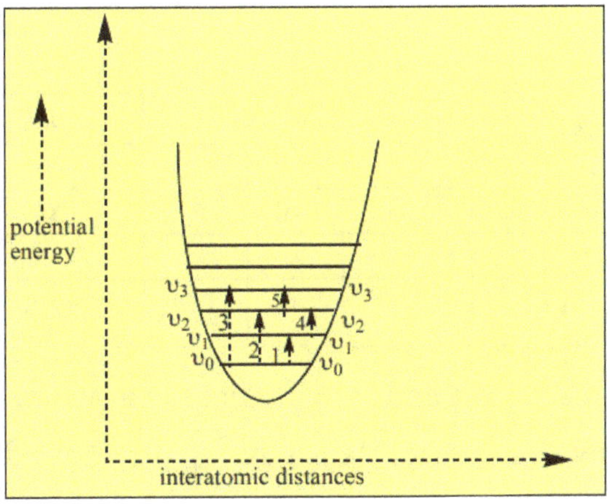

Fundamentals, overtones and hot bands in vibrational spectrum.

In the Figure transition 1 is from υ_0 to υ_1 and it is called fundamental mode. Transitions 2 and 3 i.e. υ_0 to υ_2 or υ_0 to υ_3 are called first and second overtones and transitions 4 and 5 i.e. υ_1 to υ_2 or υ_2 to υ_3 are called hot bands. This means it is possible to excite two or more vibrational modes simultaneously to give rise to overtones, combination bands, hot bands etc. i.e. excitation is from υ_0 to υ_2 or υ_0 to $(\upsilon_1+\upsilon_2)$ or υ_0 to $(\upsilon_1-\upsilon_2)$ or υ_0 to $(2\upsilon_1+\upsilon_2)$ or υ_1 to υ_2 or υ_2 to υ_3 etc. A fundamental vibrational mode that is forbidden in IR may be possibly be allowed as an overtone or combination band. Appearance of overtones or combination band is determined by symmetry selection rules similar to one we have considered earlier i.e. we have to consider the direct product and vanishing integrals involving ground state symmetry and excited state symmetry $(\upsilon_2, \upsilon_3 -----)$.

Overtones

These types of bands arise from vibrational transitions in which molecule absorbs two or more quanta of energy in same vibrational mode. If the fundamental band is found at υ_1 then the first overtone and second overtones will be generally found at $\approx 2\upsilon_1$ and $\approx 3\upsilon_1$ and so on. The difference between overtone frequency and integral multiple of fundamental frequency, is a measure of anharmonicity. For first overtone the integral multiple of fundamental is $2 x \upsilon_1 = 2\upsilon_1$ and frequency for first overtone is $\approx 2\upsilon_1$. Therefore, the difference $(2\upsilon_1)-(\approx 2\upsilon_1)$ is a measure of anharmonicity. To determine whether the overtone bands are allowed or not in IR it is necessary to take the direct product of irreducible representation of the fundamental which make up the overtone and see whether this direct product has totally symmetric representation in it or not. Let us take the example of a C_3v point group molecule NH_3 or PCl_3. The character table for C_3v point group is given in table.

Table: Character table of C_3v point group.

C_{3v}	E	$2C_3$	$3\sigma_v$	I	II
A_1	1	1	1	T_z, z	$x^2 + y^2$, z^2
A_2	1	1	−1	R_z	xy
E	2	−1	o	(T_x, T_y), (R_x, R_y)	$(x^2 - y^2 - xy)$ (xz, yz)

In Table we take A_2 mode which is IR inactive as there is no x, y, z vectors as the basis for this mode. However A_2 mode is Raman active as there is xy binary product which forms the basis for this A_2 mode. Let us find whether the first overtone of A_2 mode will be IR active or not. For this we take the direct product of A_2 mode as:

C_{3v}	E	$2C_3$	$3\sigma_v$	I	II
A_1	1	1	1	T_z, z	$x^2 + y^2$, z^2
A_2	1	1	−1	R_z	Xy
A_2	1	1	−1		
$A_2{}^{\otimes}A_2$	1	1	1	T_z, z	$x^2 + y^2$, z^2

For $A_2{}^{\otimes}A_2$ characters for E = 1, C_3 = 1 and σ_v = 1. These characters correspond to A_1 irreducible representation which is totally symmetric. Hence $A_2{}^{\otimes}A_2$ direct product has A_1 mode in it so the first overtone of A_2 mode is IR active.

Let us now see the second overtone of A_2 mode whether it is IR active or not. For this again we take the direct product of the type $A_2{}^{\otimes}A_2{}^{\otimes}A_2$.

C_{3v}	E	$2C_3$	$3\sigma_v$	I	II
A_1	1	1	1	T_z, z	$x^2 + y^2$, z^2
A_2	1	1	−1	R_z	xy

A_2	1	1	−1		
$A_2 \otimes A_2 \otimes A_2$	1	1	−1	T_z, z	$x^2 + y^2$, z^2

The $A_2 \otimes A_2 \otimes A_2$ direct product gives the characters as for E = 1, C_3 = 1 and σ_v = −1. These are the characters for A_2 mode i.e. this direct product does not contain totally symmetric representation in it. So it is inactive .Thus second overtone of A_2 mode is IR inactive.

Thus we find a mode which was IR inactive has become IR active due to overtone of it. For one dimensional representation taking this direct product and seeing the activity is straight forward .For two or three dimensional irreducible representation there is no straight forward procedure that can be used. For degenerate vibrational mode the method is more complicated. For doubly degenerate vibrational mode i.e. E mode the following equation-1 is applied for finding the characters of symmetry operations for reducible representation corresponding to nth overtones:

$$\chi_E^{n}(R) = 1/2\left[\chi_E^{n-1}(1) \cdot \chi_E(R) + \chi_E(R^n)\right]$$

Where $\chi_E^{n}(R)$ is the character of nth overtone for symmetry operation R and $\chi_E^{n-1}(R)$ is the character of (n-1)th overtone and $\chi_E(R)$ is the character of the fundamental(for which activity of its overtone is to be determined). For the first overtone n=2 and for 2nd overtone n = 3 so on $\chi_E(R^n)$ is the character corresponding to operation R performed n times in succession. For E mode of vibration and for n = 2(first overtone) the above equation can be written as equation below. Let us use equation for finding the activity of first overtone of E mode in C_{3v} point group. For first overtone n = 2.

$$\chi_E^{2}(R) = 1/2\left[\chi_E^{1}(R)\chi_E(R) + \chi_E(R^2)\right]$$

Now let us find $\chi_E(R^2)$ the character for the operation R performed 2 times in succession.

Let us first find $\chi_E(R^2)$ for each symmetry operation for C_{3v} point group.

For E representation, $\chi_E(E^2) = \chi(E) = 2$, i.e. the character of E(identity) in E-mode (here E-mode and identity should not be confused with each other), $\chi_E(C_3^{\,2}) = \chi_E(C_3^{\,-1}) = -1$, $\chi_E(\sigma_v^{\,2}) = \chi_E(E) = 2$.

Therefore, $\chi_E(R^2)$ for E 2C3 , 3σ_v = 2 −1 2 respectively.

The value of $\chi_E(R^n)$ for n up to 5 is summarized in table as:

Table: Value of $\chi_E^{n}(R)$.

C_{3v}	E	$2C_3$	$3\sigma_v$
$\chi_E(R^2)$	2	−1	2
$\chi_E(R^3)$	2	2	0

$\chi_E(R^4)$	2	−1	2
$\chi_E(R^5)$	2	−1	0

Using simplified equation-2 for n=2 i.e. for first overtone $\chi^2_E(R)$ is obtained as given in table.

Table: Value of $\chi^2_E(R)$ for E mode.

C_{3v}	E	$2C_3$	$3\sigma_v$
$\chi_E(R)$	2	−1	0
$\chi_E(R)$	2	−1	0
$\chi_E(R)^{\otimes}\chi_E(R)$	4	1	0-----(1)
$\chi_E(R^2)$	2	−1	2-----(2)
$\chi_E(R)^{\otimes}\chi_E(R)+\chi_E(R^2)$ add(1)and(2)	6	0	2
$1/2\left[\chi_E(R)^{\otimes}\chi_E(R)+\chi_E(R^2)\right]$	3	0	1

Thus the characters for the first overtone of E mode in C_{3v} point group are:

$E = 3$, $C_3 = 0$ and $\sigma_v = 1$

This is a reducible representation and can be reduced to A_1 + E irreducible representations. A_1 mode is totally symmetric hence first overtone of E mode in $C3_v$ point group is IR active.

Let us check the IR activity of second overtone of E mode in C_{3v} point group.

For n = 3 the equation $\overset{n}{\chi_E}(R)=1/2\left[\overset{n-1}{\chi_E}(1).\chi_E(R)+\chi_E(R^n)\right]$ can be simplified as equation below

$$\chi_E(R)=1/2\left[\chi^2_E(R)\,\chi_E(R)+\chi_E(R^3)\right]$$

The value of $\chi^3_E(R)$ is summarized in table:

Table: Value of $\chi^3_E(R)$ for E mode.

C_{3v}	E	$2C_3$	$3\sigma_v$
$\chi_E(R)$	2	–	0
$\chi^3_E(R)$(first overtone)	3	0	1
$x^2_E(R)^{\otimes}\chi_E(R)$	6	0	0 ----- (1)
$\chi_E(R^3)$	2	2	0 ----- (2)
$x^2_E(R)^{\otimes}\chi_E(R)+\chi_E(R^3)$add(1)and(2)	8	2	0

$1/2\left[\chi^2_E(R)^{\otimes}\chi_E(R)+\chi_E(R^3)\right]$	4	1	0
$\chi^3_E(R)$	4	0	0

Thus $\chi^3_E(R)$ for E = 4, C3 = 1, σ_v = 0. This can be reduced to $A_1 + A_2 + E$. Out of these components of A_1 is totally symmetric. Hence second overtone of E mode is also IR mode.

Combination Bands/Difference Bands

Combination bands are the types of bands arise from transitions to an excited state where the molecule has acquired two or more vibrational quanta of energy which is distributed amongst two or more modes. If two fundamentals are at υ_i and υ_j, then a binary combination will generally be found at $\approx(\upsilon_j+\upsilon_j)$. The transition energy will be slightly less than the sum of fundamentals.

Difference bands sometimes may also be detected. In this case a molecule which is already in excited state υ_n, absorbs another quanta of energy υ_m and loose the earlier quanta, υ_n, of energy. Then a difference band may be observed at $\upsilon_n - \upsilon_m$. Such bands are rare as only few molecules exist initially in excited state except at high temperature.

In order to check the activities of these combinations/difference bands it is necessary to find symmetries these bands and then check whether is are vibrational active or not Let us explain these by taking an example of tetrahedral molecule such as CH_4.

For this molecule:

$$\Gamma_{vib} = A_1 + E + 2T_2$$
$$\Gamma_{str} = A_1 + T_2$$
$$\Gamma_{bend} = E + T_2$$

And spectrum is given as:

υ_1 (A1)-----2900 cm^{-1}

υ_2 (E)-----1525 cm^{-1}

υ_3 (T_2)-----3020 cm^{-1}

υ_4 (T_2)-----1300 cm^{-1}

Let us take various combinations/differences of these frequencies and check the activities.

$$\upsilon_1+\upsilon_2 = 4425 cm^{-1},\quad \upsilon_1+\upsilon_4 = 4200\ cm^{-1}$$
$$\upsilon_1+\upsilon_3 = 5920 cm^{-1},\quad \upsilon_2+\upsilon_4 = 2825\ cm^{-1}$$
$$\upsilon_2+\upsilon_3 = 4545\ cm^{-1},\quad \upsilon_3+\upsilon_4 = 4320\ cm^{-1}$$
$$2\upsilon_3 = 6040 cm^{-1}\quad 2\upsilon_4\ cm^{-1} = 2600\ cm$$

To check the activity of $\upsilon_1 + \upsilon_2 = 4425\ cm^{-1}$ band, take the direct product of irreducible representations corresponding to these frequency i.e. of A_1 and E modes for υ_1 and υ_2 frequencies. Reduce the reducible representation if it is reducible and check whether there is A_1 mode in it or not. If A_1 mode is there then the combination band will be IR active otherwise not.

Hot Bands

These bands arise when a molecule already in excited state gains one quanta of energy and excited to higher vibrational state. These are generally found when the spectrum is recorded at high temperature. These hot bands vanish when spectrum is recorded at lower temperature. The $\upsilon_1 \rightarrow \upsilon_2$ and $\upsilon_2 \rightarrow \upsilon_3$ transitions vanish. The activity of such bands can be checked in similar manner as discussed above. At 3000 the hot band corresponds to $\upsilon_1 \rightarrow \upsilon_2$ and is nearly one tenth as intense as $\upsilon_0 \rightarrow \upsilon_1$ fundamental band.

Fermi Resonance

In complicated molecule, it may be possible that two vibrations have same frequency within experimental error; such frequencies are termed as accidentally degenerate. Thus fewer vibrational bands may be observed. When these accidentally degenerate frequencies belong to different symmetry species in the spectrum no interaction will occur between them and frequencies are nearly superimposed. But when these accidental degenerate frequencies have same symmetry these will interact with each other and Fermi resonance is observed. In Fermi resonance the lower frequency is lowered and higher frequency is raised and intensities of these are equalized i.e. weaker borrows energy from the stronger. Fermi resonance can occur between two fundamentals; one fundamental and one overtone/combination. Thus some extra band may appear in the spectrum of the molecule.

References

- Treatment-of-group-theory-in-spectroscopy: symmetry-group-theory-and-mathematical-treatment-in-chemistry: intechopen.com, Retrieved 25, June 2020
- Symmetry-and-vibrational-spectroscopy: ucsd.edu, Retrieved 04, August 2020

Permissions

All chapters in this book are published with permission under the Creative Commons Attribution Share Alike License or equivalent. Every chapter published in this book has been scrutinized by our experts. Their significance has been extensively debated. The topics covered herein carry significant information for a comprehensive understanding. They may even be implemented as practical applications or may be referred to as a beginning point for further studies.

We would like to thank the editorial team for lending their expertise to make the book truly unique. They have played a crucial role in the development of this book. Without their invaluable contributions this book wouldn't have been possible. They have made vital efforts to compile up to date information on the varied aspects of this subject to make this book a valuable addition to the collection of many professionals and students.

This book was conceptualized with the vision of imparting up-to-date and integrated information in this field. To ensure the same, a matchless editorial board was set up. Every individual on the board went through rigorous rounds of assessment to prove their worth. After which they invested a large part of their time researching and compiling the most relevant data for our readers.

The editorial board has been involved in producing this book since its inception. They have spent rigorous hours researching and exploring the diverse topics which have resulted in the successful publishing of this book. They have passed on their knowledge of decades through this book. To expedite this challenging task, the publisher supported the team at every step. A small team of assistant editors was also appointed to further simplify the editing procedure and attain best results for the readers.

Apart from the editorial board, the designing team has also invested a significant amount of their time in understanding the subject and creating the most relevant covers. They scrutinized every image to scout for the most suitable representation of the subject and create an appropriate cover for the book.

The publishing team has been an ardent support to the editorial, designing and production team. Their endless efforts to recruit the best for this project, has resulted in the accomplishment of this book. They are a veteran in the field of academics and their pool of knowledge is as vast as their experience in printing. Their expertise and guidance has proved useful at every step. Their uncompromising quality standards have made this book an exceptional effort. Their encouragement from time to time has been an inspiration for everyone.

The publisher and the editorial board hope that this book will prove to be a valuable piece of knowledge for students, practitioners and scholars across the globe.

Index

www.ingramcontent.com/pod-product-compliance
Lightning Source LLC
LaVergne TN
LVHW080904210926
848485LV00004B/588